PATTERN FORMATION

Pattern Formation
Ciliate Studies and Models

JOSEPH FRANKEL

UNIVERSITY OF IOWA

With Original Illustrations
by Jane A. Otto

New York   Oxford
OXFORD UNIVERSITY PRESS
1989

Oxford University Press

Oxford   New York   Toronto
Delhi   Bombay   Calcutta   Madras   Karachi
Petaling Jaya   Singapore   Hong Kong   Tokyo
Nairobi   Dar es Salaam   Cape Town
Melbourne   Auckland

and associated companies in
Berlin   Ibadan

Library of Congress Cataloging-in-Publication Data
Frankel, Joseph, 1935–
Pattern formation : ciliate studies and models / Joseph Frankel ;
with original illustrations by Jane A. Otto.
p. cm. Bibliography: p. Includes index.
ISBN 0-19-504890-3
1. Ciliata—Development.   2. Protozoa—Development. I. Title.
QL360.A22F73   1989
593.1'72—dc19   89-2855   CIP
9 8 7 6 5 4 3 2 1

Printed in the United States of America
on acid-free paper

To my parents, Chaja and Arie,
who helped my hobby to grow into a vocation

Preface

I have written this book because I am convinced that ciliates have a great deal to tell developmental biologists concerning mechanisms of pattern formation. This is the second book that is wholly devoted to ciliate pattern formation and is addressed to nonciliatologist readers. The first was André Lwoff's *Problems of Morphogenesis in Ciliates,* subtitled *The Kinetosomes in Development, Reproduction, and Evolution* (New York, John Wiley and Sons). Lwoff, in 1950, used descriptive studies of life cycles of ciliates that he and others had carried out during the previous three decades to draw far reaching conclusions concerning the genetic autonomy of basal bodies and the control of their development by differences in the underlying cytoplasm. These ideas had implications for both heredity and development that went far beyond the specialized studies on which they were based.

Lwoff wrote his book at the dawn of the era of cellular ultrastructure, at a time when serious experimental analysis of ciliate development was just beginning. The ultrastructural and molecular knowledge gained in the four decades since Lwoff wrote his adventurous monograph has rendered many of Lwoff's ideas concerning the autonomy of basal bodies obsolete, but the experimental studies on ciliates have strengthened the less spectacular but equally important half of Lwoff's argument, namely that differentiation of visible structures is controlled by as yet invisible spatial information in the underlying cortical cytoplasm.

During this same period, there has been a renascence of experimental embryology, much of it associated with and inspired by the concept of "positional information." Specific models, both formal (such as the polar coordinate model) and more mechanistic (especially reaction–diffusion theories) have followed hard upon the proposal of the general concept of positional information, and a great deal of effort has gone into the testing of these models.

I believe that ciliates have a special contribution to make to the new models of positional information, because they simultaneously extend and challenge these models. On the one hand, the ciliate studies support the critical idea of positional nonequivalence and invite the formulation of specific models that parallel some that have been proposed for multicellular organisms. On the other hand, these same ciliate studies challenge certain ways of thinking about positional information in multicellular organisms, in particular those views that stress cellular interactions and local genomic activation as *exclusive* means for generating and maintaining complex and extended spatial patterns.

This book summarizes the ciliate studies that are of greatest relevance to general issues of pattern formation and evaluates the contribution of these ciliate studies to our understanding of pattern formation. Its principal ideas are: first, that there exists in ciliates an intracellular hierarchy of qualitatively different systems of spatial control nested within one another; second, that these systems of spatial control can be inherited cytoplasmically; and third, that the most global level in this intra-cellular hierarchy is analogous and possibly homologous to positional information in developing embryos. The book includes a model for intracellular positional information in ciliates that is formally quite similar to some of the multicellular models, but calls for some rather different mechanisms.

This book has three broad sections: The first section (Chapters 1 to 3) is a three-part introduction, with Chapter 1 presenting the principal conceptual issues, Chapter 2 describing the relevant organizational features shared by all members of the ciliate cast to be considered in this book, and Chapter 3 introducing the individual members of the cast. The middle section (Chapters 4 to 7) describes the broad results of research on the inheritance and regulation of structural patterns in the surface region of ciliates, using as its organizing principle the major axes of ciliate clonal geometry that had been introduced and described in Chapter 2. The final third of the book (Chapters 8 to 11) concentrates on mirror-image reversals of large-scale structural arrangements, where (in my view) the existence and nature of the ciliate organizational hierarchy is best revealed. My ideas concerning the conceptual link between ciliates and embryos are made explicit in the final chapter, which is largely devoted to the *Drosophila* embryo as seen from one ciliatologist's perspective.

I should comment on the metaphor of a ciliate "cast." The analysis of pattern formation in ciliates has been carried out on several major experimental organisms, with a range of regulative capacities as great as that encountered in embryos of several different animal phyla. Selection of a single paradigmatic ciliate as a vehicle for one's ideas, therefore, might yield parochial and biased conclusions, but the opposite choice, of covering all of the important ciliates one after another, would yield something even less desirable, a ciliate equivalent of a textbook of compara-tive embryology. To solve this dilemma, I have imitated the strategy used by David Nanney in his clear and concise introduction to ciliate genetics and development, *Experimental Ciliatology* (New York, Wiley-Interscience, 1980). Nanney first introduced a cast of experimental organisms ("The Chosen Few") and then orga-nized the remainder of his book entirely by conceptual categories, drawing from members of the previously introduced cast to provide the information most perti-nent to each of these sets of scientific questions. While my subject matter is largely different from Nanney's, my organizational strategy is the same.

This book is too long to be read through at a single sitting by any but a very fast, and insomniac, reader. Some readers will seek to absorb the gist of the book without reading all of it. I have endeavored to write this book so as to facilitate two strategies of "quick reading." The fastest strategy is to read the opening sections of Chapter 11, plus the conclusions provided at the end of all but the descriptive chap-ters (2 and 3); these sections are written to be intelligible when read by themselves. A different strategy for "quick reading" is to peruse the table of contents or index and pick out the subjects of greatest interest. I have attempted to mitigate the sever-est disadvantage of entering a book in the middle, namely the lack of familiarity with what came before, by extensive cross-referencing within the text. Thus, it should be possible for a reader who does not wish to proceed from start to finish instead to alight at some internal section, and proceed forward and backward from there to obtain the desired information. I nonetheless hope that this book will not exhaust the patience of those readers who do read it through from beginning to end.

Iowa City, Ia   J.F.
December, 1988

Acknowledgments

Many colleagues helped me in the writing of this book. The entire manuscript was read by Drs. Krystina Golinska, Maria Jerka-Dziadosz, Janina Kaczanowska, Marlo Nelsen, Stephen Ng, Vance Tartar, and Norman Williams, as well as by Mr. Eric Cole, and all of these readers have provided valuable comments. In addition, I wish to acknowledge the help of Dr. Janine Beisson for Chapters 1 through 5, of Drs. Jay Mittenthal and Mark Sturtevant for Chapter 1 as well as Chapters 8 through 11, of Dr. Denis Lynn for Chapter 2, of Dr. James Berger for Chapter 5, of Drs. Gary Grimes, Xinbai Shi, and Mikio Suhama for Chapters 8 through 10, of Dr. Michel Tuffrau for Chapter 9, and of Drs. Brian Goodwin and Ms. Wendy Brandts for Chapter 11.

In addition to these helpful colleagues, I must specially acknowledge one institution and two persons without whom this book could not have been written. The institution is the U.S. National Institutes of Health, which I hope will not be offended to learn that I have employed much of two summers' worth of time supported by the salary of a research grant (HD-08485) in the writing of this book. One of the two persons is my illustrator, Jane Otto, who always understood what I wanted in illustrations and who was simultaneously highly skillful and almost unbelievably prompt in their execution. The other person is my wife, Anne Frankel, who went over drafts of each chapter and informed me bluntly of what had to be changed. She also meticulously checked the bibliography, going over every word of virtually every reference, and has provided valuable help with the proofs and index. If this book is well written and well referenced, Anne deserves a major share of the credit.

Intellectual obligations are harder to acknowledge, since it is often difficult to pin down the precise sources of one's thinking. My initial inspiration was André Lwoff's book, *Problems of Morphogenesis in Ciliates,* which I have already mentioned in the preface. I read this book in 1956, while a college senior doing a term paper in Howard Schneiderman's cell physiology course; Lwoff's views on the kinetosome (basal body) got me very excited: here was a "plasmagene" that was also an important cell organelle! The next year, as a first year graduate student at Yale University, I heard Earl Hanson give an informal talk on his studies on regeneration of locally irradiated oral structures of *Paramecium,* and his enthusiasm for this system and for the problems of intracellular development brought me into his lab as his first Ph.D. student.

There are three other major sets of intellectual influences that deserve specific acknowledgment. The first of these can be put in a dialectical form: the founder and long-time leader of ciliate genetics, the late Tracy Sonneborn, championed the thesis of cytoplasmic inheritance of cell surface patterns, while my former collaborator Klaus Heckmann, in whose laboratory I also first learned how to do ciliate genetics, insisted on the antithesis of long-term nuclear genic control. In Chapters 4, 9, and 10 of this book, I offer an attempt at a new synthesis. The second major influence was provided by those experimental ciliatologists who demonstrated the importance of large-scale relational features in ciliate cortical organization. These ideas are associated with several important names, but three are of particular significance for me: Vance Tartar, who in the mid-1950s first demonstrated that the site of cortical development in *Stentor* depended not on a specific preexisting structure but rather on a set of spatial relations; Gotram Uhlig, who was the first to apply the theoretical categories of experimental embryology to these relations; and Maria Jerka-Dziadosz, who personally introduced me to analogous relational features in ciliates very different from *Stentor*. The third influence is that of my long-time associate, Marlo Nelsen, who has always had the remarkable faculty of treasuring rather than disregarding the odd exceptions. Much of the substance and viewpoint of Chapters 8 to 10 derive from Marlo's dogged iconoclasm.

My final thanks must be given to the donor of the most valuable present that I have received. One of my father's many friends, whose name I have long since forgotten, obtained an antique but very serviceable microscope in exchange for a carton of cigarettes while traveling through destitute postwar Germany. On returning to the United States, he gave the microscope to his friend's son, who will be forever grateful.

Contents

PATTERN FORMATION

1

Spatial Order: an Analytical Framework

1.1 INTRODUCTION

During the past 20 years, there has been a major revival of interest in the generation of spatial order during development. Although the most recent manifestation of this revival is the intense ongoing work on the times and places of appearance of gene products during early development, the revival has at least two roots that lie outside of molecular biology. One is the publication and subsequent broad dissemination of the concept of positional information (Wolpert, 1969); a second is the formulation of sophisticated kinetic models of pattern formation (e.g., Gierer and Meinhardt, 1972) derived originally from Turing's demonstration of how spatial order can arise from random disorder (Turing, 1952).

As integrative ideas are being developed and refined alongside the new molecular discoveries, an old question has reemerged in various new guises (e.g., see Stent, 1985; Lawrence, 1985): are concepts that might be valid for one developing system valid for others as well? Do universal or at least broadly applicable mechanisms govern development as they are known to govern metabolism and heredity? If we believe that this question might possibly be answered in the affirmative, then we should delve as deeply as possible into mechanisms for the creation and regulation of spatial patterns in organisms selected from as wide an assortment of taxonomic groups as possible. The theoretical importance of any meaningful similarities that one detects is in direct proportion to the degree of structural difference and depth of phylogenetic separation among the organisms being compared.

This search for a unity of interpretation of pattern formation during development provides the motivating spirit for this book. Its strategy is to describe observations and experiments on pattern formation in a group of complex unicellular organisms within a framework of ideas that also apply to embryonic and postembryonic development of multicellular organisms. As pointed out by Grimstone, "Protozoa are comparable, in different contexts, with both metazoan cells and whole metazoan organisms" (Grimstone, 1961, p. 133). Among protozoa, ciliates have the special virtue of being both organizationally complex and well studied from both cellular and organismic perspectives. Although the general rule in many fields of biological research is to search for maximum simplicity, if one is considering the origin and the maintenance of spatial order, then a certain level of complexity of organization may be an asset rather than a liability. The ciliates supply this complexity in abundant measure, and therefore, allow us to ask to what degree the *intracellular* mechanisms of protozoan development resemble the mixture of intracellular and intercellular mechanisms observed in the development of multicellular organisms.

Although the goal of comparing pattern formation in ciliates with that in whole multicellular organisms requires us to examine unicellular organisms in an organismic context, if we wish to consider the actual mechanisms of development in something more than a highly schematic and abstract manner, we must also view ciliates simultaneously as cells. The dual cell/organismic focus of this book creates a practical need for a bipartite introduction: a first chapter to provide an analytical framework at the organismic level and a second to give a description of ciliates at the cellular level. Subsequent chapters concentrate primarily on ciliate observations and models, but in the final chapter we return to a multicellular embryonic system to find out whether ciliate observations and models do indeed help us achieve a more unified understanding of pattern formation.

1.2 DIVERSIFICATION AND DIFFERENTIATION

Development occurs in both space and time. This is illustrated schematically in Figure 1.1. Imagine a linear structure with four regions or units arrayed in a row. The units originally (at time 1) all appear alike. Later (at time 2) they all have become different, both from their initial condition and from each other. The process by which the units change through time as they acquire their final structural and functional states is generally termed *differentiation*. The units, however, do not differentiate at random but rather in a definite and predictable spatial order; the process that creates this spatial order has been given the apt name *diversification* by Stumpf (1967). Now it more frequently goes under the name pattern formation (Wolpert, 1971).

Diversification and differentiation are at least to some degree different processes. When we concentrate on the time line, we tend to investigate how macromolecules necessary for the expression of final specialized cell states are synthesized and then assembled into the structures that characterize these respective states. When we focus on the spatial dimension, we will still be interested in problems of gene expression and structural assembly, but we will also have to consider how such processes in one unit are influenced by many and possibly all of the other units. This question was most succinctly put by a 6-year-old boy quoted by Hörstadius (1973, p. 1): "But, daddy, how do the cells know which of them are going to form these organs?" The crucial question is how communication among developing units takes place to achieve the final coordinated result.

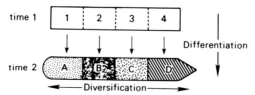

Figure 1.1   A schematic diagram illustrating the distinction between differentiation and diversification in an imaginary developing system consisting of four regions. The regions are numbered 1 to 4 and different structures are labeled A to D. In this figure and the next five, time advances from top to bottom.

1.3 REGULATION

1.3.1 Positive and Negative Regulation

The logic of communication among developing units has been analyzed in great detail by the isolation and transplantation of parts, beginning with the famous experiments of Roux, Driesch, and Boveri, and continuing through the present. The categories of results derived from such manipulative experiments can be illustrated by imagining that we have transected our linear system (of Fig. 1.1) into two equal parts. We can classify all responses into three basic categories, shown in a simplified manner in Figure 1.2. The simplest possibility (Fig. 1.2a) is that the separated portions develop into exactly the *same* structures that they would have formed had they not been separated. This outcome has usually been called mosaic development. The alternative to mosaic development is known as regulation. There are two basic kinds of regulation (Sander, 1971). Either one or both of the separated portions develop into *more* structures than they would normally have formed (positive regulation, Fig. 1.2b) or one or both of the separated portions develop into *fewer* structures than they would have formed had the system remained intact (negative regulation, Fig. 1.2c).

It is often observed that the degree and nature of regulation depends on the time as well as the place at which the operation is carried out. Thus, if operated at a sufficiently late time, virtually all developing systems will show a mosaic outcome

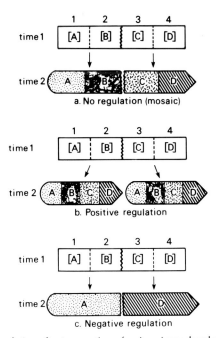

Figure 1.2   Types of regulation after transection of an imaginary developing system into two equal parts. The structures that would have been formed if the system had been left undisturbed are indicated in brackets. (**a**) no regulation (mosaic); (**b**) positive regulation; (**c**) negative regulation.

on isolation (or transplantation) of parts. Conversely, even in the most extreme "mosaic" systems, such as ascidian embryos, regulation can be demonstrated if parts are isolated early enough; in the ascidian case, complete positive regulation takes place after transection of unfertilized eggs (Reverberi and Ortolani, 1962), whereas limited negative regulation is observed after certain sets of cells are isolated at cleavage stages (Reverberi, 1961; Meedel et al., 1987). It is not surprising that all developing systems are regulative at some stage, as attainment of order implies communication, and the existence of communication implies that changes should occur after some of the communicating parts are removed.

The actual details and conditions of both positive and negative regulation in embryos are manifold, complex, and sometimes controversial. Situations that conform precisely to the simple models of positive and negative regulation, shown in Figure 1.2, in which both halves regulate symmetrically and both become reorganized globally, are relatively rare. The best examples are found in insects, in which longitudinal constrictions separating dorsal and ventral (or right and left) halves result in the formation of two normal embryos as in Figure 1.2b (Sander, 1971), whereas latitudinal constrictions separating anterior and posterior halves result in partial embryos that lack the middle segments, as in Figure 1.2c (Sander, 1976; Schubiger and Wood, 1977). Negative regulation after constriction of early *Drosophila melanogaster* embryos involves genuine respecification, not just destruction of the regions near the site of constriction (Schubiger et al., 1977).

1.3.2 Upgrading and Organizers

In most cases of embryonic development, regulation, when it occurs, tends to be asymmetrical in the sense that one-half of the system regulates (either positively or negatively), whereas the other does not. Although a review of all of these cases is beyond the scope of this book, two examples are particularly useful for thinking about how regulation might occur. The first is regulation by sequential "upgrading," illustrated schematically in Figure 1.3. Removal of unit 1, which normally forms structure A, stimulates specification (or respecification) of unit 2 to form structure A, of unit 3 to form structure B, and so forth. Such upgrading is unidirectional and appears to characterize the "equivalence groups" of cell lineages in the nematode *Caenorhabdites elegans* (Sulston and White, 1980). For example, in the extensively studied vulval equivalence group of *Caenorhabdites* hermaphrodites, if precursor cell 1 is eliminated by laser irradiation, the progeny of cell 2 can regulate to form the lineage that is normally produced by cell 1 and the progeny of cell 3 can regulate to form the lineage normally produced by the progeny of cell 2;

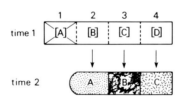

Figure 1.3 Positive regulation by upgrading. The imaginary developing system is the same as in Figures 1.1 and 1.2.

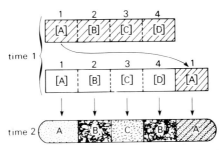

Figure 1.4 Positive regulation by induction. The source of the graft is hatched. Conventions are otherwise the same as in Figure 1.2.

if cell 2 is destroyed, the progeny of cell 1 does not regulate, whereas the progeny of cell 3 regulates to form the lineage normally produced by the progeny of cell 2; if cell 3 is destroyed, there is no regulation. (In this particular system there are only three distinguishable cell fates; the cell corresponding to unit 4 of Figure 1.3 normally forms the same structure (C) as does unit 3, so that no structure is missing after regulation is completed.) We will use this example later to evaluate the two ways of thinking about embryonic regulation that are outlined in the next section of this chapter.

In the second example, one portion of an embryo has been shown to play a leading role in stimulating specification of others. Schematically (Fig. 1.4), when unit 1 is transplanted to the opposite end of the system, it still develops into structure A, but unit 4, which normally develops into structure D, is upgraded as a consequence of the interaction with unit 1 to produce the same structure that normally is formed adjacent to A, i.e., B. The classic example of this is inductive interaction within the mesodermal mantle of amphibians (Spemann and Mangold, 1924, tr.1964). This region diversifies into notochord, somites, pronephros, and lateral mesoderm, arranged in a rough anterior/dorsal to posterior/ventral sequence. As Spemann and Mangold originally showed, this sequence can be altered by grafting the most dorsal region of an early amphibian gastrula, presumptive notochord, into the most ventral area, presumptive lateral mesoderm. Both older (Yamada, 1940) and recent (Slack and Forman, 1980; Smith and Slack, 1983; Dale and Slack, 1987) studies clearly confirmed the results of Spemann and Mangold, namely that the dorsally derived "organizer" develops as it normally would, into notochord, whereas the adjacent ventral mesodermal region that would otherwise have formed lateral mesoderm (and later blood islands) instead develops into somites (and later back muscle). Like the upgrading sequence described earlier, the inductive reorganization diagrammed in Figure 1.4 is also transitive: A ventral-to-dorsal graft (in terms of the diagram, transplanting the unit at position 4 to a location adjacent to position 1) has no effect on the adjacent host tissue (Smith and Slack, 1983).

What would happen if unit 1, the presumptive organizer, were removed from the system? One would imagine that the remainder might be able to reorganize itself by a sequential upgrading as shown in Figure 1.3. This indeed is the uniform result of classical experiments in which all or part of the organizer region was removed in amphibian blastulae or early gastrulae (Holtfreter and Hamburger, 1955, pp. 249–250; Gerhart, 1980, pp. 229–234, 256–270). Recent detailed studies on the frog, *Xenopus laevis,* however, show that this does not happen when similar operations are conducted on cleavage-stage embryos; instead the response to removal of the presumptive organizer region ranges from no regulation to negative regulation in the remainder of the mesodermal mantle (Cooke and Webber, 1985; Yamana and Kageura, 1987; Dale and Slack, 1987). The degree to which these differences are due to experimental object (salamander vs. frog), stage of development when operated (blastula/gastrula vs cleavage), and precision of assessment of experimental results (tissue vs. cell level) has not yet been sorted out fully.

In some other systems, regions that show the properties of classical organizers can indisputably be reconstituted perfectly after removal; the first and best example is that of *Hydra,* where grafting a region near the mouth into the body column converts the adjacent body column tissue into a new mouth and tentacle ring (Browne, 1909), yet this organizer region can be regenerated perfectly after the removal of the entire oral region (Webster, 1971).

1.4 SEQUENTIAL VERSUS THRESHOLD MODELS

1.4.1 The Sequential Model

Historically the first, and still influential, interpretation of positional regulation was the metabolic dominance model of C. M. Child (1915). In Child's view, organisms, in general, and developing systems, in particular, express gradients of metabolic rate. The region with the highest metabolic rate becomes a kind of developmental pacemaker, transmitting its influence (although with decrement) to the rest of the system. The metabolic "high point" maintains its metabolic advantage, and by virtue of this advantage it develops into an apical region, such as the head of a flatworm or the mouth plus tentacle ring of a hydroid. Other parts of the system are then unable to develop into the apical structure because the existing metabolic high point prevents formation of a new high point within the domain of its influence. Child's theory was thus one of sequential dominance and subordination, based ultimately on pervasive spatial differentials of metabolic rate.

Child's theory was systematically applied to embryonic development as well as to regeneration by Huxley and DeBeer (1934) under the rubric of *gradient-fields.* The gradient-field concept is a conceptual graft of Child's metabolic gradient concept onto the idea of developmental fields. A *field* was (and usually still is) defined in developmental biology as the territory in which development is subjected to a common set of coordinating influences; sometimes the term is applied to the coordinating influences themselves, in analogy to the field concept in physics. In Huxley and DeBeer's gradient-fields, the apical region both inhibits the formation of additional apical regions and induces adjacent tissues to develop into subordinate organs. This combination of inhibition and induction can then continue in a sequential manner to bring about the formation of a linear series of different structures. This idea is illustrated schematically in Figure 1.5, in which the zone that is metabolically most active (*) gives rise to structure A, the dominant apical region. This region inhibits the formation of more A elsewhere but induces the formation of structure B; B then inhibits the formation of more B elsewhere but induces the

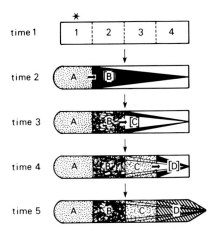

Figure 1.5 The sequential model for diversification within an imaginary developing system. Five stages are shown, with initial establishment of the metabolically most active zone [*] at time 1, specification of structure A at time 2, of structure B at time 3, of structure C at time 4, and of structure D at time 5. At each stage the next structure in the sequence is induced by the structure already present (*heavy horizontal arrows*) and a gradient of inhibition is established that prevents the structure that already is present from developing anywhere else (the inhibition established at each stage is shown as a black horizontal triangle).

formation of structure C, and so forth. In this model, positive regulation would occur after removal of the dominant region (A) because the formerly subdominant region (B) is released from inhibition and thereby can become dominant (i.e., develop as A), much as the subdominant member of a troop of baboons can become dominant after the death or removal of the dominant individual.

Development under the influence of an organizer can readily be explained by this scheme, as the transplanted dominant region would influence the neighboring territory to form the first of the transitive chain of its subordinates. This model, however, would also predict that this organizer should readily be reconstituted from the remainder of the system after it was removed. Although this prediction fits well with findings on *Hydra* and other invertebrates (Huxley and DeBeer, 1934), it does not always hold true for embryos (Section 1.3.2).

1.4.2 The Threshold Model; Positional Information

An alternative way of understanding coordination within developmental fields is to assume that there is no direct interaction of regions, but rather that the type of development at each point depends on the level of *morphogenetic potential.* A model of this type was formulated first by Dalcq and Pasteels (see Dalcq, 1938, chapter VII), and applied by Yamada (1940) to the specific case of differentiation within the mesodermal mantle of amphibians. The fundamental idea behind such a model is represented in Figure 1.6. Morphogenetic potential is smoothly graded throughout the system, and the specific structures formed in each region are determined by particular ranges of morphogenetic potential, with sharp qualitative transitions occurring at certain thresholds. Yamada's version of this idea can be rep-

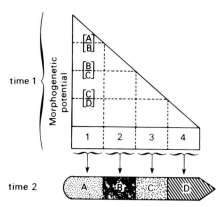

Figure 1.6 The threshold model for diversification. The level of morphogenetic potential is indicated on the vertical axis, distance along the developing system on the horizontal axis. The *dashed lines* indicate the threshold levels of morphogenetic potential (A/B, B/C, and C/D) that delimit the regions. Diversification occurs in a single step, at time 1, specifying the locations at which structures A, B, C, and D differentiate at time 2.

resented precisely in Figure 1.6, if we were to name A, B, C, and D as notochord, back muscle, pronephros, and blood islands, respectively. I have called this idea a *threshold model* because what is distinctive about it is the postulate that diversification depends on a reading of the level of a gradient in relation to critical thresholds. Although gradients may be said to be central features of both the Child–Huxley–DeBeer class of models and of the Dalcq–Pasteels–Yamada class, the conceptual uses to which they are put in the two types of model are entirely different.

Stumpf (1968) took the basic idea of the threshold models one critical step further when she demonstrated that transplanted parts of structurally different segments of the moth *Galleria* respond developmentally to particular positions within a different segment in much the same way that they respond to the corresponding positions within their own segment. Therefore, Stumpf suggested that different segments might have identical gradations of some type of morphogenetic potential. This idea was promptly generalized by Wolpert (1969), and morphogenetic potential was rechristened *positional information*.

The central idea of Wolpert's formulation was that positional information specifies *location only,* and therefore, is neutral with regard to what specific structures are formed at each location. Positional information thus is analogous to the coordinate system of a map, which allows one to find places on the map but does not by itself specify those places. The actual formation of a structure, therefore, would be determined by a separate step, called the *interpretation* of the positional information. Interpretation would correspond to the reading of the map, except that in a developing organism the specific contents of the map would be created as the coordinates were read. The actual structures formed during the process of interpretation would depend on a predetermined bias, or competence, of the territory within which a system of positional information was operative. For Wolpert, who applied this concept only to multicellular organisms, the competence would be controlled by the set of active (or potentially active) genes in the cells within a developmental field. This competence would be set by earlier interpretative events that had taken place within a more inclusive developmental field. In the cartographic analogy, the mapmaker might first set up a coordinate system over the entire globe and read it using a global index to determine that one particular region was to be France, another Italy, and so on; at a subsequent step the same coordinate grid would be applied on a finer scale to each country separately, so that the universal coordinate system would be read using a French index within the French area to fill in the details of the map of France, using an Italian index within the Italian area to make a map of Italy, and so forth. If we were dealing with an insect embryo rather than a map, the primary whole-embryo field might create separate segmental fields and simultaneously determine the "settings" of the genes of cells within these fields, presumably by creating regionally varying states of somatically inheritable genic receptivity or unreceptivity to future positional stimuli. Such stable nuclear differences would thus create unique sets of diverse potential responses (the indexes) to the reiterated identical sets of positional values that would become operative within every segment after the primary positional system had completed its work.

1.4.3 Positional Signals

Are the two classes of model depicted in Figures 1.5 and 1.6 experimentally distinguishable? In at least some cases, they are. A striking example where they have in fact been distinguished is provided by the vulval equivalence group in *Caenorhabditis*. The transitive upgrading revealed by the laser ablation studies described in Section 1.3.2 seems almost to beg for an interpretation in terms of the sequential model. A recent incisive experimental analysis by Sternberg and Horvitz (1986), however, shows that such an interpretation is insufficient. The vulval system involves three types of cell lineages arising from three linearly ordered precursor cells; in terms of our schematic diagram (see Fig. 1.1), lineage types A, B, and C arise from precursor cells at positions 1, 2, and 3. The dominance–subordination model predicts that if the units at positions 2 or 3 were physically isolated, they would be released from inhibition by units higher in the sequence and, therefore, would develop into lineage type A. What Sternberg and Horvitz (1986) instead found was that any isolated precursor cell belonging to the vulval equivalence group could develop into any of the three possible lineage types, A, B, or C, depending on its distance from the source of an external signal (which is a specific cell of the worm that belongs to a totally different lineage group). If located close to this source, the isolated precursor cell formed lineage A, if situated a little further away, it formed lineage B, and if it was far away from the source it formed lineage C. These observations indicate that the upgrading of the precursor cell at position 2 of the intact system after laser ablation of the precursor cell at position 1 does not require a relief from an inhibition normally exerted by the precursor cell at position 1. The experiments on the vulval equivalence group are thus most simply explained by the type of model shown in Figure 1.6, with the morphogenetic potential in this case being the concentration of a positional signal arising from a source external to the equivalence group itself. Although later observations showed that inhibitory interactions among the vulval precursor cells can occur (Sternberg,

1988), these may be viewed as supplementary to a primary mechanism based on interpretation of a graded positional signal.

The results described previously suggest that the positional signal (or *morphogen,* as it has often been called) in this system is a diffusible chemical substance, or set of substances. This idea has frequently been encountered (Slack 1983, Chapter 9) since it was proposed by Stumpf (1966) and Lawrence (1966) to account for diversification within insect segments; it figured prominently in Wolpert's expositions of the idea of positional information (1969, 1971). More recently, sophisticated reaction–diffusion models of the type originally proposed by Turing (1952) have been constructed by Gierer and Meinhardt (see Meinhardt, 1982). Early versions of these models were cast in terms of structure-specific dominance and inhibition, with the more slowly diffusing component of a Turing system being termed an *activator* and the more rapidly diffusing component an *inhibitor* (specifically for polar regeneration in *Hydra:* Gierer and Meinhardt, 1972). In more recent versions (e.g., Meinhardt, 1982, Chapter 8), the more rapidly diffusing component is postulated to provide the positional information that controls the diversification of the units within the system. Although this component is still called an inhibitor, in its developmental effect it is no longer an inhibitor but rather a positional signal. This is a good example of how the same formal model can have different meanings depending on the way in which the model is applied to a biological system.

When applied to the complexities of early embryonic development, diffusion-based models probably are inadequate. The difficulties of such models have been explored most thoroughly in an extensive series of isolation, transplantation, and centrifugation experiments on the frog *Xenopus* by Cooke (1981, 1985, 1987). Depending on the nature of the operation and the stage at which it was conducted, the array of structures formed within the mesoderm was either greater (Cooke, 1981) or smaller (Cooke and Webber, 1985; Cooke, 1987) than predicted by simple models based on a diffusible positional signal. Because of the lack of expected regulation in manipulated early embryos, Cooke (1985) has suggested that the positional organization of the cleavage-stage embryo "... displays durability and stability more reminiscent of a structural record than of a dynamic diffusion/metabolism maintained signalling or memory system" (Cooke, 1985, p. 79). In the extreme, this conclusion might suggest "morphogenesis without morphogens" (Goodwin, 1985). Alternatively, molecules acting as morphogens might exist, but their distribution within the embryo would be subject to major structural constraints. There is increasing evidence that such constraints, probably involving the cytoskeleton, strongly influence early patterning of the *Drosophila* embryo (Miller et al., 1985; Edgar et al., 1987; Fröhnhofer and Nüsslein-Volhard, 1987).

1.5 POSITIONAL MEMORY AND INTERCALATION

The clearest evidence for a positional organization that is *not* associated with a diffusible signal is found in the *positional memory* (Stocum, 1984) that has repeatedly been demonstrated in postembryonic systems within three very different phyla (Platyhelminths, Arthropods, and Chordates). Here one can still speak of an ordered and continuous spatial gradation of positional values, but *not* of a gradient of a diffusible substance.

Perhaps the best known demonstration of such a structural gradation was provided by the intercalation of intervening regions following grafts that created discontinuities in the proximodistal axis of segments of cockroach appendages (Bohn, 1970; Bulliére, 1971). The gradation thus demonstrated met the operational test for positional information, as grafts between disparate parts of different segments (e.g., distal femur grafted onto proximal tibia) provoked intercalary growth in the same way as did grafts between corresponding parts of the same segment (e.g., distal tibia grafted onto proximal tibia) (Bohn, 1970; Bulliére, 1971; Bart, 1988). In this case, the interactions appeared to take place at or very near the cut surfaces, and the response was localized growth originating at or near the juxtaposed surfaces, usually with both surfaces participating by providing cells for the regenerate as well as positional cues to guide what was regenerated (Bohn, 1971, 1976; Bulliére, 1971; Bart, 1988). The demonstration that similar intercalation can occur within insect segments (Nübler-Jung, 1977, 1979; Wright and Lawrence, 1981), earlier believed to regulate via a gradient of a diffusible substance (Stumpf, 1966; Lawrence, 1966), strengthens the view that regulation within postembryonic fields involves structural interactions, possibly between cell surfaces, rather than freely diffusing morphogen(s).

For our purposes, however, the flatworm *Dugesia* (formerly *Planaria*) provides the most instructive example, because it allows us to contrast the two viewpoints toward pattern regulation in a structured system that is similar in some ways to what we will encounter in ciliates. *Dugesia* is an obviously polar organism, endowed with distinctive structures in its anterior (brain and eyespots), middle (pharynx), and posterior (copulatory opening) regions (Fig. 1.7a). Figure 1.7a, modified from Wolff (1962), illustrates the sequential conception of the mechanisms coordinating structural diversification in the flatworm, complete with specific regions (given Roman numerals) that exert level-specific inductions (arrows in Fig. 1.7a, left) and inhibitions (shaded vertical triangles in Fig. 1.7a, right).

The sequential model has recently come under attack on two separate fronts. First, Chandebois (1984) performed numerous grafting operations, the results of which are inconsistent with this model. Two examples are described here. When an anterior head (anterior region I in Fig. 1.7a) was grafted on a tail (region V), the sequential model predicts that the tail piece would be remodeled under the influence of the dominant head to produce progressively more posterior regions (II, III, then IV). Chandebois documented the opposite (Fig. 1.7b): the head was remodeled under the influence of the tail, with the original eyes (E) becoming displaced anteriorly, presumably as a consequence of localized growth anterior to the graft border. Then the original eyes regressed (rE), and new eyes (nE) were formed near the new anterior end. A new pharynx (nP) developed later in the vicinity of the graft border. With slight differences in the level of the original head–tail juxtaposition, however, the new pharynx could be made to develop either anterior or posterior to the graft border. In another experiment (not shown), an entire head was grafted onto the anterior extremity of another head, so that two pairs of eyes were closely juxtaposed. Here the sequential model would predict competition between two domi-

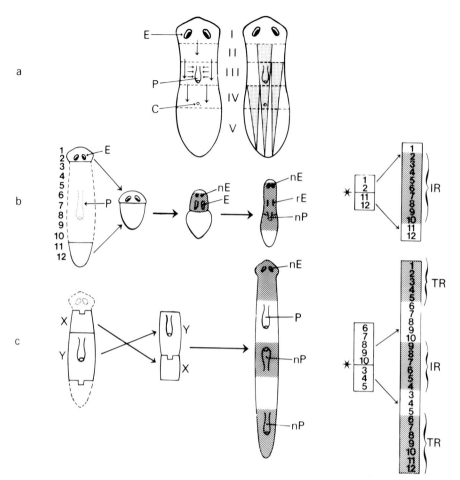

Figure 1.7 Diagrams of concepts and observations concerning diversification in flatworms. (a) Diagrams illustrating *Dugesia*, with the sequential model of E. Wolff superimposed. The labeled structures are the eyespot [E], pharynx [P], and copulatory opening [C]. Roman numerals indicate regions along the anteroposterior axis. The diagram on the left shows sequential inductions (*arrows*) and the diagram on the right shows specific inhibitions (*shaded triangles*). (b) and (c) Sequences of regeneration after transplantations, with the operation shown at the left of each sequence, the outcome in the center (with *shading* indicating the regions of growth or remodeling), and schematic interpretation in terms of 12 anteroposterior positional values on the right (with eyespots at level 2, pharynx at levels 6 to 8, * indicating a discontinuity of positional values, IR an intercalary regenerate, and TR a terminal regenerate). (b) A graft of a head on a tail. The old eyes [E], after being shifted forward, degenerated to a remnant eye [rE], whereas new eyes [nE] formed anterior to this remnant. A new pharynx [nP] developed later, near the junction between the grafted parts. (c) A graft of a piece including the pharynx [Y] onto a prepharyngeal piece [X]. A head regenerated at the anterior end of piece Y, and a posterior half-worm, including a pharynx, regenerated at the posterior end of piece X. At the junction between X and Y, an intercalary regenerate formed bearing an *inverted* pharynx. Diagram (a) is redrawn with modifications from Figure 3-17 of Wolff (1962). The left portion of diagram (b) is redrawn from Figure 1.C.1 of Chandebois (1984) and the left portion of diagram (c) from Figures 105–108 of Okada and Sugino (1937). The explanatory diagram on the right side of (c) is redrawn from Figure 41 of Chandebois (1976), and the scheme on the right side of (b) is original but follows the principles of Figure 17 of Chandebois (1984).

nant centers, with the "loser" perhaps degenerating. Chandebois reported that in this case both original pairs of eyes persisted and a third new pair of eyes was formed in a zone of intercalary growth between the two old ones.

Somewhat more recently, Saló and Baguñá (1985) addressed this problem from a different perspective. They used a chromosomal marker to investigate the cellular origin of the intercalary regenerate in a grafting experiment whose end result was similar to that shown in Figure 1.7b (although carried out by a different operative procedure). The sequential model predicts that the regenerate should arise entirely from the posterior component, under the inductive control of the anterior component. Saló and Baguñá reported that normally the material of the intercalary regenerate came equally from both components. If either component was X-irradiated, however, then the regenerate was derived entirely from the other, unirradiated, component. The anatomy of the regenerate was the same in all cases. Thus, the regenerated region can come from either component, or from both. This is very similar to what Bohn (1971, 1976) and Bart (1988) had painstakingly documented in logically similar experiments on the appendages of cockroaches and stick-insects, respectively.

Chandebois' interpretation of her results (Chandebois, 1976, 1984) is operationally similar to that derived from the insect transplantation studies. It includes only two fundamental postulates: that different anteroposterior levels of the worm express different positional values, and that a discontinuity of positional values is necessary and sufficient to stimulate intercalary regeneration to restore the intervening values. To illustrate this interpretation convincingly, we will need to describe one more flatworm experiment, carried out long ago by Okada and Sugino (1937). Here a pharyngeal piece (Fig. 1.7c, Y) was grafted onto the anterior end of a prepharyngeal piece (Fig. 1.7c, X), with no change in polarity. The result of the operation was regeneration of a new head and tail at the two respective ends of the graft complex and, most important for us, the regeneration of a *reversed* pharynx *between* the two components of the graft. This reversal was detected not only in the anatomy of the intercalated pharynx, but also in the reversed direction of the beat of the body cilia in the region that includes the reversed pharynx (Sugino, 1941). This reversed intercalary regenerate is immediately predictable from Chandebois' two basic postulates (Fig. 1.7c, right).

The results of the combination of the extreme ends of the worm shown in Figure 1.7b can also be accounted for by the intercalation model, if one assumes that some of the existing tissue (the eye region in this case) was remodeled as intercalation proceeded (Fig. 1.7b, right) (Chandebois, 1984). It is only in this remodeling aspect that Chandebois' recent results differ from the intercalation observed in insect systems. In the latter systems, localized growth prevailed when simple linear intercalation occurred (Anderson and French, 1985), although under conditions where more extensive growth was called for, some respecification of existing material might have taken place (Truby, 1986).

The ideas of structurally bound positional value and of intercalation also provide the basis for a more elaborate formal model, the *polar coordinate model,* which both in its original versions (French et al., 1976), and in various more or less radically modified forms (Bryant et al., 1981; Lewis, 1981; Mittenthal, 1981) can account for a variety of phenomena in addition to the linear intercalation described

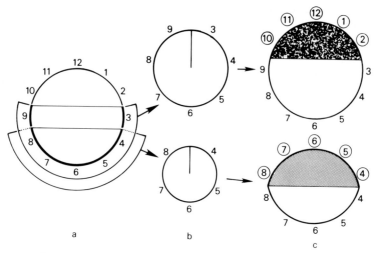

Figure 1.8 A diagram of shortest distance intercalation around a circular structure, as postulated by French et al., 1976. The structure is assumed to be a disc or a cylinder. Numbers indicate arbitrary positional values around the circumference of the disc, with 12 adjacent to 1 in the same way that it is on a clock face. (**a**) A portion of the disc is removed, with the remaining part constituting either more than half *(above)* or less than half *(below)* of the disc. (**b**) Wound healing is assumed to occur by convergence of the edges of the disc, thus creating a discontinuity in positional values, (**c**) intercalation by the shortest route produces newly formed tissue that either restores the original structure *(pebbled, above)* or mirror-duplicates the fragment *(shaded, below)*. Newly intercalated positional values are circled. Redrawn with modifications from Figure 8 of French et al. (1976), with permission.

here. Many of the details of these models are both controversial and unessential to our presentation; however, the original polar coordinate model does introduce one simple and important new idea, that when one represents positional values on a circle rather than on a straight line, intercalation may proceed in two different directions (Fig. 1.8). A basic postulate of the polar coordinate model is that intercalation always occurs by the shortest route that will restore positional continuity. This provides a consistent interpretation for the results of a large number of experiments on insect segments and appendages, some of which were paradoxical on earlier interpretations based on simple linear gradients. The physical basis for these positional values is unknown, although there are strong indications that graded cell-surface properties may be important (Nardi and Kafatos, 1976; Nardi and Stocum, 1983).

1.6 NONEQUIVALENCE

The numerous disputes that have arisen concerning the nature of positional information and the mechanisms by which it is established have tended to divert attention from the most fundamental conceptual innovation of the positional information idea. This innovation was given a name, *nonequivalence,* by Lewis and Wolpert (1976). To explain it, we return to the simple model of differentiation and diversification shown in Figure 1.1. Because differentiation involves a structural and functional specialization of cells, it would at first seem obvious that each of the different parts (A,B,C,D) of a fully differentiated system would be characterized by different cell specializations, and therefore, that before such differentiation the corresponding regions (1, 2, 3, and 4) might be "protodifferentiated" and bear these specializations in an incipient form. Lewis and Wolpert realized, however, that different portions of an organism are often not *differentially* specialized in any way that would be apparent to a cell biologist or histologist: the osteoblasts that construct a humerus do not differ in any obvious way from those that make an ulna. In terms of our simplified diagram, although the cells of both regions A and B will be highly specialized, these specializations may be the *same* in those features that a cell biologist or histologist would recognize. Regions 1 and 2, therefore, would differ not in a differentiation label but rather in a *positional label.* Nonequivalence is the condition of carrying different positional labels.

In Lewis and Wolpert's view, and that of most others who have interpreted their results under the positional information umbrella, only different *cells* can be nonequivalent. This is connected with the prevalent assumption that the state of nonequivalence is carried within the genome in the form of a set of conditions of activation or repression, or both, of specific gene loci. Long ago, however, the forerunner of the positional information idea, Stumpf (1967, p. 165), raised the possibility that different parts of the same cell may be diversified (i.e., nonequivalent), and that this diversification could have a nonchromosomal basis. This idea of intracellular nonequivalence is implicit throughout this book, and returns in explicit form in Chapters 10 and 11.

1.7 CONCLUDING REMARKS

The reader may wonder why in a book ostensibly about ciliates so much space has been devoted to systems as seemingly different from ciliates as amphibian eggs and flatworms. Do ciliates resemble either of these? The answer is that they resemble neither or possibly both, depending on whether one is considering specific organization or more general properties. Certainly, ciliates differ from any kind of egg not only in details of organization but in destiny: eggs cleave and then diversify into multicellular organisms, whereas ciliates grow and divide to form more ciliates that usually (although not always) are similar to their parents. Ciliates would thus not be expected to have an elaborate latent organization that foreshadows future development. Ciliates differ from flatworms in the obvious fact that ciliates are unicellular, whereas flatworms are multicellular. In addition, recent studies on the molecular phylogeny of eukaryotes (Sogin et al., 1986b; Field et al., 1988; Baroin et al., 1988) indicate that ciliates are not genealogically close to any multicellular animal, indeed not to any multicellular organism.

Nonetheless, there are ways in which ciliates resemble eggs (particularly insect eggs) and also flatworms. Like the former, ciliates have a complex cell surface organization, with extensive cytoskeletal elaborations and a large-scale polarity and asymmetry. Like the latter, ciliates maintain their organization while growing longitudinally and can both regenerate missing structures and clonally propagate their organization. This combination of features otherwise encountered in very diverse organisms creates certain unique possibilities for analysis, such as the ability to follow the variation in expression of a mutation in serial genetic replicas that inherit diverse preexisting structural patterns. We will see that the analysis of large-scale cell-surface patterns in ciliates allows novel approaches to problems that ciliates share with eggs, worms, and other developing systems.

2

Ciliate Organization

2.1 WHAT IS A CILIATE?

2.1.1 The Defining Characteristics of Ciliates

Ciliates express remarkably diverse variations on a few basic organizational themes. Some of these variations will be sketched when the principal experimental organisms are introduced in the next chapter. This chapter, however, is devoted to the themes themselves, with emphasis on those that relate to problems of pattern formation. In the first of four sections (Section 2.1), we summarize the special features of ciliates, starting with the nuclear aspects and then moving out to the cell surface. The second section (2.2) outlines the organization of the cell surface, with emphasis on the ciliary units that are the building blocks of cell-surface patterns in ciliates. The third section (2.3) describes, in general terms, how that organization changes through the cell cycle and regulates in response to structural or physiological disturbance. We then conclude (Section 2.4) with a brief outline of major changes brought about by nutritional shifts, including the sexual processes that are the basis of genetic analysis in ciliates.

All ciliates are characterized by a nuclear dualism, with a *micronucleus* and a *macronucleus* (Fig. 2.1a). The micronucleus is generally diploid, divides mitotically in vegetative cells and meiotically in cells undergoing conjugation or autogamy (self-fertilization), and has limited transcriptional activity (Stein-Gaven et al., 1987) and restricted but important somatic functions (reviewed by Ng, 1986). It can be considered as largely, although not exclusively, a germ-line nucleus. The macronucleus in the well-studied ciliates is made up of multiple copies of a rearranged subset of the micronuclear genome (reviewed by Blackburn and Karrer, 1986), it divides nonmitotically and imprecisely (reviewed by Berger, 1984), and it is the exclusive source of transcripts of certain specific genes (Mayo and Orias, 1985). In the ciliates considered in this book, the macronuclei divide in vegetative cells but disappear during sexual processes, when they are generated anew from division products of zygotic micronuclei. The formation of a new macronucleus involves an extraordinary genomic reorganization that has become a major subject of research in molecular biology (see papers in Gall, 1986). Most ciliates thus far analyzed also have a unique modification of the genetic code, in which UAA and UAG code for glutamine rather than peptide-chain termination (reviewed by Preer, 1986, pp. 328–329), but at least one does not (Harper and Jahn, 1989).

The surface organization of ciliates is just as distinctive as the nuclear dualism. The most general and characteristic feature of this surface organization is the con-

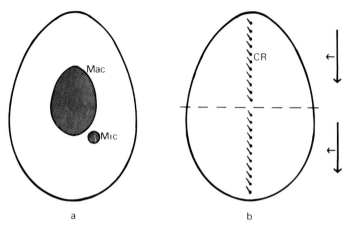

Figure 2.1 Ciliate characteristics. (**a**) Nuclear dualism, with a diploid micronucleus [Mic] and polygenomic macronucleus [Mac]. (**b**) Cell surface organization, showing one of the ciliary rows [CR] cut in two by a cleavage furrow *(dashed line)*. The *arrows* on the right indicate the polarity and asymmetry of the ciliary rows.

figuration of *ciliary units*. Ciliary units are arranged end-to-end in *ciliary rows* that commonly extend from one end of the cell to the other (Fig. 2.1b). Each ciliary unit is both polar and asymmetrical, and all units within a single row have the same polarity and asymmetry. New ciliary units are typically formed within the axis of a ciliary row, and adopt the same polarity and asymmetry as the preexisting units. At cell division, the ciliary rows are transected, usually across the cell equator (Fig. 2.1b). Thus each daughter cell gets one-half of each parental row, and the total number of rows generally remains the same. This condition is virtually unique to ciliates, and is profoundly different from that of the complex flagellates that possess multiple flagellar rows; in those flagellates, complete old flagellar rows are segregated to the two daughter cells, after which entirely new flagellar rows form in the vicinity of a particular basal body that appears to act as an organizer (Cleveland, 1960; Grell, 1973, pp. 128–131; Kubai, 1973). The absence of a unique central basal body (or of a small set of such basal bodies) as well as the scission of ciliary rows during cell division distinguishes ciliates from flagellates (Villaneuve–Brachon, 1940) and from most other cells that produce cilia or flagella at one time or another.

   These similarities make it very likely that the ciliate phylum (Phylum Ciliophora: Corliss, 1984; Small and Lynn, 1985) is a natural group in an evolutionary sense, with a presumed common ancestor shared by all members of the group. This view is consistent with the relevant molecular evidence (Sogin and Elwood, 1986; Lynn and Sogin, 1988), although this molecular evidence also indicates that there exist cases of deep divergence within the ciliate phylum (see Chapter 3).

2.1.2 The Ciliate Clonal Cylinder

The geometry of ciliate growth and division is illustrated in a highly schematic manner in Figure 2.2a. In addition to the polarized ciliary rows mentioned previously, ciliates have specialized structures located in particular regions, shown here

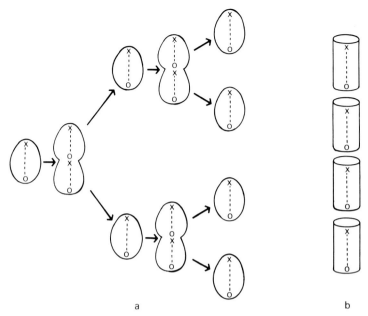

Figure 2.2 A highly schematic representation of ciliate clonal organization. (**a**) Two successive division cycles of a ciliate. The *vertical dotted line* indicates a ciliary row, x indicates a specialized structure located near the anterior end of the ciliate, o indicates a structure located near the posterior end of the ciliate. (**b**) The four-cell clone of (**a**) is represented as a clonal cylinder. For further explanation, see the text.

as an anterior structure (indicated in Fig. 2.2a by an X) and a different posterior structure (indicated in Fig. 2.2a by a 0). The ciliate grows longitudinally, and then *before* it divides it produces the new anterior and posterior structures of the future division products, in what formerly had been the midregion of the cell. An equatorial zone is thus transformed into two juxtaposed opposite poles. The cell then divides between these newly formed polar structures, generating two daughter cells that are approximate tandem replicas of the parent cell. This tandem repetition of specialized elements comes to be superimposed on the underlying continuity of ciliary rows. Ciliate division is thus akin to segmentation. This type of developmental geometry differs fundamentally from the mirror-image global organization of division products of flagellates (Chatton and Villaneuve, 1937; Cleveland, 1957) and of mammalian cells (Albrecht-Buehler, 1977).

The process of longitudinal growth, tandem subdivision, and cleavage occurs repeatedly in the presence of food. As first noted by Vance Tartar (1962), the basic topology of a ciliate clone is that of a cylinder (Fig. 2.2b). Ciliate growth consists of an elongation of the clonal cylinder, whereas ciliate division is a periodic transverse bisection of that cylinder. Reorientation of direction of growth, a normal characteristic of formation of new organ primordia in plants (Green, 1984), rarely occurs in ciliates. The closest analog to ciliate clonal topology is found not in other cells, but in certain whole multicellular organisms. In some flatworms and annelids, longitudinal growth is preceded or followed by transverse fission, with new heads and tails differentiating from former internal regions of the worm (Berrill, 1961, chapter 13). We will see in Chapter 4 that a cylindrical clonal topology permits exploration of certain developmental phenomena that are present but less easily studied in systems that have a different clonal topology.

2.2 CILIATE STATICS

2.2.1 The Surface Envelopes of Ciliates

The ciliate surface can be thought of as a two-dimensional layer richly embroidered with a complex ciliary pattern. Before considering the pattern displayed in the plane of the surface, however, we should first take a cross-sectional view and consider the nature of the surface within which this pattern is embedded.

Most ciliates possess a single outer plasma membrane underlain by two other membranes that generally are organized into more or less flattened sacs or alveoli (Fig. 2.3, Alv). Only the plasma membrane extends over the cilia. The plasma membrane includes numerous polypeptides (Adoutte et al., 1980; Williams, 1989), of which only the surface antigens that are involved in immobilization reactions are well characterized (Preer, 1986; Doerder and Berkowitz, 1986). Intramembranous particles are detected by freeze-fracture studies within all of the limiting membranes (reviewed by Allen, 1978a). The general arrangement of these particles appears to be random; however, ordered domains of intramembranous particles are observed within the plasma membrane above specific organelles (Satir et al., 1972; Beisson et al., 1976a) and in specific regions of the cell surface (Allen, 1978b; Hufnagel, 1981; Bardele, 1983).

Many ciliates have a fibrogranular layer situated directly beneath the inner alveolar membrane. This layer has been named the *epiplasm* (Fauré-Fremiet, 1962) (Fig. 2.3, E). Detergents can be used to isolate this layer and elements closely associated with it. The isolated complex retains the shape of the cell despite total dispersion of overlying membranes (Vaudaux, 1976; Collins et al., 1980). Each ciliary basal body protrudes through a transverse plate (Fig. 2.3, TP) (Hufnagel, 1969) that has the same fibrogranular consistency as the epiplasm and is closely adjoined laterally to the epiplasm (Fig. 2.3). As pointed out by Williams et al. (1979), the epiplasmic layer is topographically comparable to the membrane skeleton of other cells, such as the well-known spectrin layer of red blood cells. Even in *Tetrahymena*, however, where it exists as a continuous layer enveloping the entire cell interior, the epiplasm is structurally and chemically nonuniform (Williams et al., 1987 and personal communication). In *Paramecium*, the epiplasm appears discontinuous, forming individual plates or scales around each ciliary unit (Allen, 1971; Cohen and Beisson, 1988). The interstices between these plates are occupied by an *outer lattice* (Parducz, 1962) composed of filamentous material (Allen, 1971) with distinctive antigenic determinants (Cohen et al., 1987). One could consider the epiplasmic scales and the outer lattice as together forming a single composite layer. Such a layer has recently been isolated (Williams et al., 1989a).

Although the epiplasm is well developed in a variety of ciliates (reviewed by Pitelka, 1968; Peck, 1977), its generality in ciliates is problematic, for two reasons: first, in at least one ciliate genus, *Euplotes,* in which a well-defined epiplasmic layer was reported very early (Fauré-Fremiet and André, 1968) and a continuous fibro-

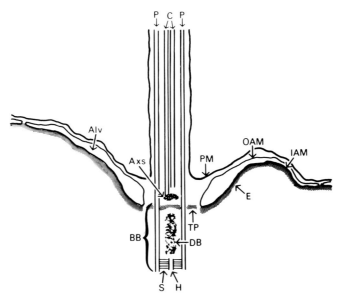

Figure 2.3 A diagram of a schematic longitudinal section of a cilium and of the adjacent cell surface of a ciliate that has a well-developed epiplasmic layer. The interior of the basal body [BB] is regionally differentiated, with the hub [H] and spokes [S] of the cartwheel structure located proximally, the RNase-sensitive dense body [DB] medially, and the transverse plate [TP], which adjoins the epiplasm [E], distally. Eighteen of the 27 microtubules of the wall of the basal body are continuous with the peripheral microtubules [P] of the cilium. One of the two central microtubules [C] of the cilium originates at the axosome [Axs] at the base of the cilium, whereas the other central microtubule originates a short distance above the axosome. The plasma membrane [PM] forms a complete envelope over the cell surface and cilium, and the outer alveolar membrane [OAM] and the inner alveolar membrane [IAM] enclose the alveoli [Alv]. The epiplasm [E] is directly adjacent to the inner alveolar membrane.

granular shell can be extracted using detergents (Jerka-Dziadosz et al., 1987; Williams et al., 1989b), careful thin-section electron microscopy has shown that a fibrogranular layer is located *within* the membrane alveoli rather than beneath them (Ruffolo, 1976a; Hausmann and Kaiser, 1979; Williams et al., 1989b). Because these alveoli form large separate plates covering the surface, the fact that the intraalveolar material [called "alveolar plates" by Hausmann and Kaiser (1979)] can be extracted as a continuous sheet enveloping the cell poses an apparent topological paradox. However, just as cytoskeletal structures can maintain continuity across cell membranes via junctional complexes (Tucker, 1981), the fibrous material of adjacent alveolar plates is joined laterally, across alveolar borders, to the dense material that Hausmann and Kaiser (1979) observed between abutting alveoli. It is this connection *across* alveolar membranes that allows an intraalveolar membrane–skeletal system to be isolated as a continuous layer (Williams et al., 1989b).

The second problem is that no fibrogranular layer is evident in appropriate cross sections of some of the larger ciliates that we will be considering, including *Stentor* (Huang and Pitelka, 1973) and Oxytrichids (Grimes and Adler, 1976; Jerka-Dziadosz, 1982), and attempts at extraction from the latter have thus far not succeeded.

The membrane–skeleton has been stressed here because it is the *only* nonmembranous surface layer that is widespread among ciliates and is thus a leading candidate for being the substratum of large-scale coordination of cell-surface pattern. Microtubules are abundant, but in most ciliates form discrete bands rather than being parts of a general layer. Actin microfilaments are associated with formation of food vacuoles (Cohen et al., 1984) but are sparse (Méténier, 1984) near the general body surface. Chemical studies also show that actin, although indubitably present, is quantitatively minor (and qualitatively aberrant) in ciliates (Hirono et al., 1987a). A contractile component of relatively low molecular weight (22 to 23 kilodaltons) is present in some ciliates, and is especially notable in *Paramecium,* where it makes up an *infraciliary lattice* at the level of the proximal ends of the basal bodies (Parducz, 1962; Allen, 1971; Garreau de Loubresse et al., 1988). The presence of bona fide intermediate filaments anywhere in any ciliate has not yet been established clearly (compare Numata and Watanabe, 1982 with Bartnik et al., 1985, p. 439; see also Williams et al., 1986).

2.2.2 Ciliary Units

The basic building blocks of ciliate cell-surface organization are the *ciliary units,*

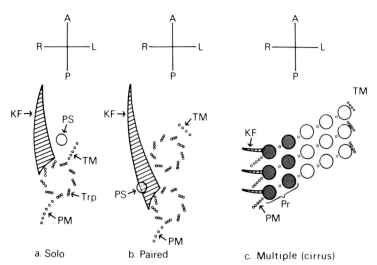

a. Solo   b. Paired   c. Multiple (cirrus)

Figure 2.4   Cross sections of the three basic kinds of ciliary units that make up ciliary rows. (**a**) A solo unit (monokinetid). The nine triplet microtubules [Trp] that make up the wall of the basal body are shown in cross section. The three adjacent fibrillar structures are the striated kinetodesmal fiber [KF], the transverse microtubule band [TM], and the postciliary microtubule band [PM]. A membrane invagination called the parasomal sac [PS] is located anterior to the ciliary unit. (**b**) A paired unit (dikinetid). The parasomal sac [PS] is located above the kinetodesmal fiber [KF]. (**c**) A multiple unit (polykinetid, in this case a cirrus) at a lower magnification. Basal bodies are drawn as circles. The basal bodies making up the pairs [Pr] that initiate formation of the cirrus are shaded. The usual orientation of these structures in the cell is shown above each diagram: A, anterior; P, posterior; R, right; L, left. (**a**) and (**b**) are redrawn, with modification, from Figures 3d and 4b, respectively, of Lynn (1981); (**c**) is redrawn from Figure 24E of Jerka-Dziadosz (1980), with addition of kinetodesmal fibers [KF] that appear transiently during cirral development. Fibrillar connectives between basal bodies are omitted from all diagrams.

or kinetids (Lynn, 1981). These units are organized around one or more ciliary basal bodies (traditionally called kinetosomes by ciliatologists), which are frequently, although not invariably, ciliated. The basal bodies and cilia have the typical eukaryotic organization (briefly described in the legend of Fig. 2.3). A striated rootlet known as a *kinetodesmal fiber* (Fig. 2.4, KF) and two sets of microtubule bands, the transverse (TM) and postciliary (PM) bands, are generally associated with the basal bodies. There are also arrays of filaments that connect the basal bodies of compound units, which are probably important in maintaining the structural stability of these units (Jerka-Dziadosz and Golinska, 1977), but these are variable in organization and, therefore, are not shown in the diagrams. A membrane invagination, the parasomal sac, which resembles the coated pits of tissue cells, is located in a definite spatial relationship to each ciliary unit (Fig. 2.4, PS).

Ciliary units are present either as solo units (monokinetids) (Fig. 2.4a), as paired units (dikinetids) (Fig. 2.4b), or as multiple units (polykinetids) (Figs. 2.4c and 2.5). In solo units, all three sets of accessory elements are typically arrayed around a single basal body and the parasomal sac is anterior to the ciliary unit (Fig. 2.4a). In most paired units, the posterior basal body bears the kinetodesmal fiber and the postciliary microtubule band, whereas the anterior basal body carries the transverse microtubule band, with the parasomal sac anterior and somewhat to the right of the posterior basal body (Fig. 2.4b). Multiple units may be thought of as sets of closely spaced and interconnected paired units (shaded in Figs. 2.4c and 2.5b) to which additional basal bodies are added.

The kinetodesmal fibers typically are directed anteriorly and to the right of each basal body. In some ciliates, notably *Paramecium,* these fibers overlap to form an ensemble that is seen in the light microscope as an apparent single fiber, the *kinetodesma.* Their invariant orientation in normal cells is the basis of the law of desmodexy (fiber on right), which was discovered before the ultrastructural nature of the fiber became known (Chatton and Lwoff, 1935).

At this point we require a clear convention for describing orientations of asymmetrical structures in the plane of the cell surface. I will follow Ng's rule: ". . . *right* and *left* are defined in terms the observer's right and left, assuming that he stands inside the animal, lines up anteroposteriorly with the animal and faces the cortical region which is being examined" (Ng, 1977, p. 233). The diagrams, however, will always be drawn as if the object were viewed from *outside* the cell, so that readers looking at the illustrations must always remember that their left is the cell's right (as in Figs. 2.4 and 2.5). Ng's rule is well suited for describing most aspects of the cell-surface organization of ciliates, as in ciliates all ciliary structures are asymmetrical and the direction of asymmetry is normally the same everywhere. Nonetheless, in some places I will be forced to violate Ng's rule; when this happens, I will explain what is being done, and why.

The location of the microtubule bands is fixed in relation to that of the kinetodesmal fiber, with the postciliary microtubule band located posteriorly and to the right of each basal body, and the transverse microtubule band anteriorly and to the left. The orientation of the transverse band in relation to the microtubules of the basal body is more variable than that of the postciliary bands (Figs. 2.4 and 2.5). Both postciliary and transverse bands typically arch upward toward the epiplasm (discussed later).

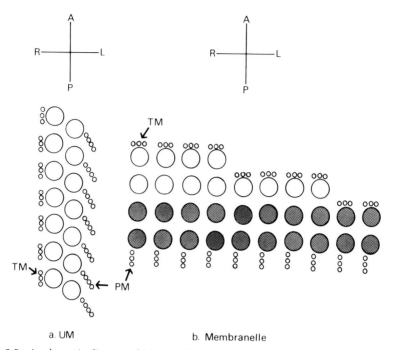

a. UM   b. Membranelle

Figure 2.5 A schematic diagram of (a) an undulating membrane [UM] and (b) a membranelle, showing basal bodies *(circles)*, transverse microtubule bands [TM], and postciliary microtubule bands [PM]. In (b), the basal bodies of the paired ciliary units of the initial membranelles (promembranelles) are shaded. A common orientation of these structures in the cell is shown above each diagram using the same conventions as in Figure 2.4.

Virtually all ciliates possess a specialized feeding region, with a distinctive and often very elaborate oral ciliature. Although oral ciliary units are quite variable in detail, most are based on elaborations of paired ciliary units. Such oral ciliary pairs exist as separate structures in a number of ciliate species. In the ciliates to be considered in detail in this book, however, the ciliary pairs are organized into two types of structure: they are compounded loosely in the *undulating membrane* (UM, Fig. 2.5a) (also called the *paroral membrane*) and more tightly in a multiple ciliary unit called a *membranelle* (Fig. 2.5b). Oral ciliary units lack striated rootlets and in some ciliates they also lack transverse microtubule bands; the postciliary bands, however, are universally present.

Although the spatial orientations of the various types of paired ciliary units are quite diverse and sometimes have been given different names on this basis (Grain, 1984), they are all rotational permutations of one another (Jerka-Dziadosz, 1981b) and in some cases are also developmentally interconvertible (Nelsen and Frankel, 1979).

2.2.3 Ciliary Organization
The nonoral cell surface

In all but a few highly specialized ciliates, most of the cell surface is covered by longitudinal arrays of ciliary units, each making up a ciliary row (called a kinety by ciliatologists). The organization of the nonoral portion of the ciliature (termed somatic by ciliate workers) tends to be conserved in ciliate evolution (Lynn, 1981). The stereoscopic diagram of parts of two adjacent ciliary rows (Fig. 2.6, from Allen, 1967), although based on one ciliate, *Tetrahymena,* therefore gives an adequate general idea of the organization of ciliary rows in most other ciliates as well (important deviations are pointed out where relevant, especially in the next chapter).

Apart from the basal bodies and cilia, three subsurface structures seen in Figure 2.6 deserve special notice, as they commonly are visible by light microscopy and are important as guideposts for organization in experimentally manipulated ciliates. We have already mentioned that the overlap of the kinetodesmal fibers (Fig. 2.6, KF) is responsible for the classic kinetodesma, always to the right of the basal-body row. The postciliary microtubule band (Fig. 2.6, PM) arches upward over the kinetodesmal fiber and is oriented at a right angle to the cell surface. This band is relatively inconspicuous in ciliates such as *Tetrahymena* and *Paramecium,* but is extremely long and well developed in some large ciliates, such as *Stentor,* in which overlapping postciliary microtubule bands replace the kinetodesma as the major microscopically visible fiber of the ciliary rows (see Chapter 3). The transverse microtubule band (Fig. 2.6, TM) arches upward transversely to the left, with its axis parallel to the cell surface.

Figure 2.6 A cutaway diagram of a portion of the cell surface of *Tetrahymena,* showing segments of two ciliary rows and their associated fibrillar elements. The surface of the plasma membrane, with protruding cilia (C) and docking sites of the mucocysts is shown in the more anterior portion *(top of diagram).* Parasomal sacs are omitted. The membrane is sectioned to show the alveoli [Alv] and one mucocyst [Mc]. Parts of the epiplasm [E], with the longitudinal microtubule bands [LM] above it, extend from the membrane-section toward the viewer. In the more posterior portion *(bottom of diagram),* both membrane and epiplasm are cut away to show the ciliary units, including basal bodies [BB], kinetodesmal fibers [KF], transverse microtubule bands [TM], and postciliary microtubule bands [PM]. Immediately beneath these are mitochondria [Mi], whose arrangement is actually more regular than shown here (Aufderheide, 1979). Slightly modified from Figure 22 of Allen (1967), with permission.

Many ciliates also possess a system of longitudinally oriented microtubules that are not closely associated with basal bodies. In *Tetrahymena* and *Paramecium*, this takes the form of a discrete longitudinal microtubule band (Fig. 2.6, LM), located between the epiplasm and the inner alveolar membrane, to the right of the ciliary rows. In *Paramecium*, the longitudinal microtubule bands appear only during division (Sundararaman and Hanson, 1976) and at certain stages after conjugation, and have been called the cytospindle (Cohen et al., 1982). In other ciliates *(Blepharisma, Euplotes)*, longitudinally oriented microtubules are found underneath much of the cell surface (Pitelka, 1968; Ruffolo, 1976a; Grim, 1982). Some of these are permanent, whereas others are transient (Jerka-Dziadosz et al., 1987).

In some ciliates, such as *Tetrahymena*, ciliary rows are made up primarily, although not entirely, of solo ciliary units. Other ciliates, like *Paramecium*, have rows that are mixed, with solo units in some regions and paired units in others. Still others, like the ciliary rows that cover the whole surface of *Stentor* and *Blepharisma*, consist entirely of paired units. Finally, the ventral surface of *Euplotes* and of the *Oxytricha–Stylonychia–Paraurostyla* group (see Chapter 3) is virtually unique in having rows made up of multiple ciliary units. These multiple units are called *cirri*. They commonly are huge and sparsely arranged over the surface, so that their homology to members of ordinary ciliary rows is obscured. An ultrastructural analysis of cirral development (Jerka-Dziadosz, 1980), however, has revealed that each diagonal file of basal bodies within a cirrus is generated from a typical basal body pair (shaded in Fig. 2.4c); the specific developmental relationship is described in the next chapter.

We can now add to our list yet another general feature that sets ciliates apart from other cells. In ciliates, the basal bodies are exclusively specialized for their role in cell surface organization. This differs from the situation in most other kinds of cells, where basal bodies tend to perform double duty. For example, in vertebrate cells centrioles may be converted into basal bodies and form flagella either normally or after drug treatment (reviewed by Fulton, 1971; Weber and Osborn, 1981). Similarly, in flagellates the flagellar basal bodies tend to become associated with nuclear division, in some cases becoming physically detached from the flagella (Johnson and Porter, 1968; Hoops and Witman, 1985) and migrating toward the cell interior to act as centrioles (Coss, 1974). In ciliates, however, micronuclear mitosis is *always* noncentriolar (summarized by Heath, 1980, 1986), and no cases of close association of basal bodies with dividing macronuclei have been reported. Indeed, instead of centrioles migrating to the interior, in certain ciliates nuclei move to the surface during division, sometimes forming loose associations with the fibrillar system, although not with specific basal bodies (Jaeckel-Williams, 1978; Tucker et al., 1980). Thus, even on morphological grounds, the cell surface organization of ciliates appears to be highly autonomous. This conclusion is buttressed experimentally in subsequent chapters.

We need a general term to describe the organized external region of a ciliate. The standard term is *cortex*. This is defined by ciliate workers as the outermost layer of the ciliate that includes the external membrane system, the membrane skeleton, the longitudinal microtubule bands when present, and the ciliary basal bodies plus all fibrillar systems associated with them (Corliss and Lom, 1985; Lynn, 1988). It should be emphasized that this definition is *not* identical to the traditional definition of the cortex as the outer gelated region of the cytoplasm (Bray et al., 1986). Thus, the relatively stationary cytoplasmic layer that is found *underneath* the visibly structured region in *Paramecium* (Aufderheide, 1977) might be considered part of the cortex according to the traditional definition, but is called subcortical by ciliatologists (Kaneda and Hanson, 1984).

The oral apparatus

Oral architecture is exceedingly diverse among different ciliates (Small and Lynn, 1985). For the purposes of this book, we can simplify matters somewhat, as most (although not all) experimental work has been done on ciliates in which the major ciliary elements of the oral structures are the undulating membranes (UMs) and the membranelles depicted in Figure 2.5. A further simplifying generalization is that UMs are always located on the right side of the oral apparatus, membranelles on the left. The number of these structures, however, is quite variable. There may be 1 to 4 UMs, and 3 to 300 membranelles. UMs are always parallel to one another and generally longitudinally oriented with respect to the cell axis, whereas membranelles typically are transverse or diagonal relative to the axis of the cell and of the UMs.

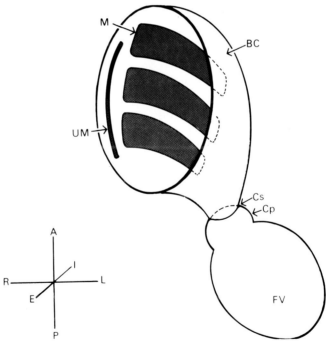

Figure 2.7   A highly schematic diagram of the regions of the oral apparatus of an imaginary ciliate. The specialized oral ciliature is shown within the buccal cavity [BC], with an undulating membrane [UM] on the right wall of the cavity near its opening, and three membranelles [M] to the left of the UM inside the cavity. The cytostome [Cs] is at the base of the buccal cavity, the cytopharynx [Cp] and the forming food vacuole [FV] are internal to the cytostome. The approximate orientation of this structure in the cell is shown at the right: A, anterior; P, posterior; R, right; L, left; I, internal; E, external.

A large number of ciliate species possess only three membranelles. In these, the entire oral ciliature typically is located within a *buccal cavity* (Fig. 2.7, BC). Other ciliates possess numerous membranelles, most of them located outside the buccal cavity. The buccal cavity terminates posteriorly in a cell mouth or *cytostome* (Fig. 2.7, Cs). This is a ring-shaped border at which the multiple membrane system that covers the general body surface ends and the single unit membrane that makes up the wall of the food vacuole begins (Nilsson and Williams, 1966; Allen, 1974). Beyond this, there is a specialized region called the *cytopharynx* (Fig. 2.7, Cp), which lacks ciliature but may have elaborate microtubular and filamentous supporting structures that connect to the forming food vacuole. The cytopharynx is characteristically rather short in ciliates, such as *Tetrahymena* (Sattler and Staehelin, 1976, 1979) or *Paramecium* (Allen, 1974), that feed on bacteria or other relatively small food, but is long and well developed in certain other ciliates, such as *Nassula,* that feed on long and tough blue-green algal filaments (Tucker, 1968).

A description of oral structure in terms only of arrangement of basal bodies plus the accessory microtubule bands, as shown in Figure 2.5, is very incomplete. Additional bands or networks of microtubules, called nematodesmata by ciliatologists (Lynn, 1981; Grain, 1984), are often associated with the proximal ends of oral basal bodies but sometimes become detached from them (Tucker, 1970). The nematodesmata may be tangential or perpendicular to the surface plane, and serve to anchor and interconnect the membranelles and UMs, and/or to reinforce the cytopharynx (reviewed by Grain, 1984). The oral apparatus is also crisscrossed by complex networks of oral filaments, which have been most studied in *Tetrahymena* (Williams and Luft, 1968; Gavin, 1977; Smith, 1982). The quantitatively predominant class of oral filament proteins in this ciliate has been analyzed both chemically (Williams et al., 1986) and structurally (Williams, 1986b), and appears to be different from both microfilaments and intermediate filaments. Recent studies have shown that one of these proteins shares an antigenic determinant with spectrin (Williams, personal communication). Localization of a putative intermediate filament protein has also been reported in the cytopharyngeal region of *Tetrahymena* (Numata et al., 1983). Microfilaments composed of actin have been detected only in a restricted region of the oral apparatus of *Tetrahymena* (Méténier, 1984) and *Paramecium* (Cohen et al., 1984), in both cases in an area closely associated with food vacuole formation.

2.2.4 Extrusomes

Extrusomes are membrane-bounded secretory organelles, which have very diverse forms in different ciliates (Hausmann, 1978). For example, *Tetrahymena* has mucocysts (Fig. 2.6, Mc), whereas *Paramecium* has the well-known trichocysts, whose secretion is a long fiber. All of these organelles are more or less elongate, and insert tip-first into "docking sites" located just underneath the cell membrane (Beisson et al., 1976a). These docking sites typically form either irregular files midway between the ciliary rows (mucocysts of *Tetrahymena:* Fig. 2.6, Mc) or alternate with cilia along the axis of the ciliary rows (trichocysts of *Paramecium*). Unlike the ciliary structures that develop near the cell surface, extrusomes mature within the internal cytoplasm and then are moved out to the docking sites at the cell surface

(Yusa, 1963). There they eject their contents when their bounding membrane fuses with the plasma membrane (reviewed by Hausmann, 1978; Allen, 1978a; Adoutte, 1988).

Remarkably, inserted trichocysts and probably also mucocysts are unnecessary for the survival of the ciliates, at least under laboratory conditions. Because of this, and because their presence or absence can readily be assayed, the *Paramecium* trichocyst has been subjected to a more detailed genetic analysis than has any other organelle found in ciliates (reviewed by Adoutte, 1988). This analysis has shown that potential trichocyst attachment sites persist at the usual locations on the cell surface in mutants in which trichocysts fail to attach, or even when they are totally absent (Beisson et al., 1976a; Pouphile et al., 1986). Although these unoccupied attachment sites may sometimes be abnormal, the organization of the cell surface pattern of attachment sites is clearly independent of the pathway of trichocyst development.

2.2.5 Unique Exit Sites: Cytoproct and Contractile Vacuole Pore

Whereas extrusomes liberate their contents from multiple sites dispersed over the cell membrane, there are two other classes of membrane-bound inclusions that dump their respective contents at more localized sites. These are, respectively, the food vacuoles, which evacuate their contents at a unique *cytoproct,* and the contractile vacuole(s), which make use of *contractile vacuole pore(s)* to expel excess fluid.

The cytoproct is an elongated region of specialized membrane and underlying filamentous material (reviewed by Allen, 1978a; Allen and Wolf, 1979) that can also be identified as an irregular slit in cells suitably prepared for light microscopy. Although, like the cytostome, the cytoproct is functionally dynamic, it is formed at a definite location in the cell (see Chapter 3), making it a useful marker for studies on pattern formation.

The contractile vacuole pores (CVPs) are elaborate structures with both circular (actually spiral) and radial microtubules plus associated intramemembranous specializations (McKanna, 1973; Allen, 1978a). In the light microscope, the CVPs appear as deceptively simple rings. These structures, although themselves quite distinct from ciliary basal bodies, develop near basal bodies at particular locations in the cell. They are even more useful than the cytoproct as markers of both local and global cell-surface organization.

2.3 CILIATE DYNAMICS

2.3.1 The Perpetuation of Ciliary Rows

Ciliary rows are perpetuated in a similar manner in all ciliates. This is accomplished by formation of new ciliary units at a well-defined position and orientation relative to preexisting ciliary units, as illustrated for solo ciliary units in Figure 2.8.

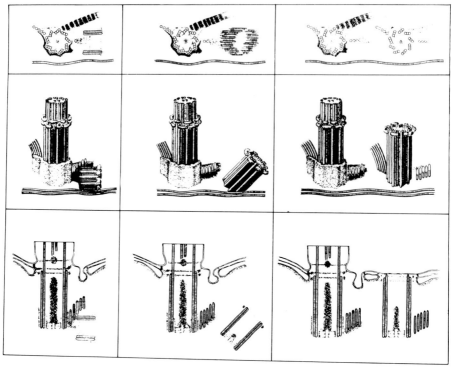

Figure 2.8 Diagrams summarizing the maturation of probasal bodies within ciliary rows. All diagrams are oriented so that the anterior of the cell is to the right. The top row shows cross sections near the base of mature basal bodies, the middle row shows a three-dimensional reconstruction that includes only the basal bodies and adjacent fibrillar structures, and the bottom row shows median longitudinal sections that include the cell membranes as well (with a parasomal sac immediately anterior to each mature basal body). The columns show successive stages of maturation of probasal bodies. For further description, see the text. From Figure 22 of Allen (1969), with permission.

A new probasal body develops anterior to the proximal region of an old mature basal body. The probasal body initially is oriented with its long axis perpendicular to that of the old basal body and parallel to the cell surface (Fig. 2.8, left). Then, as it elongates, the nascent basal body rotates through 90° to end up parallel to the old basal body and perpendicular to the cell surface (Fig. 2.8, center and right). The accessory microtubule bands and kinetodesmal fiber begin to form near the proximal end of the new basal body during rotation and complete their development after the rotation has ended (Fig. 2.8, right). The maturing basal body becomes integrated into the epiplasm by a probable aggregation of fibrous material around its distal end [this is seen more clearly in *Nassula* (Tucker, 1971b)]. The distal ends of the basal body microtubules have by this stage somehow penetrated through the alveolar layers, and made contact with the plasma membrane. Subsequently, although often after a considerable delay (Nanney, 1975), a cilium grows out from the distal end of the new basal body.

Fundamentally similar processes are involved in the perpetuation of ciliary rows made up of paired basal bodies (Jurand and Selman, 1969, plate 38; Bohatier,

1979) and of multiple basal bodies (Jerka-Dziadosz, 1980). When many new basal bodies are formed in a short time, probasal bodies may form next to older basal bodies that are still immature, so that a chain of several new basal bodies may be seen anterior to a mature basal body, with the basal body closest to the mature basal body the most completely developed and the one farthest away the most immature (Dippell, 1968; Allen, 1969; Tucker, 1970; Jerka-Dziadosz, 1980).

We have not yet considered how a probasal body (Fig. 2.8, left) first arises. This was described by Dippell (1968) for *Paramecium*. Before any microtubules appear in the probasal body, a *generative disc* of dense material forms anterior to the proximal region of a mature basal body (Fig. 2.9a, GD). This disc is the foundation of a new probasal body. The innermost (A) microtubules of each of the nine triplets appear first, and become connected to one another by temporary dense links (Fig. 2.9b). The second (B) microtubules begin to form before the ring of A microtubules is complete (Fig. 2.9b, right). Finally, the outermost (C) microtubules appear in the same manner (not shown). As the microtubules form, the central portion of the generative disc becomes hollowed out. The "hub and spoke" structure of the "cart-

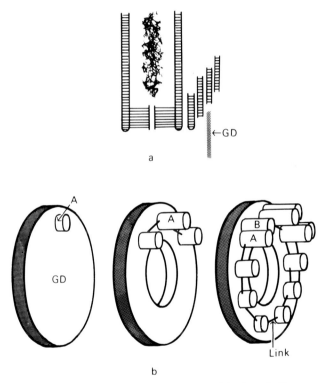

Figure 2.9 Diagrams of early stages of the formation of basal bodies of ciliary rows in *Paramecium*. (**a**) The approximate location of the generative disc [GD], oriented similarly to the bottom row in Figure 2.8 (anterior to the right). (**b**) Progressive stages in the development of microtubules of the basal body from the generative disc, assuming an absence of the cartwheel at early stages. For further explanation, see the text. (**a**) drawn from the description in Dippell (1968), (**b**) redrawn with modifications from Figure 5 of Tucker (1977), with permission, and reproduced by permission from *Nature*, Vol. **266**, pp. 22–26, copyright © 1977 Macmillan Magazines Ltd.

wheel" found at the base of the mature basal body (see Fig. 2.3) is not seen clearly until the probasal body is well formed, at the stage shown in the left diagram of Figure 2.8. There are hints, however, of its possible presence at an earlier stage [see Figs. 9 and 10 in Dippell (1968)]. In a variety of other organisms, ranging from symbiotic flagellates that inhabit termite guts to mammals, the cartwheel was observed to appear before the first microtubules (Kalnins and Porter, 1969; Anderson and Brenner, 1971; Tamm and Tamm, 1980, 1988). The question of whether a cartwheel might actually form at an early stage within the generative disc in *Paramecium* and other ciliates deserves reinvestigation.

This morphological sequence leaves open two crucial questions. One is whether the generative disc actually generates the organization of the basal body, that is, does it act as a "microtubule-nucleating-template" (Tucker, 1977) that specifies the arrangement of microtubules in basal body? An alternative possibility is that this disc may act as a favorable site for microtubule formation but that the specific arrangement of microtubules is determined by some other means, such as a sequential determination of sites of microtubule nucleation via a selective positioning of intermicrotubule links (Roth et al., 1970; Bardele, 1977; Tucker, 1979). There is only scattered evidence that bears on this question. The absence of microtubule triplets in some otherwise normal basal bodies of a mutant of *Paramecium* may argue for specification of the organization of the basal body by a "prepattern" rather than by sequential nucleation (Ruiz et al., 1987). The fact that the cartwheel complex has been shown capable of reassembly in vitro (Gavin, 1984) and may be present at early stages in vivo (see above) suggests that it might possibly provide or be associated with such a prepattern.

The other, related question is whether the old basal body in some sense specifies the organization of the new one. Although the classical idea of direct division of basal bodies (Lwoff, 1950) was disproven by the ultrastructural observations, the related notion of "genetic continuity" (Lwoff, 1950) has lived on through the repeated claims that basal bodies contain DNA (reviewed by Fulton, 1971). As pointed out in Fulton's review, however, most of this evidence was equivocal; the best evidence was a yellow-green fluorescence with acridine orange (Smith-Sonneborn and Plaut, 1967) that was later shown to be caused by RNA (Hartman et al., 1974). Detailed digestion studies by Dippell (1976) showed that although DNase had no effect on the ultrastructure of basal bodies, RNase removed the central dense material (Fig. 2.3, DB). The strong likelihood that RNA is present inside basal bodies does not in itself answer the question of whether and how the old basal body contributes to the organization of the new one, as no one yet knows what the basal body RNA is nor what it does, although there is evidence in nonciliate systems that centriolar RNA is essential for nucleation of microtubules (Heidemann et al., 1977; Petersen and Berns, 1978). Nonetheless, the minimal possibility that the old basal body provides only a specialized local environment near its proximal anterior surface within which generative discs can assemble readily has not been ruled out. The fact that new basal bodies sometimes assemble in the absence of any ultrastructurally distinguishable old basal bodies, both in nonciliate systems (Fulton, 1971) and in ciliate cysts (Grimes, 1973a,b) does not resolve this issue: basal body RNA might become detached from basal bodies, whereas a specialized local environment near old basal bodies may promote basal body assembly without being indispensable.

2.3.2 Development of the Oral Apparatus

Formation

In every ciliate division cycle, new specialized cell surface structures are formed before division, and then passed on to the daughter cells. The most complex of these newly formed structures is the oral apparatus (OA).

The details of formation of the OA vary immensely in different ciliates (Tuffrau, 1984; Lynn, 1990), and it would be a serious error to think that one generalized description applies to all or even most ciliates. Nonetheless, the course of oral development in the seven ciliate genera that are considered in detail in this book (Chapter 3) can be summarized in the form of three generalizations, using a diagram of oral development in *Tetrahymena* (Fig. 2.10) as a pictorial aid. Papers by Jerka-Dziadosz (1981a,b), Bernard and Bohatier (1981), and Bakowska et al. (1982b) provide a more detailed review of the observations that are summarized briefly in this section.

The first generalization is that oral development begins with the formation of an *anarchic field* (Fig. 2.10, AF) of basal bodies, in which the ciliary units are oriented in a variable, possibly random, manner. Any initial orderliness is not preserved as the anarchic field develops (phase A in Fig. 2.10). This loss of order is demonstrated ultrastructurally by a highly variable orientation of postciliary microtubule bands (Bernard and Bohatier, 1981) and by an equally variable location of the parasomal sacs in relation to the nearby basal bodies (Hufnagel, 1983). This variable orientation may result, at least in part, from a considerable spatial irregularity in the generation of new basal bodies in the anarchic field (Grimes, 1973c; Williams and Frankel, 1973), which contrasts to the order observed when new basal bodies are formed within ciliary rows. Thus, the morphological evidence

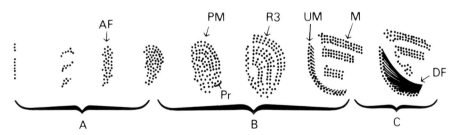

Figure 2.10 Oral development in *Tetrahymena*. **Phase A:** An anarchic field [AF] is produced by proliferation of basal bodies. **Phase B:** Promembranelles [PM] develop by alignment of basal body pairs [Pr], a third row of basal bodies [R3] is added to each promembranelle to convert it into a membranelle [M]. The undulating membrane [UM] develops to the right of the membranelles. **Phase C:** Basal bodies at the right end of the membranelles shift their location, a deep fiber [DF] and other fibrillar systems, as well as the buccal cavity, cytostome, and cytopharynx (not shown) develop. For further explanation, see the text. Redrawn, with modifications, from Lansing et al. (1985).

suggests that the spatial order of the later stages of oral development is newly created, rather than being carried over from any preexisting order (Grain and Bohatier, 1977). Experimental support for this conclusion is described in Chapter 7.

The second generalization, foreshadowed in our earlier description of oral anatomy (Fig. 2.5b, shaded basal bodies), is that membranelles always begin as double rows of basal bodies. These initial *promembranelles* (Fig. 2.10, PM) are formed by the side-by-side alignment of pairs of basal bodies (phase B in Fig. 2.10) that had been generated near the end of the anarchic field stage. A corollary to this generalization is that additional rows of basal bodies (Fig. 2.5b, unshaded basal bodies) are produced within the nascent membranelles by the formation of new basal bodies on one side of the initial pairs (Fig. 2.10, R3).

This second generalization may not apply quite as generally to the formation of the undulating membrane. Alignment of initially disordered basal-body pairs to make a UM has clearly been demonstrated in *Stentor* (Bernard and Bohatier, 1981) and in *Paraurostyla* (Jerka-Dziadosz, 1981b). In *Tetrahymena*, however, the UM appears to form by alignment of solo units (Bakowska et al., 1982b).

The third generalization is that the extensive membranous and fibrillar differentiation associated with the formation of the oral cavity, cytostome, and cytopharynx, as well as certain distinctive modifications of the basic membranellar order found in those ciliates that possess only three membranelles, all take place *after* the elaboration of the ciliary structures of the new OA (phase C in Fig. 2.10). This is evident chemically as well as morphologically; for example, the major oral filament proteins of the *Tetrahymena* OA appear in substantial amounts only *after* both membranelles and the UM are fully formed (Williams et al., 1986).

In general, phases (A) and (B) precede cell division, whereas phase (C) is coincident with it. The formation of the fibrillar structures of the OA thus roughly coincides with the time at which a filament ring is detected underneath the division furrow. Such a ring has been detected in four diverse ciliates (*Nassula:* Tucker, 1971a; *Tetrahymena:* Yasuda et al., 1980; Jerka-Dziadosz, 1981c; *Paramecium:* Cohen et al., 1984; *Stentor:* Diener et al., 1983) and is likely to be present in all dividing ciliates. In animal cells the filaments are actin (Schroeder, 1975; Mabuchi, 1986). Heavy meromyosin, however, fails to decorate filaments of the division furrow in *Paramecium* (Cohen et al., 1984) and in *Tetrahymena* (Méténier, 1984). These findings apparently conflict with the more recent detection of actin in the filament ring of *Tetrahymena* by light-microscopic immunochemical methods (Hirono et al., 1987b). Possibly actin is a minor component of the filament ring, whereas the nature of the major component(s) remains unknown (see Grain, 1986, pp. 234–236).

Remodeling

Cytoplasmic microtubular systems in both animal and plant cells tend to be highly labile in two different senses: they are physicochemically labile, as exhibited by depolymerization at high pressure and low temperature (Behnke, 1970) and in the presence of drugs, such as colchicine (reviewed by Dustin, 1984), and they are developmentally labile, as manifested by a radical remodeling during mitosis and cell division (McIntosh, 1979; Weber and Osborn, 1981; Tiwari et al., 1984).

Microtubular systems found on the cell surface of ciliates show a remarkable phys-icochemical stability, as with few exceptions they are not directly depolymerized by low temperature and chemical agents (Williams, 1975; Cohen and Beisson, 1988). For the most part, they show developmental stability as well. In general, preexisting basal bodies persist through cell division, and most but not all of the fibrillar systems of the general body surface also persist. In the OA, however, fibril-lar structures are resorbed and then formed anew, and some parts of the OA such as the UM are reorganized drastically (Nelsen 1981; Bakowska et al., 1982).

When a ciliate divides, the new OA passes to the posterior division product, whereas the anterior division product retains the old OA (see figures in Chapter 3). The membranelles of the anterior OA remain intact, but shortly before the onset of division the UM undergoes a substantial reorganization while the accessory fibrillar structures are resorbed. The old anterior OA, therefore, reverts to a con-dition close to that observed in developing OAs at the end of phase B. The cell thus becomes temporarily unable to feed. Then the new fibrillar structures reform within the old OA at the same time that they are developing in the new OA, so that old and new oral systems go through phase C of oral development together. This loss and reformation of fibrils can be detected not only by standard microscopic techniques in a variety of ciliates, but also by immunofluorescent studies of the proteins that make up some of these fibrils (Numata et al., 1983; Williams et al., 1986).

Although the remodeling of the anterior OA is quite impressive, especially in ciliates with elaborate cytopharyngeal specializations, we must not forget that dur-ing this entire process the organized arrangement of basal bodies of the oral mem-branelles remains intact. Furthermore, in some ciliates oral proteins, probably including tubulins, show a high degree of conservation within preexisting membra-nelle bands (Ruffolo, 1970; Grimes and Gavin, 1987). Thus, when new fibrillar structures form in old mouths, they get hung onto a preexisting scaffold that does *not* dedifferentiate.

2.3.3 "Morphogenesis as a Common Denominator"

Although the OA persists essentially intact through division, there are situations in which dedifferentiation of old oral structures with partial or total replacement by a newly formed set of structures does occur. This happens when an old OA is dam-aged, when it is the wrong size for the cell, or when the ciliate is going through certain forms of starvation-mediated development.

But why make a new set? One could imagine that when an OA is damaged or when a cell becomes substantially larger, the preexisting OA could grow by addition of ciliary units, and that when a cell gets smaller it could selectively resorb parts of its oral system. Both of these processes actually occur in *Dileptus,* a ciliate in which the OA has neither membranelles nor UMs, but rather a ring of ciliary units, some paired and others solo (Grain and Golinska, 1969). An OA of *Dileptus* can grow by addition of ciliary units within the portion of the oral ring that contains solo ciliary units (Kink, 1976) and can shrink by resorption of ciliary units (Golinska and

Kink, 1977). The OA of *Dileptus* is thus a "steady-state structure" (Golinska, 1986), a ring that enlarges or shrinks to meet the exigencies of this highly adaptable ciliate's existence.

In those ciliates in which the oral ciliature is made up of multiple ciliary units (including all of those to be considered in subsequent chapters of this book), the OA for the most part is *not* a steady-state structure. Although such an OA can reform its fibrillar systems in situ during cell division, and can sometimes become smaller while remaining differentiated (Tartar, 1961, p. 123–127; Buhse, 1966a), it has never been observed to become larger or to restore missing parts by addition of new ciliary units to old oral structures that retain their previous organization. As pointed out most clearly by Jerka-Dziadosz and Golinska (1977), increase in structural complexity of ciliary structures is associated with a loss of capacity to proliferate. This loss is either irrevocable (as is typical of membranelles) or can be reversed only by a partial dedifferentiation of the structural system (as is sometimes observed in UMs as well as cirri). In this sense, differentiation and growth do conflict in ciliates, but at the level of structural assemblies rather than of whole cells.

Most ciliates with complex oral structures have a standard way of evading this conflict. They develop an entirely *new* OA, not only at cell division, but also whenever a need exists to replace old oral structures. If this happens after damage to or loss of a preexisting OA it is called *regeneration,* and if it happens for any of a variety of other reasons it is called *reorganization* or *oral replacement.* But whatever the stimulus or the name, the process itself is always the same. For any ciliate species that has an OA made up of multiple ciliary units, one could draw a *single* diagram of oral development in that species (such as Fig. 2.10 for *Tetrahymena*) that would, except for details of the location and mobilization of the original anarchic field, be equally accurate for *all* types of oral development carried out by that species. This simplifies the problem of analysis of oral development because a single underlying physiological state of activation is likely to operate for all cases of oral development (Tartar, 1961, Chapter VIII), and a single set of instructions might be used to construct an OA irrespective of circumstances.

The activation of oral development characteristically stimulates other processes in addition to formation of a new OA. This is obvious in cell division, in which oral development is integrated with formation of a new cytoproct and new contractile vacuole pores as well as of the division furrow itself (see figures in Chapter 3). Some, although not all, of these additional changes occur when oral primordia form under other circumstances: micronuclei commonly divide (Schwartz, 1935; Parker and Giese, 1966; Jerka-Dziadosz and Frankel, 1970) and macronuclei reorganize (Parker and Giese, 1966; Jerka-Dziadosz and Frankel, 1970; de Terra 1975). Even when nuclei appear largely unchanged, cells may develop according to the same time schedule and also undergo shape changes similar to those observed during cell division (Tamura et al., 1966; Frankel, 1970). There are thus both morphological and physiological grounds for considering reorganization (oral replacement) to be a type of "pseudo-division" (Tartar, 1966b) that takes place when some but not all of the conditions for cell division have been satisfied (de Terra, 1969). It is, therefore, reasonable to think of "morphogenesis as a common denominator" (Tartar, 1967b, pp. 34–37) with a high degree of integration of both oral and nonoral developmental processes.

2.4 STARVATION-MEDIATED DEVELOPMENT

2.4.1 Ciliate Sexuality

When sexually mature ciliates begin to starve in the presence of other sexually mature ciliates of the same biological species but another mating type, they unite and exchange gametic nuclei. This process, called *conjugation,* is very similar in all ciliates. The nuclear events are shown diagrammatically in Figure 2.11. The essen-

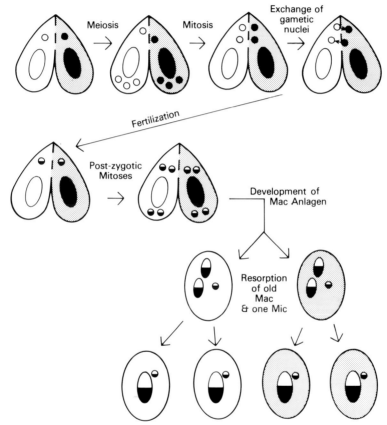

Figure 2.11   A schematic diagram of conjugation in *Tetrahymena.* The two conjugating cells enter meiosis synchronously. Three of the four products of meiosis are resorbed, whereas the remaining one divides mitotically to produce two pronuclei (gametic nuclei). These are exchanged reciprocally. Each gametic nucleus from one cell fuses with a gametic nucleus from the other cell, to form zygotic nuclei that are genically identical in the two cells. The zygotic nuclei then undergo two postzygotic divisions. Of the resulting four nuclei in each conjugant, two form macronuclear [Mac] anlagen that develop into new macronuclei, one forms the new micronucleus, and one micronucleus [Mic] is resorbed. The old macronucleus also is resorbed. The two new macronuclei of each exconjugant are segregated to two daughter cells while the micronucleus divides. In these diagrams, the nuclei of the two partners are marked in black and white to indicate formation of a heterozygote, and the cell bodies are shaded and unshaded to indicate the propagation of a cortical difference between the conjugating cells. Redrawn, with modifications, from Figure 1 of Bruns and Brussard (1974), with permission.

tial sexual process is a reciprocal exchange of mitotic products of a single meiotic nucleus, resulting in two genically identical zygotic nuclei. The products of the zygotic nuclei form both the micronuclei and the macronuclei of the cells of the next sexual generation, whereas the old macronuclei degenerate. This process allows genetic analysis in those ciliates in which the mating system has been brought under control, notably species of *Tetrahymena, Paramecium,* and *Euplotes.* Isolation of new recessive mutations is facilitated in ciliates (*Paramecium* and *Tetrahymena*) for which one can obtain homozygotes in one step by self-fertilization.

Apart from creating an opportunity for conventional genetic analysis, ciliate conjugation has one special advantage that makes it extremely useful for a less usual type of analysis of inheritance of cell surface patterns. Conjugating cells exchange nuclei and also exchange some internal cytoplasm (McDonald, 1966; Berger, 1976), but their cell surfaces remain separate. Although oral replacement takes place during conjugation (Ng and Newman, 1984a), the ciliary rows remain intact. Therefore, if two clones capable of interbreeding manifest any consistent difference in number or organization of ciliary rows, one can find out whether or not the cortical difference is based on a difference in nuclear genes simply by inducing conjugation between the two clones. If the cortical difference is based on a difference in nuclear genes then it should disappear during the cell divisions taking place after conjugation; if, however, the cortical difference is based on a difference in some self-perpetuating cortical pattern, then it should persist for a substantial number of cell divisions after conjugation. This test has been used to excellent advantage, notably by T. M. Sonneborn, in assessing the extent of cortical autonomy (see Chapter 4).

2.4.2 Cystment

Conjugation is only one of several developmental responses possible during starvation. A few ciliates are capable of "escaping" from the consequences of starvation by forming resting cysts, from which they can emerge when food is again present (reviewed by Bussers, 1984). The great significance of cystment to analysis of cell surface patterns in ciliates is that it offers another crucial test of continuity of ciliary patterns. During conjugation the nonoral ciliature generally persists unchanged as the nuclear apparatus undergoes sexual reorganization, whereas during cystment in *Oxytricha* and related ciliates the nuclear genome probably remains unaltered but the ciliature totally disappears (Grimes, 1973b). This, then, provides a different approach, complementary to the sexual one, of asking to what degree and in what manner preexisting ciliary organization is heritable.

2.4.3 Other Transformations

If starving cells cannot find mates and cannot or do not encyst, they may undergo any of a variety of other changes. The hypotrich *Stylonychia* can carry out repeated cycles of ciliary replacement while undergoing progressive diminution in size (Dembowska, 1938). In some species of *Tetrahymena,* oral replacement in starving cells is accompanied by other changes, including increase in number of cilia, formation of a long caudal cilium, and general elongation to form a "rapid swimmer" phenotype (Nelsen, 1978; Nelsen and DeBault, 1978) that is never observed in cells in nutrient medium. Other species of *Tetrahymena* are even more versatile. Under starvation conditions, *Tetrahymena patula* uses oral replacement as part of a more elaborate developmental sequence of cleavage within a reproductive cyst (Gabe and Williams, 1982), whereas in the presence of potential prey, *T. patula* and other *Tetrahymena* species *(T. vorax, T. paravorax)* go through a different type of oral replacement to form a specialized large-mouthed (macrostome) phenotype (Williams, 1960; Buhse, 1966a,b; Méténier and Groliere, 1979). Certain parasitic *Tetrahymena* species can undergo dramatic changes in number of ciliary rows associated with different life cycle stages in their hosts (Kozloff, 1956; Batson, 1983). These alternative phenotypes are useful in analysis of oral patterning, cell proportioning, and readjustments of cell surface patterns; they allow us to ask what aspects of cell-surface patterns change and what aspects remain the same when a ciliate is challenged to make a maximum adjustment to drastically altered conditions.

3

The Experimental Cast

3.1 INTRODUCTION

3.1.1 Why Is the Cast So Large?

This chapter introduces seven ciliates. A nonciliatologist reader is likely to ask why a book whose purpose is to search for unity must present so much diversity. Although a part of any honest answer to this question is that investigators in this field have tended to spread their efforts out over many organisms rather than concentrate on a few, there are also serious intellectual reasons for taking diversity seriously, especially as we search for unity. The most important reason for examining several ciliates is that when there are major differences among organisms within a large taxonomic group, a comparative approach is essential to ascertain which rules and mechanisms are most widespread and hence most likely to be of general importance. For example, the two ciliate genera best suited for genetic analysis, *Paramecium* and *Tetrahymena,* differ greatly in regulative capacity: the former is a ciliate counterpart to a mosaic embryo, whereas the latter is more like a regulative embryo. Thus it becomes a matter of great importance to ascertain whether phenomena of local structural determination discovered in *Paramecium* are also observable in *Tetrahymena,* and conversely whether any counterpart to the global regulation found in *Tetrahymena* also exists in *Paramecium.*

A second justification for considering several ciliates relates to the differing experimental opportunities provided by different ciliates. *Paramecium, Tetrahymena,* and *Euplotes* are currently the most suitable objects for genetic analysis but offer limited opportunities for microsurgical study. The large ciliates, *Stentor* and *Blepharisma,* are the best organisms for microsurgical analysis, but have not yet been used in genetic studies of pattern formation. Finally, *Oxytricha/Stylonychia* and *Paraurostyla* are the best suited for a combination of precise cytological analysis with microsurgery and other ways of altering ciliary patterns. We can learn something important from each of these groups of ciliates that we cannot now learn from the others.

Despite the diversity of characters in our story, this book will nonetheless be organized by theme and not by character. To make this approach feasible without cumbersome descriptive digressions, the major members of the cast are introduced at the outset, concentrating on those traits of each member that are important in the subsequent narrative.

3.1.2 A Note on Phyletic Distances

Our interpretation of biological differences among experimental organisms is affected by ideas concerning the closeness of their phyletic relationships. In two important ciliate cases, persuasive new evidence has shown that traditional presumptions are incorrect.

The first such case involves the two best-known ciliates, *Paramecium* and *Tetrahymena*. From 1950 to 1980, these two genera were considered to be members of the same ciliate order, the Hymenostomatida (Corliss, 1979). If this close taxonomic relationship truly reflects a relatively recent common ancestry, then the existence of major differences between these two organisms would imply that developmental mechanisms might be quite labile (in an evolutionary sense) in ciliates. Recently, however, the taxonomy of ciliates has undergone a major reconstruction (Small and Lynn, 1981) based on the postulate of an inverse relationship between the level of complexity of cell-surface structural configurations and their evolutionary conservatism (Lynn, 1981). In this reconstructed taxonomy, *Paramecium* and *Tetrahymena* are placed in different classes, sharing only a subphylum in common (Small and Lynn, 1981, 1985). Small and Lynn's conclusion of a great phyletic distance between *Paramecium* and *Tetrahymena* has recently received strong support from the molecular phylogeny of ribosomal RNA genes (Sogin and Elwood, 1986; Baroin et al., 1988). This parallelism of conclusions founded on altogether different types of evidence suggests that the biological differences between these two genera might reflect ancient and possibly quite stable divergences.

The other taxonomic and phyletic divorce that is important for us involves *Euplotes* and two other well-known ciliates, *Oxytricha* and *Stylonychia*. This trio had traditionally been placed together in the same order, the Hypotrichida, indeed since 1961 in the same suborder, the Sporadotrichina (Corliss, 1979). As the structure and development of these ciliates became better known, it became increasingly clear that *Euplotes* is very different from the other two, although different taxonomists have made widely different assessments concerning the taxonomic level at which the Euplotids should be separated from the other hypotrichs, from suborder (Fleury et al., 1985; Tuffrau, 1987) to subphylum (Small and Lynn, 1985). The decision to split off *Euplotes,* while keeping *Oxytricha* and *Stylonychia* together, has been supported strongly by the subsequent and conceptually independent molecular analysis of ribosomal RNA genes (Sogin et al., 1986a). The most recent molecular diagnosis of ciliate relationships supports a phyletic separation that is very deep, although not quite as profound as originally envisaged in the Small–Lynn taxonomic scheme (Lynn and Sogin, 1988). This recognition justifies introducing *Euplotes* separately from the other hypotrichs.

Sogin and Elwood conclude that the "... deep branching pattern within the ciliate line of descent indicates that the major classes of the Ciliophora arose during a relatively early radiative period of eukaryotic evolution" (Sogin and Elwood, 1986, p. 59). Thus, we should not be surprised to find major differences among our representative ciliates. Recognition of the depth of phyletic branching increases the significance of any widespread developmental mechanisms, because it strongly suggests that such mechanisms themselves are ancient.

3.2 THE GENETICAL TRIO: *TETRAHYMENA, PARAMECIUM,* AND *EUPLOTES*

The three ciliates that I introduce first are united only by the methods used for their investigation. Most studies on patterning of these organisms have made use of genetics, either in the strict sense of assessment of results of sexual crosses or in the broader sense of analysis of inheritance of differences maintained by clonal lineages. Results have been assessed using methods of silver-staining, which at a minimum allow ascertainment of the locations of basal bodies situated at the cell surface (the Chatton–Lwoff wet-silver procedure) and at a maximum permit visualization of the asymmetrical organization of ciliary rows (the protein-silver or protargol technique). These methods have been supplemented by transmission and scanning electron microscopy, and, recently, by immunofluorescence (Iftode et al., 1989), and by high-resolution interference-contrast light microscopy (Aufderheide, 1986). This combination of techniques has been used to approach the whole range of problems of pattern formation in ciliates; hence these organisms (especially *Tetrahymena* and *Paramecium*) will make frequent appearances in subsequent chapters.

3.2.1 *Tetrahymena*

When speaking of genetic analysis of development, *"Tetrahymena"* means *Tetrahymena thermophila.* However, *T. thermophila* is a part of a complex assemblage of sibling and nonsibling *Tetrahymena* species (Nanney and McCoy, 1976). Although most of the comparative considerations bearing on the *Tetrahymena* assemblage are beyond the scope of this book, it is instructive to note that during a probable evolutionary history of 30 to 40 million years (Van Bell, 1985) cortical patterns have tended to remain conservative despite considerable changes in cortical proteins (Williams, 1984, 1986a). Therefore, the following description, although focused on *T. thermophila,* actually applies to the organization of all bacteria-feeding *Tetrahymena* forms.

Statics

Figure 3.1 gives a composite portrait of the surface of *T. thermophila,* as seen after staining with protein silver. This ciliate is about 50 micrometers long, and is somewhat pear-shaped. The oral apparatus (OA) is located near the anterior end of the cell. The remainder of the cell surface typically is covered by 17 to 21 longitudinal ciliary rows. The basal bodies, cilia, and transverse (TM) and longitudinal (LM) microtubule bands are easily distinguished in these preparations. Most of the ciliary rows extend from the posterior end of the cell to the apical crown (Fig. 3.1b, AC) or to the short anterior suture (Fig. 3.1b, AS). The exceptions, two ciliary rows that terminate at the posterior end of the OA, are known as the postoral ciliary rows. They are labeled "1" and "n," following a traditional numbering system in which the right postoral ciliary row is assigned the number 1 and numbering proceeds in an ascending order from the cell's left to right (clockwise as viewed from the anterior end), ending at the left postoral or *n*th ciliary row.

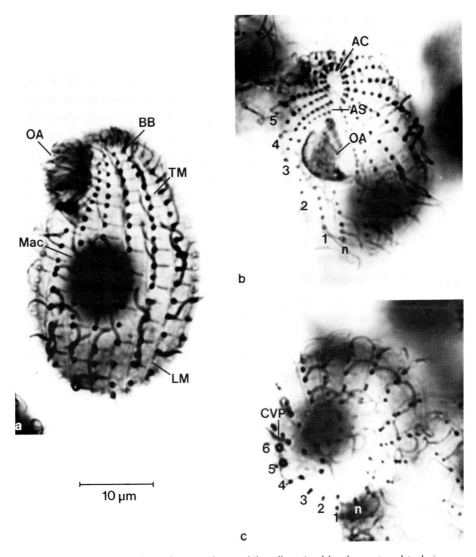

Figure 3.1 Three views of *Tetrahymena thermophila* cells stained by the protargol technique. Basal bodies [BB], cilia, and accessory microtubule bands are stained. (**a**) A view of the left side, showing the ciliature of the oral apparatus [OA] in profile. Six ciliary rows are in focus. Transverse microtubule bands [TM] and longitudinal microtubule bands [LM] are clearly visible. The macronucleus [Mac], located in the center of the cell, is out of focus. (**b**) An anteroventral view. The oral apparatus [OA] and the two postoral ciliary rows, 1 and *n*, are in clear view. Anterior to the oral apparatus some ciliary rows converge on a short anterior suture [AS]. An apical crown [AC] of basal body pairs encircles the anterior end of the cell. (**c**) A posterior view, focused through the cell. The ciliary rows terminate at varying distances from the posterior end of the cell. Contractile vacuole pores [CVP] are located at the posterior ends of ciliary rows 5 and 6. The scale bar represents 10 μm.

Although most of the units of the ciliary rows of *Tetrahymena* are solo ciliary units (see Fig. 2.6), paired ciliary units are found at the anterior ends of most ciliary rows (Jerka-Dziadosz, 1981d). These paired units are aligned laterally to form a conspicuous apical crown (Fig. 3.1b, AC), which characteristically extends from row 5 to row *n*-2 (McCoy, 1974). The other ciliary rows (the two postoral rows plus rows 2 to 4 plus row *n*-1) have solo basal bodies at their anterior ends. The crown of paired ciliary units thus is asymmetrical with respect to the OA, a feature that will be important in Chapters 8 and 9, when we consider symmetry reversals in *Tetrahymena.*

The total number of basal bodies in all of the ciliary rows in *T. thermophila* is near 400 at the start of the cell cycle, with a nearly even distribution of basal bodies among rows, except for a lesser number in the postoral rows (Nanney et al., 1978). The average number of ciliary units per row is negatively correlated with the number of ciliary rows, so that the total number of ciliary units in all of the ciliary rows maintains a rough constancy irrespective of the number of rows (Nanney, 1971b).

The ciliature of the OA consists of the four compound ciliary structures from which *Tetrahymena* gets its name, the undulating membrane (Fig. 3.2, UM) and the three membranelles (Fig. 3.2, M1, M2, M3). The membranelles are located within a relatively shallow buccal cavity, whereas the UM is perched on the right margin of that cavity. Consideration of the detailed organization of these structures is postponed to Chapter 7, where the normal organization is compared to that observed in mutants.

The cytoproct is an elongated slit located near the posterior end of row number 1 (Fig. 3.2, Cyp). The single contractile vacuole is situated just beneath the surface layer near the posterior end of the cell, and typically opens through two round contractile vacuole pores (CVPs), which in 19-rowed cells are typically located near the posterior ends of rows 5 and 6 (Fig. 3.1c).

Dynamics

In *Tetrahymena,* new basal bodies are formed anterior to old ones within the ciliary rows (see Fig. 2.8 and accompanying text). Most of this basal body proliferation occurs at the same time as oral development (Nelsen et al., 1981). Basal bodies are produced along most of the length of the two postoral ciliary rows and along approximately the posterior two thirds of the other rows (Nanney, 1975), with a peak in intensity of proliferation in the midbody region (Kaczanowski, 1978). Although the total number of basal bodies in the ciliary rows doubles in each cell cycle, and the two daughter cells get approximately equal numbers of basal bodies (Nanney et al., 1978), this equality clearly is *not* achieved by each individual old basal body generating one new one per cell cycle (Nanney, 1975). Instead, some anterior basal bodies are not involved in the formation of new basal bodies, whereas other posterior basal bodies generate more than one new basal body (Nanney, 1975). The two daughter cells thus get a varying mix of "old" and "new" basal bodies, with a greater proportion of "old" ones in the anterior division product and of "new" ones in the posterior division product (Frankel et al., 1981).

All of the prominent fibrillar systems of the cell cortex (see Fig. 2.6), including the longitudinal microtubule bands (Ng, 1979a) and the kinetodesmal fibers (Nelsen, personal communication; Buzanska, personal communication), remain intact through all stages of the cell cycle.

When *Tetrahymena* cells prepare to divide, new oral structures, new cytoprocts, and new CVPs develop at their final locations (Fig. 3.2). A new OA is produced from a field of basal bodies that forms adjacent to the equatorial region of a ciliary row, usually the right postoral row (row 1). Membranelles and the UM become organized within this field (see Fig. 2.10 and accompanying text). A fission zone then develops as an equatorial ring of discontinuities in the ciliary rows (Fig. 3.2c, FZ). The new contractile vacuole pores (Fig. 3.2c, nCVP) and the new cytoproct (Fig. 3.2d, nCyp) form just anterior to the fission zone, usually along the same ciliary rows as the corresponding old structures. New contractile vacuoles appear internal to their pores at about the same time as the pores develop at the surface.

In *Tetrahymena,* these new structures all develop at a considerable distance from old structures of the same kind, so that after constriction is completed they will be in their proper places in the daughter cells, with no need for further migration or displacement. The anterior division product will then have acquired the old OA plus the newly formed cytoproct and CVPs, whereas the posterior division product will have obtained the newly formed OA together with the presumed old cytoproct and old CVPs.

The micronucleus divides during the period of alignment of oral membranelles, before cell division begins, whereas the macronucleus divides later, during cell con-

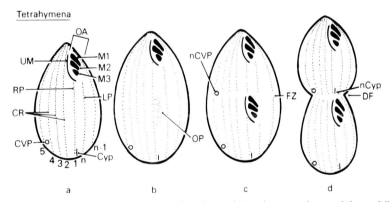

Figure 3.2 Structural features of the ventral surface of *Tetrahymena thermophila* at different developmental stages. (**a**) A nondeveloping cell. Seven longitudinal ciliary rows [CR] are shown, including the right postoral [RP] and left postoral [LP] rows. The numbering scheme for the ciliary rows corresponds to that illustrated in Figure 3.1 The three membranelles [M1, M2, M3] plus the undulating membrane [UM] of the oral apparatus [OA] are located near the anterior end of the cell, and the cytoproct [Cyp] and the two contractile vacuole pores [CVP] are near the posterior end (only one of the two CVPs is visible in the diagram). (**b**) The anarchic field stage of oral development. An oral primordium [OP] appears at midbody adjacent to the RP ciliary row. (**c**) The stage of completion of the membranelles and UM of the new OA (corresponding to the end of phase B in Fig. 2.10). The fission zone [FZ] appears as an equatorial discontinuity in the ciliary rows, just anterior to the new OA. New CVPs [nCVP] are formed just anterior to the fission zone. (**d**) Cell division. The division furrow [DF] constricts along the site of the fission zone. The new cytoproct [nCyp] becomes apparent at the posterior end of the RP ciliary row of the anterior division product. Redrawn with modifications from Figure 1 of Nelsen and Frankel (1986).

striction. Both nuclei are located near the center of the cell during most of the cell cycle, but the micronucleus moves out to the cell surface at the beginning of mitosis and maintains a temporary association with ciliary rows as it completes its mitotic division (Jaeckel–Williams, 1978).

As was mentioned in Section 2.4.3, oral development can occur in *Tetrahymena* even when cell division does not. Typically, a composite anarchic field is formed, in part near the anterior end of the right postoral ciliary row, in part from the disaggregation of the basal bodies of the UM of the old OA (Frankel, 1969a; Kaczanowski, 1976; Nelsen, 1978). The membranelles of the old OA are resorbed while the membranelles and UM of the new OA are becoming organized. In *T. thermophila*, oral replacement often is integrated into special developmental sequences triggered by starvation, such as conjugation or transformation into a thin "rapid swimmer" form. In conjugation, resorption of old oral structures, including the UM, sometimes *precedes* the appearance of a new oral primordium, in which case an anarchic field is formed de novo near the anterior end of a ciliary row (Tsunemoto et al., 1988, Fig. 28a). This also occurs in reproductive cysts of *T. patula* (Gabe and Williams, 1982). In these reproductive cysts, the temporal relationship between resorption of old structures and development of new ones is known to be variable (Gabe and Williams, 1982).

With the exception of the parasitic *T. rostrata* (Stout, 1954), no *Tetrahymena* species is known to "escape" from starvation by forming resting cysts.

3.2.2 *Paramecium*

Paramecium, like *Tetrahymena,* is a genus that includes a large assemblage of sibling and nonsibling species. The members of the *"Paramecium aurelia"* sibling-species swarm (Sonneborn, 1975b) were the first to be domesticated genetically. Our description, therefore, will be of these organisms, and of *P. tetraurelia* in particular, with occasional borrowing of data from other *Paramecium* species.

Statics

Paramecium is both larger and more complex than *Tetrahymena*. The average length of nondividing *Paramecium tetraurelia* cells is somewhat over 100 micrometers (Kaneda and Hanson, 1974). The general arrangement of structures on the cell surface is shown in Figure 3.3. The compound oral ciliature is located in a deeply invaginated buccal cavity, situated in the midregion of the cell. Only its opening is shown in Figure 3.3; the internal details are drawn in Figure 3.4. The buccal cavity opens into a shallower but broader concavity of the ventral cell surface.

The ciliary rows that cover the nonoral surface of *Paramecium tetraurelia* are more numerous [about 75 (Chen-Shan, 1970)] and also have a larger number of ciliary units [about 70 for a middorsal ciliary row (Kaneda and Hanson, 1974)] than in *Tetrahymena*. Whereas the ciliary rows of the dorsal surface extend longitudinally (Fig. 3.3b), those on the ventral surface have curved paths, with the curvature greatest close to the opening of the buccal cavity (Fig. 3.3a). The majority of the ciliary rows terminate at two sutures, an anterior suture (Fig. 3.3, AS) and a

Paramecium

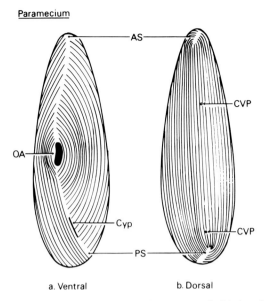

a. Ventral   b. Dorsal

Figure 3.3   The surfaces of *Paramecium tetraurelia,* (**a**) ventral, (**b**) dorsal. The ciliary rows are shown as lines, which abut on the anterior suture [AS] and posterior suture [PS]. The cytoproct [Cyp] is situated within the posterior suture, and the contractile vacuole pores [CVP] are located between ciliary rows in the mid-dorsal surface of the cell. Only the opening of the oral apparatus [OA] is shown. Redrawn with slight modification from Figures 3B and 3D of Kaneda and Hanson (1974), with permission.

posterior suture (PS), which extend anteriorly and posteriorly from the opening of the buccal cavity (Fig. 3.3a), and bend over the cell apices to terminate on the dorsal surface (Fig. 3.3b).

Most ciliary rows contain both solo and paired ciliary units. In general, the ventral and anterior ciliary units are paired, the dorsal and posterior ones solo (Sonneborn, 1970; Iftode et al., 1989). The ultrastructural organization of the surface of *Paramecium* is considerably more complex and elaborate than that of *Tetrahymena,* as in addition to all of the "standard" cortical elements *Paramecium* has a regular outer lattice of fine fibers located just beneath the cell membrane along cortical ridges, a more irregular infraciliary lattice situated at the base of the cortex (Parducz, 1962; Allen, 1971; Garreau de Loubresse et al., 1988), and a system of microtubules attached to the nonoral basal bodies (Plattner et al., 1982) that extends deep into the interior of the cell (Cohen and Beisson, 1988). The first of these systems is probably associated with the exceptional structural rigidity of the surface of this cell, the second with its limited contractility, and the third with interactions between the cell surface and structures of internal origin (Cohen and Beisson, 1988).

The compound oral ciliature of *Paramecium* is topologically similar to that of *Tetrahymena,* with an undulating membrane (UM) at the right margin of the buccal cavity and three membranelles within the buccal cavity to the left of the UM (Fig. 3.4). The buccal cavity is much deeper, however; also, the membranelles are considerably longer than in *Tetrahymena,* and the four constituent basal-body rows of membranelle 3 are unusually widely spaced. The UM of *Paramecium* is short and inconspicuous but organized in the same way as the *Tetrahymena* UM (Patterson, 1981; Grain, 1984, pp. 128–129). *Paramecium* workers have developed a terminology of their own for the oral ciliary structures of this organism, calling membranelles 1 and 2 "peniculi", membranelle 3 a "quadrulus", and the UM an

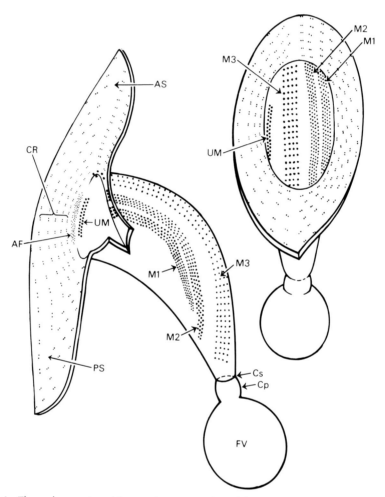

Figure 3.4   The oral apparatus of *Paramecium tetraurelia* and the nonoral cell surface immediately to its right. (**a**) Side view looking toward the right side of the buccal cavity and the surrounding depression of the body surface. The undulating membrane [UM] is on the right wall of the buccal cavity near its opening, whereas the three membranelles [M1, M2, and part of M3] are on the left wall of the buccal cavity. The cytostome [Cs], the cytopharynx [Cp], and a forming food vacuole [FV] are also shown. A permanent anarchic field [AF] is located at the right opening of the buccal cavity, between the UM and the nonoral ciliary rows [CR] immediately to its right. The ciliary rows adjacent to the right side of the buccal cavity terminate on the anterior suture [AS] and posterior suture [PS]. (**b**) Frontal view into the opening of the buccal cavity, showing parts of the three membranelles [M1, M2, M3] as well as the undulating membrane [UM].

"endoral membrane"; however, ultrastructural and developmental studies have removed the justification for a unique terminology. The one major special feature in the organization of the oral apparatus (OA) of *Paramecium* is the presence of a *permanent* anarchic field of unciliated basal bodies, located between the UM and the nonoral ciliary row immediately to its right (Hufnagel, 1969; Jones, 1976; Shi, 1980a); the reality and location of this field was verified by thin-section electron microscopy (Patterson, 1981).

The cytoproct of *Paramecium* is located posterior to the OA, within the posterior suture (Fig. 3.3a, Cyp). There are two contractile vacuoles, with their pores (Fig. 3.3b, CVP), located in anterior and posterior regions between two middorsal ciliary rows, not always the same two rows (Kaneda and Hanson, 1974).

Dynamics

In *Paramecium,* as in *Tetrahymena,* new basal bodies are formed anterior to old ones within the axis of the ciliary rows. All new ciliary units are produced in the late part of the cell cycle, coincident with nuclear divisions and oral development (Kaneda and Hanson, 1974). All old ciliary units participate in proliferation except for some located near the poles (Suhama, 1971; Iftode et al., 1989). Proliferation occurs in two waves, the first progressing from the site of the fission zone toward the poles, the second moving from anterior to posterior in both division products. The first wave generates new ciliary units, whereas the second produces the second (anterior) basal body within those of the new units that are to become paired (Iftode et al., 1989). The net result is that the basal bodies near the poles do not proliferate at all, whereas those close to the equator triplicate or quadruplicate (Gillies and Hanson, 1968; Suhama, 1971; Iftode et al., 1989). It is even clearer in *Paramecium* than in *Tetrahymena* that the nearly equal distribution of ciliary units to the two division products is an outcome of a globally regulated process, not of a simple one-for-one replication of basal bodies (Suhama, 1971; Iftode et al., 1989).

Postciliary and transverse microtubule bands persist intact throughout the cell cycle (Cohen et al., 1982), whereas the longitudinal microtubule bands, which are absent during the interfission period, appear transiently while cells are elongating during fission (Sundararaman and Hanson, 1976; Cohen et al., 1982). At the same time, all old kinetodesmal fibers are resorbed and entirely new ones are formed (Fernández-Galiano, 1978; Iftode et al., 1989). The meshes of the outer lattice elongate and become subdivided (Cohen et al., 1987) simultaneously with the subdivision of the epiplasmic scales (Iftode et al., 1989) and reorganization of the infraciliary lattice (Garreau de Loubresse et al., 1988). All of these events proceed in waves from the fission zone to the ends of the cell, but, except for the subdivision of the epiplasm and outer lattice, the progression of the different waves are not closely coordinated with each other (Iftode et al., 1989).

In *Paramecium,* the oral ciliature of the posterior division product is generated from the anarchic field located at the right margin of the old OA (Fig. 3.4, AF). The field enlarges with the addition of new basal bodies, after which the membranelles become organized much as in *Tetrahymena.* The UM and anarchic field of the posterior oral structures are formed from a portion of this same field at a late stage (Jones, 1976; Shi, 1980b; Perez-Paniagua et al., 1988). The new ciliary structures become enclosed in a new buccal cavity that forms by an inpocketing of the right

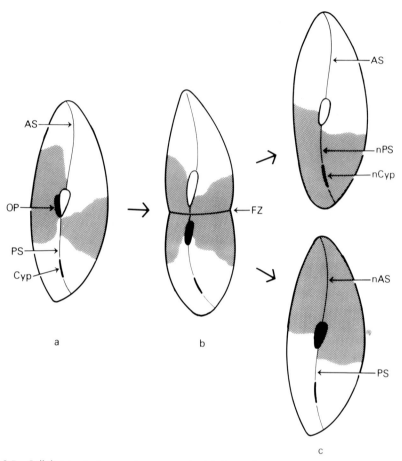

Figure 3.5 Cell division in *Paramecium tetraurelia*. (**a**) The prefission cell, with an oral primordium [OP] located in an evagination *(blackened)* of the right wall of the old buccal cavity. The anterior suture [AS] and posterior suture [PS] are indicated as lines, with the cytoproct [Cyp] situated in the posterior suture. The *shaded* region represents that portion of the cell destined to become the future posterior half of anterior fission product and anterior half of the posterior fission product. (**b**) The separation of the old *(white)* and new *(black)* buccal cavities and formation of the fission zone [FZ]. (**c**) The division products, showing the newly formed posterior suture [nPS] and new cytoproct [nCyp] of the anterior division product and the newly formed anterior suture [nAS] of the posterior division product. Redrawn, with modifications, from Figure 18 of Iftode et al. (1989), with permission.

·wall of the old one, and this new cavity then buds off from the old one and moves in a posterior direction (Yusa, 1957; Roque, 1961).

The OA of the anterior division product retains the intact membranelles of the preexisting OA, whereas the UM and anarchic field of this OA develop from an initial row of basal bodies "which emerges on the site where the old endoral (= undulating) membrane has existed" (Shi, 1980b). The origin of this initial row is not completely clear, but it appears to be continuous with the old UM, most of which has been resorbed before separation of the two sets of oral structures (Perez-Paniagua et al., 1988).

We could think of the OA of *Paramecium* as a self-reproducing structure, as both the anterior and posterior OAs may arise from different parts of the preexisting OA; however, we cannot be completely confident about this conclusion, as the sources and fates of the anarchic field and undulating membrane are very hard to follow.

New contractile vacuoles and their pores appear at about the same time that oral development begins, a short distance anterior to the corresponding old structures (King, 1954). The new cytoproct forms much later, during cytokinesis, within the growing posterior suture of the anterior division product (Kaneda and Hanson, 1974). At the same time, the old cytoproct becomes remodeled extensively, possibly being replaced in situ (Dubielecka and Kaczanowska, 1984).

Cell division in *Paramecium* is partially a process of growth. The cell shortens just before cell division begins and then elongates dramatically, with growth occurring most rapidly in the midregion (Fig. 3.5a, stippled zone) (Kaneda and Hanson, 1974; Iftode et al., 1989). A fission zone spreads laterally in both directions from the place where the buccal openings are separating (Fig. 3.5b). As division constriction occurs, the new posterior and anterior sutures form between the anterior and posterior OAs (Fig. 3.5c). Thus, in *Paramecium,* the posterior half of the anterior fission product and the anterior half of the posterior division product are formed during the division process (Schwartz, 1963); however, the cortical structures of these regions are a mosaic of interspersed old and new elements.

Although the separation of OAs during cell division is largely a consequence of growth of the intervening region, contractile vacuoles and their pores migrate to their final positions. As initially deduced by King (1954) and subsequently confirmed by Suhama (1973), the major increase in distance between the CVPs that takes place during cell division is a result of a movement relative to other surface structures, mostly a posterior movement of the posterior pores in both nascent daughter cells (Suhama, 1973).

Because actual movement of a structure within the plane of an apparently rigid cell surface may appear surprising to many readers, a brief digression is warranted. In the ciliate *Nassula* (which is not described in detail), a posterior migration of the CVP over a similar distance is associated with the transient appearance of a fibrous strand, composed of both microtubules and filaments, which is anchored to a basal body at its posterior end and to the migrating CVP at its anterior end (Tucker, 1971b). The relevant micrograph (Fig. 30 in Tucker, 1971b) gives the strong impression that these fibers might actually be pulling the CVP posteriorly. Perhaps something similar is also taking place in dividing *Paramecium* cells.

Whereas in *Tetrahymena* the micronucleus moves to the surface of the cell before cell division while the macronucleus remains centrally located, in *Paramecium* it is the macronucleus that makes the most striking migration. The macronucleus, originally centrally located, migrates toward the anterodorsal region of the cell cortex, elongates dramatically just before the onset of cytokinesis, and then divides while cell constriction is occurring (Tucker et al., 1980). During the entire process of its elongation and division, the macronucleus is situated about 4 micrometers beneath the cell surface, at the base of the layer of inserted trichocysts. The interaction of trichocysts with the cell surface is necessary for proper segregation of macronuclei to division products. All mutants that affect attachment of trichocysts to the cell surface also prevent positioning of the macronucleus under the cell cortex and cause subsequent misdivision of macronuclei (Cohen and Beisson, 1980; Cohen et al., 1980). Mutants that merely prevent discharge of normally positioned trichocysts do not affect macronuclear positioning or division (Cohen and Beisson, 1980). Recent studies indicate that the absence of trichocysts alters the spatial organization of a microtubule meshwork that usually extends perpendicular to the cell surface; when inserted trichocysts are lacking, this meshwork is oriented predominantly parallel to the surface (Cohen and Beisson, 1988). Thus it is likely that this microtubule meshwork performs some essential function in the normal division of the macronucleus in *Paramecium.*

When starving, *Paramecium tetraurelia* may either conjugate or undergo self-fertilization (autogamy). During both processes, oral replacement occurs. Its early stages are remarkably similar to those observed before cell division, although later on the old membranelles become resorbed as the new ones are formed (Ng and Newman, 1984a). This oral replacement during sexual reorganization requires the activity of the prezygotic macronucleus (Tam and Ng, 1987) and micronuclei (Ng and Mikami, 1981; Ng and Newman, 1984b; Jurand and Ng, 1988), but not of the postzygotic macronuclear anlagen (Ng and Newman, 1985, 1987).

The process of conjugation includes some remarkable cytoplasmic localizations that affect nuclear destinies. First, after completion of the meiotic divisions, a *paroral cone* formed near the OA "rescues" one of the four haploid products of meiosis that otherwise would degenerate (Sonneborn, 1954; Yanagi and Hiwatashi, 1985; Yanagi, 1987). Much later, two of the mitotic products of the zygotic nucleus are determined to differentiate into new macronuclei under the influence of transiently localized posterior cytoplasm (Mikami, 1980; Grandchamp and Beisson, 1981). These localizations are probably dependent on regional differences in the cell surface.

Although *Paramecium* has great sexual capabilities, it has no other way of coping morphogenetically with starvation. Not only does it fail to form cysts, but it also does not undergo oral replacement under nonsexual circumstances. This implies that in a vegetative pedigree the line of successive anterior division products should inherit the same membranelles, with at best limited possibilities for reconstruction and none for replacement (Hanson, 1962). Siegel (1970) has observed that lineages of successive anterior division products have significantly longer and more variable generation times than lineages of successive posterior division products, and suggested unrepaired damage to the OA as a possible cause.

3.2.3 *Euplotes*

Euplotes is a dorsoventrally flattened ciliate with a dramatic difference in the ciliature of the dorsal and ventral surfaces. The ventral surface has a highly specialized compound ciliature adapted to crawling on the substratum, whereas the dorsal and lateral surfaces are sparsely covered by rows of stiff cilia, of presumed sensory function (Görtz, 1982). Because the analyses of *Euplotes* to be considered in this book relate primarily to the dorsal ciliature of *Euplotes minuta,* we restrict our description to this one system.

Euplotes

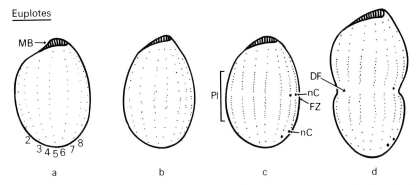

Figure 3.6 Development on the dorsal surface of *Euplotes minuta,* in a 9-rowed cell. Each ciliary unit is represented by a dot. The structure at the anterior end is a portion of the membranelle band [MB], most of which is on the ventral surface (not shown). (**a**) A stage before proliferation of ciliary units. (**b**) An early stage of unit proliferation. Some of the ciliary units have duplicated. (**c**) A stage near the end of proliferation. The region of proliferation [Pl] is clearly demarcated by the closely spaced ciliary units. The fission zone [FZ] appears near the center of this region as an equatorial band of discontinuities in the ciliary rows. New caudal cirri [nC] develop adjacent to the posterior ends of rows 7 and 8. (**d**) Cell division. The division furrow [DF] constricts along the fission zone. Redrawn from Figure 1 of Frankel, 1973a.

Euplotes minuta is, as its name implies, the smallest *Euplotes,* with an average length of about 50 micrometers (Machelon et al., 1984; Valbonesi et al., 1988), and thus is similar in size to *T. thermophila.* The dorsal ciliature of *E. minuta* consists of 7 to 10 ciliary rows (Frankel, 1973a; Valbonesi et al., 1988) with 8 to 16 ciliary units per row (Fig. 3.6a), a total of about 100 units in all (Heckmann and Frankel, 1968). Although each unit of the *Euplotes* ciliary rows appears single in the light microscope, ultrastructural studies show it to be paired, with only one of the two basal bodies bearing a cilium (Ruffolo, 1976a; Görtz, 1982). The ciliary units are apportioned among ciliary rows according to a precise spatial pattern (Frankel, 1975a).

All new basal bodies within the ciliary rows of *Euplotes* develop just before cell division in an equatorial zone located on both sides of the future division site (Fig. 3.6b–d) (Ruffolo, 1976b and references cited therein). This zone is similar to that of *Paramecium,* but is more restricted and more clearly demarcated. In *Euplotes minuta,* about 5 old ciliary units within each row participate in the formation of 2 to 3 new units apiece, to make a final complement of approximately 18 units (Fig. 3.6c) (Frankel, 1973b). The units from the zone of proliferation are then distributed somewhat unequally to the two division products, balancing reciprocal inequalities in the number of nonproliferating ciliary units at the anterior and posterior ends (Frankel, 1975a). The equal distribution of basal bodies to division products is thus achieved by a compensating process that once again implies a global integration.

3.3 THE MICROSURGICAL DUO: *STENTOR* AND *BLEPHARISMA*

The ciliates of our second group, unlike the first, are closely related. *Stentor* and *Blepharisma* both belong to the same suborder, the Heterotrichina (Small and

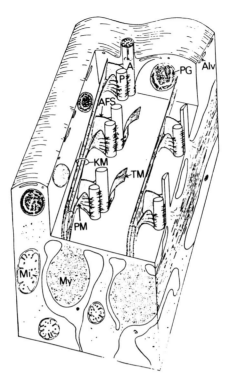

Figure 3.7 A portion of the nonoral cell surface of *Stentor coeruleus*, showing segments of two ciliary rows and their adjacent fibrillar elements. The anterior direction is toward the top of the diagram. The cell surface is underlain by scattered alveoli [Alv]. The ciliary units consist of a ciliated anterior basal body [A] and a nonciliated posterior basal body [P]. Associated with each ciliary unit are an anterior fiber sheet [AFS], a transverse microtubule band [TM], and a postciliary microtubule band [PM]. The postciliary microtubule bands overlap to form the KM fiber. Other elements shown are the contractile myonemes [My] located beneath the ciliary row, the pigment granules [PG], and the mitochondria [Mi]. Slightly modified from Figure 12 of Huang and Pitelka (1973), with permission.

Lynn, 1985). These ciliates are both large and eminently operable, healing easily and capable of regenerating even after drastic surgery. On the other hand, they tend to be difficult objects for cytological study, in large part because the abundant pigment granules located near the cell surface interfere with visualization of ciliary structures in silver-stained specimens. Most observations, therefore, have been made by light microscopy of living specimens.

I introduce *Stentor* and *Blepharisma* together. The reason for this joint introduction is that these two heterotrichs are rather similar both topologically and developmentally despite a striking difference in cell shape and not very recent evolutionary divergence (Baroin et al., 1988). The more familiar of the two, *Stentor*, is also geometrically the more deviant; its peculiar geometry can be understood better if its description is coupled with that of the more conventionally organized *Blepharisma*.

Statics

The overall organization of the nonoral ciliary rows of *Stentor* and *Blepharisma* differs from that of *Tetrahymena* (Fig. 2.6) in two respects. First, in the heterotrichs *all* of the ciliary units are paired, with only the anterior basal body of each pair ciliated. Second, striated kinetodesmal fibers are absent [although a fibrillar sheet (Fig. 3.7, AFS) is present that might be its homolog]. The postciliary microtubule bands (Fig. 3.7, PM) are very long and overlap to form a composite microtubular structure called a KM-fiber (Fig. 3.7, KM).

Two other important features of the heterotrich cell surface are the myonemes (My) and pigment granules (PG). *Stentor* is highly contractile and *Blepharisma* slightly so, and the myonemes (Fig. 3.7, My) are strongly suspected to be the active contractile structures, whereas the KM system may be involved in relaxation (Huang and Pitelka, 1973). The pigment granules (Fig. 3.7, PG) are localized in the cell surface layer. These granules are most concentrated in the spaces *between* the ciliary rows. In living cells these interrow stripes of concentrated pigment are highly visible, whereas the ciliary rows are seen as "clear stripes" between the pigmented stripes. The ciliary row–pattern is, therefore, a negative of the stripe pattern.

The largest species of *Blepharisma* and *Stentor*, *B. japonicus* and *S. coeruleus*, respectively, are understandably the favorites for microsurgery. The former ranges from 250 to 500 micrometers in length, the latter from 500 micrometers to a millimeter. The overall organization of the ventral surfaces of *Blepharisma* and *Stentor* are illustrated in Figures 3.8a and 3.8e. In both of these ciliates, the oral apparatus is very large and has many membranelles, each with two or three rows of cilia plus a single long undulating membrane (UM). In *Blepharisma*, the membranelles (Fig. 3.8a, MB) are transverse and on the left side of the oral apparatus whereas the UM is longitudinal and on the right (Jenkins, 1973; Suzuki, 1973a). The posterior end of the membranelle band spirals into a funnel-shaped buccal cavity, with a cytostome at its base. The ciliary rows on the nonoral cell surface are essentially longitudinal and parallel to one another. They are unbranched except in a ramifying zone (Fig. 3.8a, RZ) located just posterior to the oral apparatus (Suzuki, 1973a).

The oral apparatus of *Stentor* is topologically identical to that of *Blepharisma* but topographically very different. It is perched on the rim of the broad end of an inverted cone. The membranelle band (Fig. 3.8e, MB) is located on the outer edge, whereas the UM is internal to it (Pelvat, 1985). Enclosed within this encircling oral apparatus is a portion of the nonoral ciliature, the frontal field (Fig. 3.8e, FF), with the path of its ciliary rows distorted into sweeping arcs. The remainder of the nonoral ciliary rows are situated along the side of the cone and oriented more or less longitudinally, but show two special features that are absent in *Belpharisma*. First, the ciliary rows are not uniformly spaced. The width of spacing of the ciliary rows grades imperceptibly around the circumference of the cell from the most widely spaced rows (separated by the broadest pigment stripes) situated to the left of the buccal cavity, to the most closely spaced ciliary rows (separated by the narrowest pigment stripes) located immediately posterior to the buccal cavity. Second, along the midventral meridian the most closely spaced rows abut on the most widely spaced rows at an acute angle, forming a suture. This suture is called the *contrast*

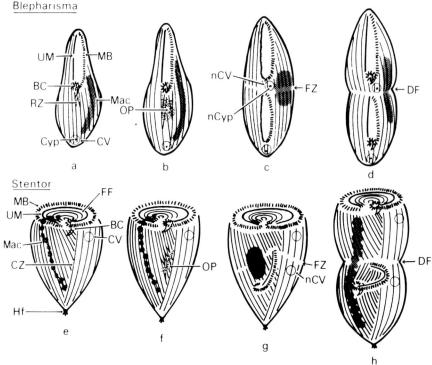

Figure 3.8 Predivision development of *Blepharisma (top)* and of *Stentor (bottom)*. (**a**) The ventral surface of a nondeveloping *Blepharisma*, showing the membranelle band [MB], the undulating membrane [UM], and the buccal cavity [BC] of the oral apparatus, the ciliary rows *(vertical lines)* with the ramifying zone [RZ] in the postoral region, and the contractile vacuole [CV] and the cytoproct [Cyp] near the posterior end of the cell. The macronucleus [Mac] is located in the cell interior. (**b**) The development of an oral primordium [OP] as multiple small fields of basal bodies adjacent to ciliary rows of the ramifying zone. (**c**) The stage of completion of ciliary structures of the new oral apparatus. The old oral apparatus has temporarily lost its buccal cavity. The fission zone [FZ] is formed, and a new contractile vacuole [nCV] and new cytoproct [nCyp] appears just anterior to it. The macronucleus has condensed into a sphere. (**d**) Cell division, with the constricting division furrow [DF]. The macronucleus is reelongating. (**e**) The ventral surface of a nondeveloping *Stentor*, with structures corresponding to those shown for *Blepharisma*, plus the frontal field [FF] of nonoral ciliary rows, the contrast zone [CZ] separating the most closely spaced ciliary rows from the most widely spaced rows, and the holdfast [HF] at the posterior end of the cell. (**f**) A developing *Stentor*, with a coalesced anarchic-field oral primordium [OP]. (**g**) The stage of completion of the ciliary structures of the oral apparatus and the beginning of spiraling of its posterior end. A fission zone [FZ] forms and a new contractile vacuole [nCV] appears posterior to it. (**h**) Cell division. The posterior division product is beginning to form the characteristic apical disc and frontal field posterior to the division furrow [DF], while the macronucleus is elongating.

zone (Fig. 3.8e, CZ) and is easily seen in living stentors as a "locus of stripe contrast," where narrow pigment stripes meet broad ones (Tartar, 1956c). This site is of great significance in the experimental analysis of pattern formation in *Stentor*.

   Both *Stentor* and *Blepharisma* have a single large contractile vacuole (CV), but its location differs in the two organisms. The CV in *Blepharisma* is near the posterior pole, whereas the one in *Stentor* is in the anterior region, among the widely spaced ciliary rows a few rows to the left of the contrast zone. The cytoproct is located close to the contractile vacuole in both of these ciliates. It has been observed more often in *Blepharisma,* where it is localized in the cortex immediately above the CV (Fig. 3.8a, Cyp).

Stentor has a special structure near the posterior end, called the *holdfast* (Tartar, 1961, pp. 37–40). This structure, which is pseudopodial and adhesive, is used for anchorage to the substratum when feeding. *Blepharisma* feeds while swimming and has no holdfast.

Dynamics

The proliferation of ciliary units of nonoral ciliary rows has not been studied in *Stentor* or *Blepharisma.* [In the related heterotrich *Condylostoma,* the entire increase in number of nonoral basal bodies takes place during cell division (de Terra, 1972); the *Condylostoma* paired ciliary unit proliferates by separation of the anterior and posterior basal bodies of each unit, followed by the formation of a new basal body anterior to each (Bohatier, 1979).]

In ciliates such as *Tetrahymena* and *Paramecium* it is obvious that new membranelles are not added to existing oral structures. The matter is not as clear in the heterotrichs, where the number of membranelles is large (over 200 in *Stentor*) and accurate counts are difficult. Membranelle counts, however, made on *Stentor* (de Terra, 1966) and *Condylostoma* (de Terra, 1972) at different times after division are consistent with the conclusion that in situ membranelle addition does not occur.

New oral structures are formed in *Stentor* and *Blepharisma,* as in *Tetrahymena,* adjacent to ciliary rows posterior to the old buccal cavity. Several ciliary rows, however, are involved rather than just one. Furthermore, immediately before and during formation of oral primordia the ciliary rows involved in this process become branched. These events occur in the ramifying zone of *Blepharisma* (Fig. 3.8b) (Suzuki, 1973c; Sawyer and Jenkins, 1977), and at the contrast zone in *Stentor,* mostly among the closely spaced ciliary rows (Uhlig, 1960; Paulin and Bussey, 1971; Pelvat and de Haller, 1979). Shortly after their formation, the multiple anarchic fields coalesce into one (Fig. 3.8f). The field later splits longitudinally, with membranelles forming in the left half and the UM in the right (Fig. 3.8c,g). This is followed by the clockwise spiraling of the posterior end of the membranelle band, and the formation of the buccal cavity and the cytostome (Fig. 3.8d,h).

During the later phases of oral development, the anterior oral apparatus of *Blepharisma* undergoes an extensive remodeling, with a loss and subsequent reformation of the buccal cavity, a reorganization of the UM, and even a partial reorganization of the membranelles (Sawyer and Jenkins, 1977). This matter has not been examined as closely in *Stentor,* but the degree of reorganization appears to be less than in *Blepharisma* (Tartar, 1961, p. 75).

New contractile vacuoles appear far from the old ones just before the onset of division, at the posterior pole of the anterior division product in *Blepharisma* (Fig. 3.8c,nCV), and in the anterior region of the posterior division product in *Stentor* (Fig. 3.8g,nCV). New cytoprocts appear at about the same time.

In *Blepharisma,* as in *Tetrahymena,* the anterior division product obtains the reorganized old oral structures plus the new contractile vacuole and cytoproct, and the posterior division product gets the new oral structures plus the old contractile vacuole and cytoproct. There are no major dislocations of structures during division, although the cell grows extensively during and after division, reconstructing the ramifying zone in the postoral ciliary rows in both daughters (Fig. 3.8d). In *Stentor*, cell division is topographically more complex. The new oral apparatus of the posterior division product bends during fission as the new apical disc is formed, cutting off some of the ciliary rows to its right to form a new frontal field (Fig. 3.8h). At the same time, the contrast zone is reconstituted in both division products by extensive growth of the narrowly spaced ciliary rows.

Both *Blepharisma* and *Stentor* have multiple micronuclei, which divide mitotically before cell division. The macronuclei in both organisms are elongated during interphase, being rod-shaped in *Blepharisma* (Fig. 3.8a) and nodulated in *Stentor* (Fig. 3.8e). In both organisms, the macronucleus condenses to a sphere just before the onset of cytokinesis (Fig. 3.8c,g), and then reelongates during cell division (Fig. 3.8d,h). In *Stentor*, the macronucleus is located close to the cell surface, beneath the fairly closely spaced ciliary rows to the right of the contrast-meridian (Fig. 3.8e).

Oral replacement occurs readily under a variety of circumstances in both *Blepharisma* and *Stentor*. In both ciliates, the oral replacement primordium forms at virtually the same site and develops in the same way as the predivision oral primordium. The oral replacement primordium, however, is somewhat smaller than the predivision oral primordium, and it does not replace the entire membranelle band; only the posterior portion of the old oral apparatus regresses, whereas the anterior membranelles persist unchanged. The new oral structures formed from the oral replacement primordium then join with the remaining old membranelles, producing a composite oral apparatus that is partly old and partly new. During this process, micronuclei divide mitotically (Schwartz, 1935; Suzuki, 1973b) while the macronucleus condenses and reextends as it does during cell division but does not divide while it is reextending (Tartar, 1961, pp. 91–93; Suzuki, 1973b). The same cortical and nuclear events take place after the old oral structures are partially or completely removed (Pelvat and de Haller, 1979).

Conjugation has been described both in *Blepharisma* (Suzuki, 1973b) and in *Stentor* (Tartar, 1961, pp. 323–332). The control of the process is well understood in *Blepharisma* (Miyake, 1981), but it is unexplored in *Stentor*.

Blepharisma readily forms resting cysts, in which the ciliature appears to dedifferentiate (Giese, 1973). This feature, however, has not yet been exploited for studies in pattern formation, as has been done with oxytrichids (discussed later). *Stentor* has been known to encyst but does so rarely (Tartar, 1961, p. 26).

3.4 THE VERSATILE OXYTRICHIDS: *OXYTRICHA/STYLONYCHIA* AND *PARAUROSTYLA*

The final group of ciliates to be introduced has perhaps the greatest potential for analysis of global patterning. These ciliates have two opposed surfaces with a completely different ciliature. Both surfaces are rich in locational markers. Nucleated pieces heal and regenerate readily after surgical transection. There are no pigment granules to interfere with silver staining, and operated cell fragments can be prepared for cytological examination. Thus, although the types of microsurgical experiments conducted on these ciliates have been more limited than with *Blepharisma* or *Stentor*, the analyses of the outcomes have been more precise. Microsurgical and other manipulations also have been combined with assessment of pattern regulation through the events of encystment and excystment, in which the ciliature first regresses and then is reconstructed. Finally, mating types have been described in some of these ciliates, and a start has been made in analyzing the genetics of pattern formation.

Although several of these ciliates have been studied, I introduce only three, *Oxytricha, Stylonychia,* and *Paraurostyla,* which are quite similar to one another. *Oxytricha* and *Stylonychia* are very much alike, both in their ciliary patterns (Wirnsberger et al., 1985) and in their rDNA sequences (Elwood et al., 1985); indeed the rDNA sequences of *O. "nova"* and *S. pustulata* are as closely related as those of morphologically indistinguishable *Tetrahymena* sibling species (Sogin et al., 1986c). Ammermann (1985) observed that isoenzyme patterns of different *Oxytricha* and *Stylonychia* species did not reveal a distinct clustering of the species into the two genera, and also showed that the classical morphological distinction between the two genera was based on an observational error. In his judgment, *Oxytricha* and *Stylonychia* should really be one genus, which he calls *Oxytricha/Stylonychia;* I shall do the same when referring to the genus as a whole. *Paraurostyla* differs somewhat from *Oxytricha/Stylonychia* in its cortical anatomy, but is sufficiently similar to it in its developmental patterns that most specialists place them in the same taxonomic family, the Oxytrichidae (Borror, 1972; Wirnsberger et al., 1986; Tuffrau, 1987).

Statics

Both of the oxytrichid genera considered here are dorsoventrally flattened and roughly oval in contour. The three most commonly used experimental species differ, however, in size: *O. fallax* is the smallest, 100 to 150 micrometers in length, *S. mytilus* is intermediate at 150 to 220 micrometers, and *P. weissei* the largest at 160 to 240 micrometers. We use *Oxytricha/Stylonychia* as our model for the description of ciliary patterns and their development. The ultrastructural analysis is most detailed for *Paraurostyla*, so we switch to this genus at one point in our narrative.

The organization of the ventral and dorsal surfaces of *Oxytricha/Stylonychia* during preparation for cell division is shown in Figure 3.9. The oral apparatus is located near the anterior end of the ventral surface, and is similar in general organization to that of *Blepharisma* except for the presence of two UMs rather than one. The remainder of the ventral surface, however, is covered exclusively by multiple ciliary units known as cirri (Fig. 3.10a). The cirri of the ventral surface consist of three major sets: the right and left marginal cirral rows (Fig. 3.9a, RM, LM) and the frontal–ventral–transverse (F-V-T) cirri (Fig. 3.9a, F, V, T). Eight frontal cirri are located to the right of the UMs, four ventral cirri are in the midregion, and six transverse cirri make up an asymmetrical "V" near the posterior end of the cell. In *Paraurostyla* (not shown), the marginal cirral rows are similar to those of *Oxytricha/Stylonychia* but the ventral surface is more generously provided with four

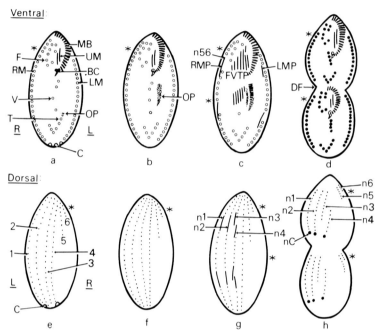

Oxytricha

Figure 3.9 Predivision development of *Oxytricha fallax*, of both ventral (**a–d**) and dorsal (**e–h**) surfaces. The asterisk in all of these diagrams marks a particular site on the anterior right edge of the cell (see text). R, right; L, left. (**a**) The stage of initiation of the oral primordium. The membranelle band [MB], two undulating membranes [UM], and the buccal cavity [BC] of the old oral apparatus are shown. Cirri are represented as *open circles*, with right marginal [RM], left marginal [LM], frontal [F], ventral [V], transverse [T], and caudal [C] sets. The oral primordium [OP], located just anterior to the left-most transverse cirrus, is at a very early stage of its development. (**b**) The late anarchic field stage. The oral primordium [OP] has enlarged and migrated to its definitive position. (**c**) The stage at which membranelles and the UMs are differentiating within the new oral primordium, while the ciliary streaks of the left marginal cirral primordium [LMP], right marginal cirral primordium [RMP], and frontal-ventral-transverse cirral primordium [FVTP] have been formed (only the anterior set is labeled). The primordium for the new dorsal rows 5 and 6 [n56] is located at the anterior end of the RMP. (**d**) Cell division, with a division furrow [DF]. The cirral primordia have developed into new cirri *(shaded)*. The old cirri have been resorbed or are destined to be resorbed and, therefore, are not shown. (**e**) The dorsal surface at a stage corresponding to (**a**). The ciliary rows are numbered from left to right. Caudal cirri [CC] are present near the posterior ends of ciliary rows 1, 2, and 4. (**f**) The dorsal surface at a stage corresponding to (**b**). Development has not yet begun. (**g**) The dorsal surface at a stage corresponding to (**c**). Streaks of new cirral rows [n1, n2, n3, n4] have formed within old rows 1, 2, and 3. (**h**) The dorsal surface at a stage corresponding to (**d**). The ciliary units of the new dorsal rows are spreading apart, while those of the old rows are regressing (not shown). The new rows 5 and 6 [n5, n6], derived from the anterior end of the right marginal streaks, have moved around to the dorsal surface and are in the process of assuming their final positions. New caudal cirri [nC] are present near the posterior ends of new rows 1, 2, and 4.

longitudinal rows of ventral cirri. The sparse ventral cirri of *Oxytricha/Stylonychia* can be considered as derived by reduction from the full set of *Paraurostyla* (Martin, 1982; Wirnsberger et al., 1986).

The ciliature of the dorsal surface of *Oxytricha/Stylonychia* resembles that of *Euplotes* in being composed mostly of longitudinal rows of widely separated paired ciliary units (Grimes and Adler, 1976; Jerka-Dziadosz, 1982). The rows are numbered from the viewer's left to right, a convention opposite to that used in *Tetrahymena* (compare Fig. 3.9 with Fig. 3.1). The lengths of these rows are unequal, with three or four full-length rows on the left side of the cell (Fig. 3.9e, rows 1–4), and two much shorter rows on the right side (Fig. 3.9e, rows 5 and 6). The only multiple ciliary units on the dorsal surface are the three caudal cirri, located near the posterior ends of rows 1, 2, and 4, respectively (Fig. 3.9e, C). This description applies equally to *Paraurostyla,* except for some unimportant differences in details.

The single contractile vacuole of *O. fallax* is located on the dorsal surface, near the middle of the third ciliary row (Grimes, personal communication). The outlet pores do not stain with the silver-staining method most commonly used in these organisms (protein silver), and thus virtually nothing is known about them.

I must now pause to confess that when describing the geometry of the dorsal surface, I have deliberately violated Ng's rule, in which right and left are always designated from the viewpoint of an observer standing inside the cell and facing out toward the surface (see Section 2.2.2). The reason for this violation is that oxytrichids, like people, have a very pronounced dorsoventral differentiation, and also have a restricted number of structures located at definite places. If you look someone in the face, that person's right ear is on your left; if you then look at the same person's head from behind, the right ear is now on your right. Consider, now, the site marked by the asterisk in the diagrams of Figure 3.9. A point near the anterior end of the right marginal cirral row is on your left as you view the ventral surface. Flip the cell over to view the dorsal surface, and that same point is now on your right.

Dynamics

In oxytrichids, all ciliary structures are formed in a coordinated manner during predivision development. Once produced, the number of ciliary elements remains constant, as has been demonstrated in detail for cirri and membranelles of *Paraurostyla* (Jerka-Dziadosz, 1976). Cortical development in oxytrichids always involves reorganization or replacement of the entire ciliature, excepting only the old membranelles. These cells cannot produce one new structure without simultaneously replacing virtually all of the others.

The developmental process begins with formation of an oral primordium, at a site that differs in different species of *Oxytricha/Stylonychia* (Foissner and Adam, 1983; Wirnsberger et al., 1985). In the two species most commonly used in experimental work, *O. fallax* (Grimes, 1972) and *S. mytilus* (Tuffrau, 1969; Wirnsberger et al., 1986), the oral primordium is normally initiated by formation of a small group of basal bodies just anterior to the transverse cirrus at the anterior end of the left arm of the "V" (Fig. 3.9a). The oral primordium then migrates anteriorly and grows by addition of new basal bodies (Fig. 3.9b). Once it is at its final position,

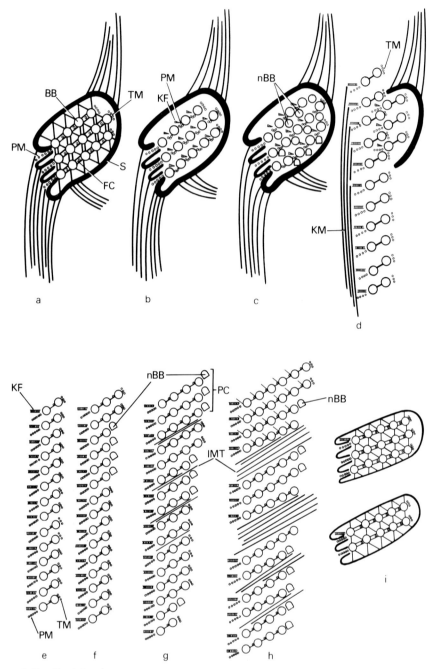

Figure 3.10 Cirral development in *Paraurostyla weissei*, including formation of a streak from an old cirrus (**a–e**) and the subsequent differentiation of new cirri from the streak (**f–i**). (**a**) An intact mature cirrus, with basal bodies [BB] connected by fibrillar connectives [FC]. Transverse microtubule bands [TM] and postciliary microtubule bands [PM] are located at opposite ends of the cirrus. The cirrus is surrounded by a fibrous sheath [S]. (**b**) Dedifferentiation of the cirrus. The fibrillar connectives disappear, but a postciliary microtubule band [PM] and a kinetodesmal fiber [KF] form next to each basal body, transforming each basal body of the cirrus into a solo ciliary unit. (**c**) Development of a new basal body [nBB] adjacent to each old basal body, converting the

membranelles and UMs differentiate much as they do in *Blepharisma*. The simultaneous reorganization of the anterior oral apparatus also resembles the corresponding process in *Blepharisma*.

New cirri begin to develop while new membranelles are differentiating within the oral primordium. During this process, some old cirri contribute basal bodies that will eventually become parts of the new cirri, as was first described for *Paraurostyla* (Jerka-Dziadosz and Frankel, 1969) and subsequently observed in the other oxytrichids (*O. fallax:* Grimes, 1972; *S. mytilus:* Wirnsberger et al., 1986). In the marginal cirral rows, two or three cirri within the anterior and posterior portions disaggregate to form longitudinal rows of very closely spaced paired ciliary units (Fig. 3.9c, RMP, LMP). These rows are called *streaks* because they are most easily seen as longitudinal streaks of closely spaced cilia. The progress of disaggregation of an old cirrus to form these streaks, described in most detail for *Paraurostyla*, involves the dissolution of the fibrillar connections between the basal bodies of the cirri (Fig. 3.10b), the formation of new basal bodies near old ones to make basal body pairs (Fig. 3.10c), and the longitudinal displacement of these newly generated pairs to form a row of very closely spaced basal body pairs (Fig. 3.10d,e). At this stage these streaks are similar to the ciliary rows of heterotrichs (see Fig. 3.7), except that the paired ciliary units are much closer together, oriented diagonally, and both basal bodies of each pair are ciliated. The later differentiation of these streaks into new cirri involves a lateral addition of three or four more basal bodies adjacent to each of the primordial pairs (Fig. 3.10f,g), and concomitant formation of microtubular partitions between sets of three or four of these diagonal rows of basal bodies (Fig. 3.10g,h). These sets are the *procirri*, from which mature cirri later differentiate by formation of fibrillar connectives and accessory microtubule bands (Fig. 3.10h,i). A true appreciation of these amazing structural transformations can be gained only by consulting the original ultrastructural description (Jerka-Dziadosz, 1980).

All new marginal cirri of both anterior and posterior division products arise from the marginal ciliary streaks, whereas those old cirri that did not participate in the formation of the streaks are resorbed during or shortly after cell division.

The new frontal, ventral, and transverse cirri arise from the FVT streaks (Fig. 3.9c). Twelve such streaks form, six in the future anterior division product and six

---

←

solo units into paired units. (**d**) Rearrangement of the paired ciliary units to form a longitudinal streak. The postciliary microtubule bands of the posterior-right basal body of each pair elongate and overlap to form a KM fiber, while a transverse microtubule band [TM] is now present next to the anterior-left member. The sheath and elongated accessory microtubule bands of the old cirrus are being resorbed. (**e**) The completed streak, made up exclusively of paired ciliary units, each possessing a kinetodesmal fiber [KF], a postciliary microtubule band [PM], and a transverse microtubule band [TM]. (**f**) Onset of proliferation of new basal bodies [nBB] adjacent to the anterior-left basal body of the basal body pairs. (**g**) Continuation of proliferation of new basal bodies [nBB], with simultaneous formation of intra-streak microtubules [IMt] that separate procirri [PC]. (**h**) Completion of proliferation of new basal bodies and further elaboration of intrastreak microtubules. (**i**) Completion of differentiation of cirri, with formation of new fibrous connectives between basal bodies, and the elaboration of a new sheath. The long accessory microtubule bands (not shown) also are formed at this time. Redrawn with slight modifications from Figures 4 and 24 of Jerka-Dziadosz (1980), with permission.

in the future posterior division product. These are composite in their derivation. In *Oxytricha/Stylonychia,* some arise from the disaggregation of certain old cirri, and others might arise from portions of the oral primordium (Grimes, 1972; Foissner and Adam, 1983; Wirnsberger et al., 1986). In *Paraurostyla,* most or all of the FVT streaks arise from the disaggregation of preexisting frontal and ventral cirri, although, as will be seen in Chapter 6, this mode of origin is not obligatory (Jerka-Dziadosz and Frankel, 1969). In all oxytrichids, the two sets of FVT streaks subsequently differentiate into the new frontal, ventral, and transverse cirri of the two daughter cells, whereas all of the remaining old cirri of these groups are resorbed.

Events on the dorsal surface are the same in all oxytrichids, and in many ways resemble those on the ventral surface. On the left half of the dorsal surface, certain ciliary units within the old ciliary rows proliferate repeatedly to form the streaks for the new rows 1, 2, 3, and 4 (Fig. 3.9g). These streaks subsequently expand to cover the entire length of the daughter cells (Fig. 3.9h), whereas the remaining old cilia are resorbed (Grimes and Adler, 1976; Jerka-Dziadosz, 1982). Note that this sequence is very different from the events on the dorsal surface of *Euplotes,* where all old cilia are conserved (Fig. 3.6). Dorsal development in oxytrichids differs from ventral development primarily by the omission of the final steps of formation of multiple ciliary units from paired units everywhere except at the posterior ends of new dorsal rows 1, 2, and 4, where three new caudal cirri form (Fig. 3.9h, nC). Dorsal development in oxytrichids shows us that a cirrus is an elaboration of the fundamental unit of ciliate cortical architecture, the basal body pair.

The knowledge that cirri are derived from ordinary basal body pairs allows us to appreciate the most peculiar event in ciliary development in the oxytrichids. Although the new dorsal ciliary rows 1 to 4 are generated from streaks arising from within the old rows, new ciliary rows 5 and 6 normally arise from a different source, the anterior end of the right marginal cirral streaks. Most of the paired ciliary units of each of these streaks undergo additional basal body proliferation to form new cirri, but about ten basal body pairs at the anterior end of each of these streaks (Fig. 3.9c, n56) do not do so. These anterior segments break off from the remainder of the streak and move around the edge of the cell to end up on the anterior-right portion of the dorsal surface (the region marked by the asterisks in Fig. 3.9g,h). There, each anterior streak-segment breaks up a second time into two pieces that become laterally displaced relative to one another and then elongate to form the new fifth and sixth dorsal ciliary rows; the old fifth and sixth dorsal rows are simultaneously resorbed (Jerka-Dziadosz and Frankel, 1969; Grimes, 1972).

Nuclear events during cell division (not shown) include mitosis of micronuclei followed by a coalescence and division of the macronucleus, much as in heterotrichs. The nuclei do not appear to be closely associated with the cortex at any stage.

Oral replacement and regeneration in oxytrichids bear a close relationship to the predivision events. During all three processes, the nonoral ciliature is totally replaced by corresponding new structures derived from ciliary streaks, whereas the old membranelle band is replaced either partially (in oral replacement) or totally (in regeneration after removal of the old oral structures).

Although micronuclei divide and macronuclei elongate during the regeneration process, the attendant DNA synthetic events are uncoupled in a uniform way in three different oxytrichids (*Paraurostyla, Stylonychia,* and *Laurentiella*): micro-

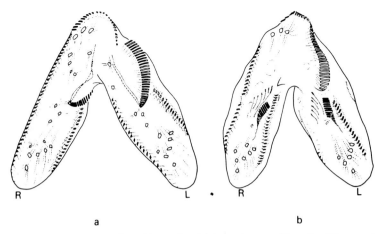

Figure 3.11 Conjugation in *Stylonychia mytilus*. (**a**) An early stage. Note the difference in relative position and organization of the anterior ends of the right [R] and left [L] conjugant. (**b**) A later stage. The anterior ends of the two conjugants have formed an integrated membranelle band, which is now in the process of being replaced by membranelles belonging to the first reorganization primordia. For further explanation, see the text. From Figures 1 and 4 of Tuffrau et al. (1981), with permission.

nuclear DNA synthesis occurs on schedule, just before micronuclear division, but macronuclear DNA synthesis, which normally precedes micronuclear DNA replication, is arrested and postponed until after the completion of regeneration (Jerka-Dziadosz and Frankel, 1970; Sapra and Ammermann, 1974; Torres et al., 1980).

Conjugation is well known and readily inducible in all three of the oxytrichids considered here. The cortical events, however, are uniquely asymmetrical. In *Oxytricha/Stylonychia*, the two partners become distinguishable by their relative orientations and behavior as they mate, and then each partner resorbs *different* preexisting structures to achieve a partially integrated ciliary organization of the anteriorly fused pair (Hammersmith, 1976a; Ricci, 1981; Tuffrau et al., 1981; Yano, 1985a). The left partner retains the posterior half of its membranelle band, whereas the right partner retains the anterior half (Fig. 3.11a), with the complementary portions being resorbed; the two partners then temporarily appear to share a composite membranelle band (Fig. 3.11b). The choice of whether to be a left or right partner is apparently made at random shortly before or during actual pairing (Ricci, 1981), but thereafter is irreversible (Yano, 1985b). Cortical development involves *three* successive rounds of replacement of ciliary structures, the first during conjugation (Fig. 3.11b) and the last two after the left and right partners separate (Tuffrau et al., 1981).

In *Paraurostyla*, this asymmetry of conjugation is carried further: the left partner is actually absorbed by the right partner, and the fused cells then form a zygotic cyst (Heckmann, 1965; Jerka-Dziadosz and Janus, 1975). The first of the three rounds of ciliary replacement takes place during the fusion process, before dedifferentiation of the ciliature during the formation of the zygotic cyst. The second replacement takes place during the excystment process and the third afterward, finally resulting in the formation of a single vegetative exconjugant (Jerka-Dziadosz and Janus, 1975). The nuclear events are only a slight variation of the "typical" ones shown in Figure 2.11, and conventional genetic transmission through the zygocyst has been demonstrated (Jerka-Dziadosz and Dubielecka, 1985).

Permanent cell fusion also occurs during *paraconjugation* in some strains of *Oxytricha/Stylonychia*. In this unique process, macronuclei rather than micronuclei undergo fusion (Esposito et al., 1978; Banchetti et al., 1980; Yano, 1985a), with an active suppression of micronuclear meiosis (Ricci et al., 1980; Yano, 1985b). This condition is extremely interesting for analysis of patterning because the cell surfaces of the fused partners may become integrated without proceeding through a dedifferentiated zygocyst (Hammersmith and Grimes, 1981).

Finally, vegetative oxytrichids readily form resting cysts. Cortical events of encystment have been investigated most intensively in *O. fallax,* in which the total disappearance of the preexisting ciliature was first described by light microscopy (Hashimoto, 1962) and then confirmed by serial-section electron microscopy (Grimes, 1973a). Thus, when the ciliature reappears on excystment, it must form de novo. The redevelopment of ciliary structures during excystment has been studied in several oxytrichids. Basal bodies first reappear in small clusters (Jareno and Tuffrau, 1979; Calvo et al., 1988). In one oxytrichid, *Onychodromus acuminatus,* the scattered initial basal bodies appear on the cell surface in a specific spatial relationship to "positioning fibers" arising from the macronucleus (Jareno and Tuffrau, 1979; Jareno, 1984); such fibers have not been observed in *Oxytricha* (Hammersmith, personal communication) or in the closely related oxytrichid *Histriculus similis* (Calvo et al., 1988). In the oxytrichid *Gastrostyla steini,* "paired kinetosomes (basal bodies) initially appear assembled deep in the endoplasm" (Walker et al., 1980). One wonders if this may be true of other excysting oxytrichids as well; if so, this is an exception to the general rule that basal body formation is restricted to the cell surface.

Once on the surface, the initial basal bodies apparently become consolidated to form an oral primordium (Hashimoto, 1963; Grimes, 1973b). Other ventral ciliary primordia develop somewhat later in close spatial relationship to the oral primordium, with the left marginal cirral row in particular developing in direct continuity with the oral primordium, while the dorsal ciliary primordia arise as separate rows from the start (Calvo et al., 1988; Hammersmith, personal communication). Despite the unconventional origin of the ciliary primordia, the later steps of formation of ciliary structures are the same as in vegetative cells. The pattern of differentiation of the ciliary primordia is more conservative than the modes of origin of these primordia.

3.5 CILIATE ASYMMETRY

From these brief descriptions, we see that the organization of the ciliate surface is asymmetrical at every level of structural organization: at the most local level of the geometry of the ciliary units, at the intermediate level of the organization of the oral apparatus, at the most global level of the configuration of the ciliature as a whole, and sometimes even at a multicellular level in the configuration of conjugating cells. Such asymmetry is the normal condition in all ciliates. As we will see in Chapters 8 to 10, cases of experimentally or mutationally induced symmetry are both exceptional and illuminating.

4

Structural Inheritance

4.1 INTRODUCTION

4.1.1 *Omnis forma ex DNA?*

Beginning with Weissman's germ-plasm hypothesis, biologists have become accustomed to thinking of the characters of organisms as expressions of determinants that are separate from the characters themselves. Most present-day biologists probably subscribe to the doctrine of *Omnis forma ex DNA*—All form from DNA—and, therefore, are likely to think of the problem of pattern formation as one of how a one-dimensional molecular sequence specifies three-dimensional structural organization.

The sufficiency of this biological world view has been challenged from two different perspectives. The first of these emphasizes the continuity of structural organization and the indispensibility of preexisting structure for the genesis of new structure. This perspective is "genetic" in the broadest sense of the term, as it focuses on inheritance of phenotypic differences, albeit at a structural rather than a molecular level (Nanney, 1968b; Sonneborn, 1970). The second perspective concentrates on physical rules that may constrain and to some degree determine biological organization. This viewpoint is "generative" rather than genetic, as it is preoccupied not with the direct continuity of form but rather with the physical processes that might generate particular classes of form while excluding other classes (Goodwin, 1985; Green, 1987). Although these approaches to the problem of elaboration of structure differ substantially, they share an insistence that research following the *Omnis forma ex DNA* idea is incapable of providing a complete explanation for biological forms and patterns, and that structural organization must be taken seriously in its own right.

Much of the significance of information concerning pattern formation in ciliates lies in the demonstration of the importance of higher-order organization in inheritance and development. The genetic aspect of this view is strongly associated with ciliates, due in large part to T. M. Sonneborn's celebrated demonstration of inheritance of structural organization in the ciliate cortex. This idea provides our starting point for an analytical inquiry into pattern formation in ciliates.

4.1.2 Implications of the Clonal Cylinder

The cylindrical topology of the ciliate clone (see Section 2.1.2) makes possible the indefinite propagation and, therefore, the inheritance of any structural system that

retains its individuality while growing longitudinally. Because the patterned organization of ciliates is most conspicuous in the surface region (see Chapter 2), the
structural systems that are known to be inherited in ciliates are localized or at least
expressed within the ciliate cortex. We may ask, for *each* such structural system, to
what extent is its inheritance independent of the genotype, how enduring is it, and
what is the scope of its cellular influence? We also may ask a second type of question, namely *how many* such systems of structural inheritance might there be and,
if there is more than one, do they coexist side-by-side or are they nested within one
another? When this second question is asked, the demonstration of structural
inheritance becomes not only an end in itself but also a means for dissecting the
levels of organization within the cell.

   This chapter begins with a detailed analysis of the classic example of the inheritance of the preexisting number and organization of ciliary rows. Then it will proceed to a consideration of the inheritance of structural configurations of both a
lower and a higher degree of complexity than the ciliary rows. This leads us to the
conclusion that we are dealing not with one unique system of heritable structural
patterns but rather with several systems, some of which might be encapsulated
within each other. This conclusion sets the stage for a more detailed analysis of
these levels of structural organization in subsequent chapters.

   I use the expression "structural inheritance" for the examples that follow. The
term structure is used here in the broad sense of an organization that is maintained
at a level above that of individual macromolecules.

4.2 INHERITANCE OF CILIARY ROWS

4.2.1 Inheritance of Differences in the Number of Ciliary Rows

In the simplest interpretation of the ciliate clonal cylinder (see Section 2.1.2), each
ciliary row undergoes indefinite longitudinal extension and, therefore, is in principle capable of self-perpetuation. The reality of such self-perpetuation has been demonstrated by proving that differences in both *number* and spatial *orientation* of ciliary rows are clonally inherited. We begin with inheritance of number of rows, and
in the next section consider the inheritance of their orientation.

The demonstration of clonal inheritance

Clonal inheritance of differences in the number of ciliary rows was demonstrated
first in *Tetrahymena thermophila*, later in *Euplotes minuta*. I begin by presenting
an illustrative example from the latter species. Cells of the nonautogamous subspe-
cies of *E. minuta* commonly have 8 or 9 dorsolateral ciliary rows (see Section
3.2.3). Suppose that we begin with a stock in which some cells have 8 rows and
others have 9. If there were no true inheritance of number of ciliary rows from
generation to generation (i.e., zero fidelity), then all clones selected from that stock,
whether initiated by an 8-rowed cell or by a 9-rowed cell, should manifest a similar
mixture of 8- and 9-rowed cells (Table 4.1, column A). On the other hand, if individual ciliary rows were inherited with perfect fidelity, then the members of clones
initiated by an 8-rowed cell should all have 8 rows, whereas members of clones
initiated by a 9-rowed cell should all have 9 rows (Table 4.1, column B). Twelve

Table 4.1 Expected and Observed Arrays of Numbers of Ciliary Rows in Subclones of *Euplotes minuta*

	Presumed number of rows in "founder" cell		Distribution of number of ciliary rows						
			Expected on the basis of:				Observed at 30 fissions		
			A. Zero fidelity[a]		B. Perfect fidelity		C.		
			8	9	8	9	8	9	p
	8	→	16	24	40	0	40	0	<0.001
	8	→	21	19	40	0	40	0	<0.001
Starting	8	→	24	16	40	0	39	1	<0.001
clone	8	→	16	24	40	0	37	3	<0.001
———	9	→	24	16	0	40	21	19	>0.2
36%—8	9	→	18	22	0	40	19	21	>0.2
64%—9	9	→	18	22	0	40	18	22	>0.2
	9	→	16	24	0	40	6	34	<0.001
	9	→	13	27	0	40	5	35	<0.001
	9	→	17	23	0	40	3	37	<0.001
	9	→	13	27	0	40	2	38	<0.001
	9	→	20	20	0	40	2	38	<0.001
							232	248	

[a]Random samples from a binomial distribution in which the respective proportions of 8-rowed and 9-rowed cells is the same as the observed *overall* frequency of 8-rowed and 9-rowed cells at 30 fissions (derived from the sums shown at the bottom of column C). For further explanation, see the text. From Table 1 of Frankel 1975b, with permission.

cells were selected from a clonal culture of *E. minuta,* and used to initiate 12 subclones. The result observed from scoring 40 randomly chosen cells in samples fixed 30 fissions after initiation of these clones (Table 4.1, column C) is sharply at variance with the hypothesis of zero fidelity but also falls short of perfect fidelity. Nine of the 12 subclones had a preponderance of 8-rowed or of 9-rowed cells, strongly suggesting a substantial, although not total, inheritance of the preexisting number of rows (Frankel, 1973b, 1975b).

How long are such differences maintained? The answer was sought by serially propagating all 12 clones (without selection) and periodically examining stained samples over a period of about 240 fissions. Although the clones that were "high" (mostly 9-rowed) at 30 fissions tended to remain the highest, and the clones that were "low" (mostly 8-rowed) at 30 fissions tended to remain the lowest, there was a gradual convergence, so that by 240 fissions after establishment of the subclones the differences, although still statistically significant, had become small (Frankel, 1973b).

Similar investigations of clonal inheritance of the preexisting number of ciliary rows have been carried out in *Tetrahymena thermophila* (Nanney, 1966b, c, 1968a; Frankel, 1980), with comparable results. For various reasons, it was possible to obtain a greater diversity of ciliary–row phenotypes in *T. thermophila* than in *E. minuta,* with *corticotypes* (numbers of ciliary rows) ranging from 14 to 35 (Nanney, 1966b,c). Nanney showed that when cells of very different corticotypes were cloned under optimal growth conditions, the variability observed within each clone after 20 fissions of growth was related to the corticotype in a manner that strongly sug-

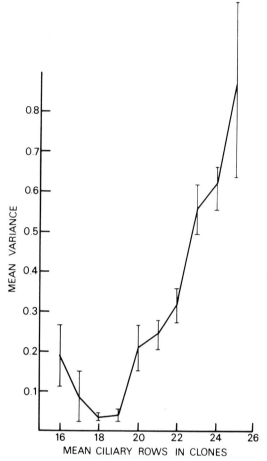

Figure 4.1 The mean intraclonal variance of 20-fission clones of *Tetrahymena thermophila (ordinate)* plotted against the mean number of ciliary rows of these clones *(abscissa)*. The vertical bars give the standard errors of the estimates of the variance. The higher the mean variance, the greater the instability of propagation of ciliary rows. Redrawn from Figure 4 of Nanney (1968b), with permission (copyright © 1968 by the AAAS).

gested convergence to a "stability center" of 18 to 19 rows, at a rate proportional to the initial distance from the stability center (Fig. 4.1) (Nanney, 1966c, 1968a). A subsequent longitudinal study confirmed the general conclusions of Nanney's cross-sectional analysis, although it indicated a stability center somewhat broader than that shown in Figure 4.1, encompassing a range from 18 to 20 (or 21) ciliary rows (Frankel, 1980). Within the stability center, differences in the number of ciliary rows can be inherited for a long time, possibly indefinitely, whereas outside of it the degree of inheritance is more limited but still is sufficient to "brake" the "fall" into the stability center.

Genic influences

What can be said concerning the effects of nuclear genes on the clonal inheritance of number of ciliary rows? Three things. First, the short-term clonal inheritance of the preexisting number of rows is not altered by conjugation between cells that possess different numbers of ciliary rows (*T. thermophila:* Nanney, 1966c; *E. minuta:* Heckmann and Frankel, 1968). This is not surprising in view of the fact that both *T. thermophila* and *E. minuta* retain their ciliary rows unaltered through conjugation. Second, the stability center is under genic control. This was originally demonstrated in *E. minuta* by a conventional genetic analysis of two strains that differed in their stability centers (Heckmann and Frankel, 1968), and later confirmed by a similar analysis in the closely related *E. vannus* (Laloë, 1979). The control of these naturally occurring differences probably involves alleles at more than one gene locus (Frankel, 1973b; Laloë, 1979). Third, single-gene differences are known that affect both the number of ciliary rows and the stability of inheritance of these rows; this has been documented in detail for *E. minuta* (Frankel, 1973a) and also has been found in *T. thermophila* (Frankel and Jenkins, unpublished observations). It is likely that the genotypes of some parasitic *Tetrahymena* species prescribe a high fidelity of transmission of corticotype under some conditions and a low fidelity under others (Batson, 1983). Thus the dualistic view originally stated by Nanney (1968b) is almost certainly correct; genes control the "Platonic pattern," which includes the location of the center of the "stability profile" and the shape of its contours, whereas the preexisting organization directly and strongly influences the "permutations on the pattern," such as the precise number of ciliary rows that a particular cell possesses at any particular time.

Developmental mechanisms

Allelic differences that diminish the fidelity of inheritance of differences in number of ciliary rows help in the analysis of the structural basis of this fidelity. Clones of

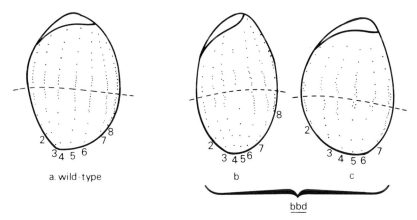

Figure 4.2   Camera-lucida drawings of the dorsal surfaces of protargol-stained wild-type (**a**) and *bbd* (**b**, **c**) cells of *Euplotes minuta* at a stage shortly before the beginning of cell division. The ciliary units are shown as dots. The *dashed lines* indicate the fission zone. The *closely-spaced dots* near the fission zone represent proliferating ciliary units. In the wild-type cell (**a**), a zone of proliferation is present in all ciliary rows both anterior and posterior to the fission zone. In *bbd* cells (**b**, **c**), proliferation sometimes fails to occur where it should [posterior moiety of row 5 and anterior moiety of row 8 in diagram (**b**)], or may be misaligned relative to the anteroposterior axis of the cell [posterior moiety of rows 5 and 6 in diagram (**c**)].

cells homozygous for the *basal body deficient (bbd)* allele of *E. minuta* have a distribution of row numbers not far from the zero-fidelity prediction shown in column A of Table 4.1 (Frankel, 1973a). Examination of protein-silver preparations of cells preparing for division shows the probable reason for this loss of fidelity. While in dividing wild-type *E. minuta* cells the basal bodies in the zone of proliferation are virtually always lined up longitudinally (Figs. 3.6 and 4.2a), in cells homozygous for the *bbd* allele proliferation in a particular row–segment may be lacking (Fig. 4.2b) or may be directed laterally rather than longitudinally (Fig. 4.2c) (Frankel, 1973a). The former aberration would eventually result in loss of a ciliary row in some progeny, whereas the latter might eventually create a new row where there had been none before (Heckmann and Frankel, 1968). The high frequency of such cortical aberrations in *bbd* clones can account for their loss of cortical fidelity. Therefore, the greater but not perfect fidelity of wild-type clones might be caused by a less frequent occurrence of these same aberrations. This analysis suggests that inheritance of number of ciliary rows is based on formation of new ciliary units at the correct time, in the correct regions of the cell, and, above all, in the correct (longitudinally aligned) orientation relative to the preexisting ciliary units.

The existence of a genically controlled stability center implies that a cell must somehow assess its number of ciliary rows and make appropriate adjustments in the probabilities of loss or gain of ciliary rows during cell division. Such adjustments could be made either by fine-tuning the likelihood of "errors" in the time, place, or orientation of formation of ciliary units, or by calling into play some additional mechanism(s) that might lead to a more rapid gain of a ciliary row by insertion or to loss by resorption. One mechanism of rapid gain has been documented in *T. thermophila*, in which an entirely new postoral ciliary row is produced from the base of the new undulating membrane (UM) during transformation to the "rapid-swimmer" form in 18- and 19-rowed cells but is not produced in 20- and 21-rowed cells (Nelsen and Frankel, 1979). In this case, a seemingly local cellular mechanism is dependent on some form of global assessment.

4.2.2 Inheritance of Inversions of Ciliary Rows

Organization, generation, and inheritance of inverted ciliary rows

The most spectacular demonstration of the autonomy of ciliary rows is the proof that 180-degree rotated configurations of ciliary rows can be inherited. This was shown first in *Paramecium tetraurelia* (Beisson and Sonneborn, 1965), and later confirmed for ciliary rows in *Tetrahymena thermophila* (Ng and Frankel, 1977) and for marginal cirral rows in *Stylonychia mytilus* (Grimes et al., 1981) and *Paraurostyla weissei* (Jerka-Dziadosz, 1985). In all of these cases, one or more ciliary (or cirral) row(s) maintained a 180-degree rotated ("inverted") permutation of the normal configuration (Fig. 4.3, ciliary rows marked "I").

The key to understanding the nature of the inversion is the fact that the ciliary units of inverted ciliary rows are intrinsically entirely normal structurally, functionally, and developmentally. In thin sections viewed by transmission electron microscopy, 180-degree rotated ciliary rows themselves are indistinguishable from normally oriented ciliary rows; indeed, when the two are side-by-side in the absence of external markers of cellular polarity one cannot tell which ciliary row is normally

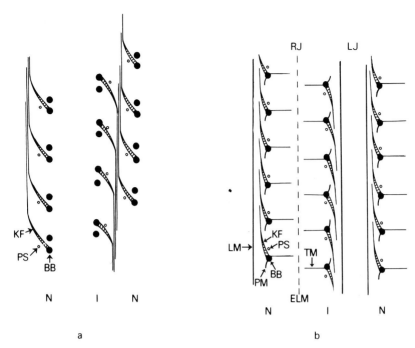

Figure 4.3 Normal [N] and inverted [I] ciliary rows of *Paramecium tetraurelia* (**a**) and *Tetrahymena thermophila* (**b**). Basal bodies [BB], parasomal sacs [PS], and kinetodesmal fibers [KF] are shown in both diagrams; whereas transverse [TM], postciliary [PM], and longitudinal [LM] microtubule bands are drawn only in diagram **b.** In (**b**), the left junction [LJ] and the right junction [RJ] between inverted and adjacent normally oriented rows are labeled. The *dashed vertical line* in the middle of the right junction indicates the location of the extra longitudinal microtubule band [ELM] that sometimes is present. (**a**) is redrawn from Figure 13 of Sonneborn (1970), with the outer lattice omitted; (**b**) is redrawn from Figure 6 of Aufderheide et al. (1980).

oriented and which is inverted (Ng and Williams, 1977). The direction of ciliary beat is unchanged in relation to the intrinsic polarity of inverted ciliary rows, and thus is opposite to that of the adjacent normally oriented ciliary rows (Tamm et al., 1975); this disharmony at the cellular level generates the "twisty" swimming patterns that makes possible selection of cells bearing inverted ciliary row(s). Finally, new basal bodies form at the normal position in 180-degree rotated ciliary units, meaning that in terms of the cell's polarity they develop posterior to old basal bodies rather than anterior to old basal bodies as in the adjacent normally oriented ciliary rows (Ng and Frankel, 1977). There is thus no intrinsic difference between a normal and an inverted ciliary row.

Both the origin and the genetics of the row inversion support the conclusion that the ciliary units of normally oriented and inverted ciliary rows differ *only* in their cellular orientation. Clones bearing inverted row(s) were isolated from belatedly separating conjugating cells in *P. tetraurelia* and from transiently arrested dividing cells in *T. thermophila*. In both cases, the conjoined cells tended to twist into a heteropolar orientation. During the period of union, one or more elongating ciliary rows would occasionally "cross over" from one cell into the other. If this occurred after the two cells had twisted into a heteropolar orientation, then the invading ciliary rows would be in an inverted orientation relative to the native ciliary rows of the attached host cell (Fig. 4.4a). As the two oppositely oriented cells divided (Fig. 4.4b), singlet progeny cells would arise with full-length inverted ciliary rows (Fig. 4.4c). Such cells could then be selected on the basis of their twisty swimming. The existence, number, and location of inverted rows could then be checked cytologically.

This mode of origin is itself sufficient to indicate that the "mutation" that generated the inverted ciliary rows is one of cell–surface configuration rather than of nuclear genes. In *P. tetraurelia,* the nongenic basis of the configurational change was proven by the unaltered clonal perpetuation of that change after conjugation with normal cells, whereas the cortical site of the alteration was demonstrated by its clonal persistence even after conjugation under conditions in which the nuclear events were accompanied by a massive exchange of internal cytoplasm (Beisson and Sonneborn, 1965).

Ciliary-row inversions have been inherited for about 800 fissions in clones of *P. tetraurelia* (Sonneborn, 1970) and 1500 fissions in *T. thermophila* (Ng and Frankel, 1977). Hence, it has been demonstrated beyond reasonable doubt that in these

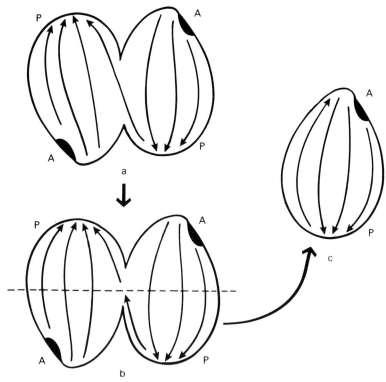

Figure 4.4 A highly schematic representation of the way in which inverted ciliary rows may be generated. (**a**) Two cells fused in a heteropolar orientation (A, anterior; P, posterior), with a ciliary row invading from one cell into the other. (**b**) Division of the fused cells. (**c**) A singlet progeny cell with an inverted ciliary row. The ciliary rows are indicated by lines, with their polarity represented by the *arrowheads*. The *dashed horizontal line* in (**b**) is the fission zone.

cases a major difference in a cellular phenotype is based on an inherited difference in spatial arrangement of cellular structures.

The basis of configurational heredity of ciliary rows

Sonneborn gave a strong interpretation to the observations of continuity of inversions of ciliary rows: "There is no escape from the conclusion that the site of initiation of basal body assembly, its path of migration to the surface of the cell, and the orientation of associated structures around it are indeed determined by the molecular geography *within* the unit territory and *not* by any other outside influence, either molecular or cellular" (Sonneborn, 1970, p. 353). The last part of this statement can be taken as a challenge to find an "outside influence" that *does* affect the site of initiation of new basal bodies, its subsequent path to the surface, or the orientation of associated structures. Such an outside influence might be provided by a genic mutation.

Genic mutations are known, both in *Paramecium* and *Tetrahymena,* that affect the orientation of ciliary units. The best studied of these mutations, *kin241* in *P. tetraurelia* (Beisson et al., 1976b) and *disA1* in *T. thermophila* (Frankel, 1979), both result in appearance of erratic ciliary units lying outside of the ciliary rows or in the total disorganization of these rows. The units themselves, however, always preserve the normal relative positions of accessory structures, as seen by light microscopy (Aufderheide, 1980; Cohen and Beisson, 1988) and (in *disA1*) by electron microscopy (Jaeckel-Williams, personal communication). These observations strengthen rather than weaken Sonneborn's assertion, as they show that even when the anchorage of ciliary units is disrupted (presumably due to some molecular defect in the epiplasm or the cell membrane), the units retain their rigid *internal* organization. This implies that there exists a profound spatial asymmetry around the circumference of the basal body, so that a ciliary unit is as unavoidably anisotropic at the structural level as an L-amino acid is at the molecular level.

The inheritance of inversions of ciliary rows, therefore, is based on the inherent structural asymmetry of ciliary units together with a tendency to maintain longitudinal order in the ciliate cell surface. The *kin241* and *disA1* mutations (and possibly the *bbd* mutation of *E. minuta* described earlier) disrupt the latter without affecting the former. To what extent this longitudinal order is maintained by features of the units themselves (for example, by the extensive overlap of kinetodesmal fibers in *Paramecium*) and to what extent by a cellular polarity extrinsic to the rows (see Section 4.2.4) is an intriguing question for future research.

4.2.3 Inheritance of Longitudinal Microtubule Bands

The inheritance of the longitudinal microtubule bands of *Tetrahymena* is frequently associated with that of ciliary rows. Inverted ciliary rows of *T. thermophila* have longitudinal microtubule bands on their *left,* opposite to their usual location to the right of normally oriented ciliary rows (Fig. 4.3b) (Ng and Frankel, 1977). The same is true for the transient longitudinal microtubule bands of *P. tetraurelia* (Cohen et al., 1982). These observations imply that ciliary units control the location of the physically separate (see Fig. 2.6) longitudinal microtubule bands. This

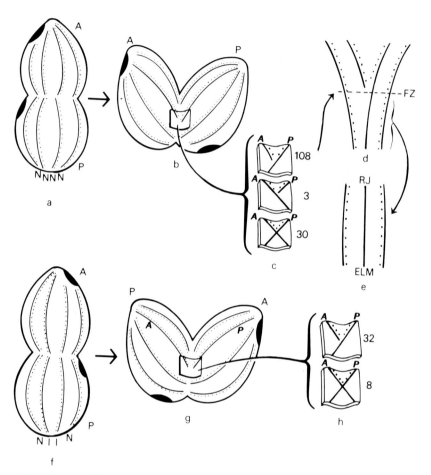

Figure 4.5 Diagrams illustrating the formation of extra longitudinal microtubule bands [ELM] in *Tetrahymena thermophila* and the preferential direction of growth of these bands. In all of these diagrams, "A" and "P" indicate anterior and posterior ends, either of the cell *(normal letters)* or of the ciliary rows *(heavy-italicized letters)*. **(a)** The left side of a dividing *Tetrahymena* cell showing four normally oriented rows of basal bodies *(dots)* plus their adjacent longitudinal microtubule bands *(lines)*. **(b)** A cell arrested in division, with an incipient right junction. **(c)** The boxed region of diagram **(b)**, enlarged and shown at a slightly later stage, with the posterior *(upper)*, anterior *(middle)*, or both *(lower)* portions of the longitudinal microtubule band growing into the right junction. The numbers next to each sketch indicate the number of cells scored with each configuration. **(d)** Extension of the extra microtubule band to the posterior end of the heteropolar complex. **(e)** The eventual configuration in a cell derived from this complex, with an extra longitudinal microtubule band [ELM] in the right junction [RJ]. **(f)** The *right* side of a cell bearing two inverted [I] ciliary rows flanked by normally oriented [N] ciliary rows. **(g)** Such a cell arrested in division, with an incipient right junction between the two original inverted ciliary rows. **(h)** The boxed region of diagram **(g)**, enlarged, with outgrowth of the microtubule band predominantly from the intrinsic posterior portion of the inverted ciliary row *(upper)*, sometimes from both portions *(lower)*. Data and design of diagrams **(a)** to **(e)** from Ng (1978), **(f)** to **(g)** from Ng (1979c), with permission.

does not, however, rule out the alternative possibility that the longitudinal microtubule bands also are capable of independent propagation.

Ng (1979a) has demonstrated that microtubule bands of *T. thermophila* can be inherited independently of the ciliary rows. This demonstration had three parts. First, Ng found that in cells bearing inverted ciliary rows an *extra* longitudinal microtubule band (Fig. 4.3b, ELM) was sometimes present approximately midway between an inverted ciliary row and a normally oriented ciliary row located to its right, i.e., in a "right junction" (Fig. 4.3b, RJ); such extra longitudinal microtubule bands (ELMs) were never found in left junctions (Fig. 4.3b, LJ). Second, Ng observed that the ELMs had a strong tendency to be inherited; cells of clones originally possessing such extra bands tended to retain them, whereas most cells of clones that initially lacked ELMs continued to lack them (Ng, 1979a). Third, Ng found evidence of a mechanism for the origin of the extra bands by introduction of longitudinal microtubule bands into right junctions between inverted and normally oriented rows. This process is illustrated diagrammatically in Figure 4.5. In the heteropolar stage that preceded the insertion of ciliary rows into oppositely oriented cells, some ciliary rows bent into a V-shape (Fig. 4.5b). At such sites, the adjacent longitudinal microtubule band was often visibly broken. Ng (1978) observed that instead of a "V" configuration, there was often a "Y"; one of the two ends of the broken band was beginning to grow down into the future right junction between normal and inverted ciliary rows (Fig. 4.5c). Eventually, this would form an ELM midway between the normal and inverted ciliary rows in a right junction (Fig. 4.5d). This never occurred in left junctions, where the longitudinal microtubule bands were separated from the junction by the ciliary row itself (Ng, 1978).

In addition to accounting for the origin of the ELMs, Ng showed that there was a preferential direction of growth of these bands. In heteropolar cells in which the ciliary rows in the affected region were normally oriented, the "Y" usually formed by extension of the band from posterior to anterior (Fig. 4.5c), although occasionally an opposite "Y" was seen, and more commonly an "X" in which *both* microtubule bands began to undergo extension (Ng, 1978). To test the conclusion of a preferential direction of extension, Ng took cells that already carried inverted ciliary rows through the process of forming new heteropolar cells, and studied the formation of ELMs in a nascent junction formed between previously *inverted* rows (Fig. 4.5f,b). Now the preferential direction of growth of the band was reversed (anterior to posterior) with respect to the cell's polarity, and therefore, unchanged with respect to the inverted ciliary row's intrinsic polarity (Fig. 4.5h) (Ng, 1979c).

Studies by Ng thus showed not only that longitudinal microtubule bands could be inherited separately from the adjacent ciliary rows, but also that they could maintain a preferential direction of outgrowth from free ends, reminiscent of that observed in vitro in other microtubular systems (e.g., Horio and Hotani, 1986). The interpretation of this finding is complicated somewhat by the fact that individual microtubules do not run the entire length of the band, so that preferential nucleation as well as outgrowth of microtubules must be involved (see discussion and references in Ng, 1979c). Nonetheless, the results clearly imply that the preformed assembled microtubular structure has a strong influence on the localized geometry of the newly assembled microtubules.

4.2.4 Can Ciliary Units Count?

Our analysis thus far has concerned structures that form uniformly along the entire ciliary row. A different problem is presented by structures that are generated within or close to ciliary units but only in particular regions of the cell. One can then ask whether ciliary rows carry information about regional location as well as directional orientation. We consider here two cases for which the observations are somewhat different but the fundamental answer is the same.

The first case is that of the contractile vacuole pore (CVP) of *T. thermophila*. As described earlier (see Section 3.2.1), typically two CVPs form near the posterior ends of two adjacent ciliary rows at about the same time that the fission zone appears. Ng (1979b) showed that at the time of its first appearance in normally oriented ciliary rows, the new CVP is located to the posterior-left of a basal body. In inverted ciliary rows, however, CVPs are formed to the anterior-right of basal bodies (Fig. 4.6) (Ng, 1979b). Thus, the "fine positioning" of the CVPs (Ng, 1977) is under the exclusive control of the internal geometry of the ciliary units that make up the ciliary row. This conclusion was subsequently confirmed by a detailed ultrastructural analysis of CVP formation in another ciliate, *Chilodonella steini* (Kaczanowska and Moraczewski, 1981).

But what decides that CVPs must appear only at the *posterior* end of ciliary rows? Two opposing predictions can be made, depending on one's interpretation of the polarity of ciliary rows. If successive units in a ciliary row carry successively different anterior-to-posterior positional values, counting off their positions like people counting off along a line, then new CVPs should form at the cell's *anterior* end of inverted ciliary rows. Alternatively, if each ciliary unit has an intrinsic direction and nothing more, then new CVPs should form at the normal *posterior* loca-

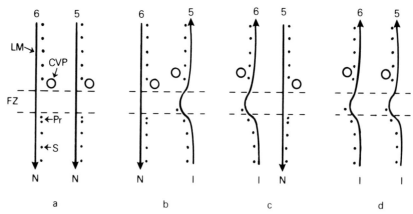

Figure 4.6 The region around the fission zone [FZ] in two CVP-rows shown in the four possible combinations of normally oriented [N] and inverted [I] configurations. The anterior cellular direction is upward on the page. The structures shown are the solo basal bodies [S], basal body pairs [Pr], longitudinal microtubule bands [LM], and contractile vacuole pores [CVP]. The *vertical arrows* indicate the differences in intrinsic polarity of the ciliary rows, and the *dashed horizontal lines* show the anterior and posterior limits of the fission zone gap as it is seen in normally oriented ciliary rows.

tion in inverted ciliary rows. In fact, with rare exceptions (Ng, 1979d), new CVPs appear near the cell's *posterior* ends of inverted as well as normally oriented ciliary rows (Fig. 4.6) (Ng, 1977, 1979b). Therefore, there is no anterior-to-posterior gradation intrinsic to the organization of a ciliary row, or, if there is, it is overridden by a cellular polarity that is extrinsic to the ciliary rows.

Although CVPs are distinctive markers of the posterior ends of certain ciliary rows of *Tetrahymena,* paired ciliary units mark the anterior ends of most ciliary rows (see Section 3.2.1). Therefore, one can ask whether inverted ciliary rows determine their own anterior ends. If they do, then the paired ciliary units must form at the (cellular) posterior end of inverted ciliary rows, whereas if they do not the pairs should form at the anterior ends of inverted as well as normally oriented ciliary rows. In fact, inverted ciliary rows bear no paired ciliary units anywhere along their length (Fig. 4.6) (Ng and Williams, 1977; Frankel et al., 1981). In addition, fission zone gaps tend to form belatedly and often somewhat abnormally within inverted ciliary rows in *T. thermophila* (Ng, 1979d); after cell division the anterior ends of such rows extend beyond the apical crown (Ng and Williams, 1977).

In *Paramecium,* paired ciliary units are more widely distributed over the cell than in *Tetrahymena,* and basal body pairs do form within inverted ciliary rows, in the same anteroposterior cellular regions as in the adjacent normally oriented rows (Sonneborn, 1970; Suhama, 1975). Inverted ciliary rows, however, extend abnormally into the anterior suture, beyond the location where all other ciliary rows have their anterior termination (Sonneborn, 1975a, 1977). When many adjacent ciliary rows are inverted, the fission zone fails to extend across the zone of inversion (Sonneborn, 1977). Thus, in *Paramecium* as in *Tetrahymena,* inverted ciliary rows cannot demarcate their anterior ends.

A reasonable conclusion from the results on both organisms is that "the two ends of a cortical unit of structure interact differently with the anterior suture and with the plane of division" (Sonneborn, 1977, p. 845). Possibly a posteriorly directed cellular signal originating from the fission zone must interact with the *anterior* end of a ciliary unit to create a ciliary-row terminus. In this example as in the previous one, we see that "To form and position an organelle, the cell relies on two sets of information: a local structural reference in the immediate vicinity of the organelle that is to be formed and a specification which involves consideration of other often remote body structures or parameters" (Ng, 1979b, p. 309). In the examples described earlier, the relevant parameter appears to be a cellular polarity that is extrinsic to the ciliary rows themselves.

4.3 INHERITANCE OF THE DOUBLET CONFIGURATION

4.3.1 The Doublet Biotype

The most complete genetic proof for the inheritance of cell surface organization was provided by Sonneborn (1963) for the homopolar doublet biotype of *Paramecium tetraurelia.* Because there are good reasons for believing that the conclusions of Sonneborn's analysis apply not only to *Paramecium* but to ciliate doublets in general, we first describe the doublet condition in general terms, and then consider specific features of doublets in two very different ciliates, *Paramecium* and *Oxytricha/Stylonychia.*

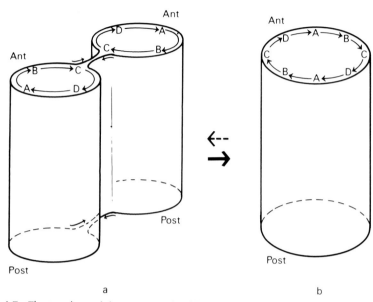

Figure 4.7 The topology of the majority of stable homopolar doublets in ciliates. The cells are represented as cylinders, with their anterior [ant] and posterior [post] ends lined up. The letters indicate positions around the circumference of the cells. (**a**) The lateral union of two cells at corresponding circumferential positions. (**b**) The resulting completely fused homopolar doublet.

The topology of the majority of hereditary ciliate doublets is illustrated in Figure 4.7. Imagine two cylinders lined up side by side with their anteroposterior axes aligned (homopolar), and turned so that equivalent longitudes (Fig. 4.7a, c) are lying next to each other. Imagine that one makes a slit along these longitudes in both cylinders, and then joins the new free edges of one opened-out cylinder to the adjacent edges of the other cylinder, putting the two cylinders in parabiosis. The "siamese twin" pair of cylinders produced after the adjacent edges heal is topologically equivalent to a single composite cylinder shown in Figure 4.7b. In this composite doublet–cylinder, the sequence of longitudes around the circumference (A-B-C-D) is repeated (A-B-C-D-A-B-C-D). Such doublet–cylinders express a twofold rotational symmetry around an imaginary central axis. One important feature of this organization is that *every* point on the doublet–cylinder of Figure 4.7b has the *same* set of neighboring points as does each singlet of Figure 4.7a. We will see later (Section 4.3.3) that the "rule of normal neighbors," originally stated for predicting stability of pattern juxtapositions in animal systems (Mittenthal, 1981), is an equally important criterion for stability of ciliate doublets.

How are doublets formed in real ciliate life? There are three ways: One is by a failure of separation of conjugating cells, a second by spatial adjustments that follow arrest of cell division, and a third by responses to microsurgically created fusions or transplantations. We have already seen that in the first two circumstances the two joined cells often bend back on each other, as was shown in Figures 4.4 and 4.5, to form heteropolar (head-to-tail) doublets. Although such heteropolar doublets are of immense value in generating inverted ciliary rows in budded-off singlet cells (see Section 4.2.2), the heteropolar doublets themselves cannot perpetuate their organization and tend to become irregular monsters. Permanently joined conjugating cells, however, may also maintain a homopolar orientation rather than bending into a heteropolar configuration, and cells that are arrested in division sometimes slide and then twist into a homopolar alignment rather than folding to make a heteropolar complex. As first shown systematically by Fauré-Fremiet (1945), doublets generated after division arrest typically are stable and self-perpetuating if (and only if) they attain the balanced homopolar condition indicated schematically in Figure 4.7b. The same applies to microsurgically created doublets (Tartar, 1961, Chapter XII).

Beginning with studies by Dawson (1920) on *Oxytricha* and by Chatton (1921) on *Glaucoma* (a close relative of *Tetrahymena*), inherited homopolar doublets have been generated in many ciliate species (reviewed by Sonneborn, 1963). In all doublets, the duplicate sets of surface structures surround a common endoplasm with no trace of an internal partition. The term *doublet,* therefore, implies a doubled cortical organization, *not* two separate cells.

The doublet condition may be highly stable as long as one selects doublets for continued propagation (Dawson, 1920; Fauré-Fremiet, 1945; Sonneborn, 1963). Such selection is necessary for two reasons: first, doublets can revert to singlets and second, singlets often (although not always) multiply more rapidly than doublets. The reversion from doublet to singlet may occur in two ways: either by a separation of the two units, directly reversing the fusion that originally produced the doublet, or by an internal reorganization of the doublet that eliminates one of the two organizational entities. We will have more to say about the latter mode of reversion in Chapters 9 and 10, but here it is more important to stress that the doublet configuration *can* be inherited, sometimes for hundreds of generations. The next two sections show two different ways in which this may happen.

4.3.2 Structural Continuity in *Paramecium* doublets

Sonneborn obtained homopolar doublets in *Paramecium tetraurelia* by preventing separation of conjugating cells (Sonneborn, 1963). The organization of these doublets is shown schematically in Figure 4.8b, with the corresponding singlet in Figure 4.8a (for more details on the cortical organization of *Paramecium,* see Section 3.2.2). The doublets possess two oral apparatuses (OAs) with associated anterior (AS) and posterior (PS) sutures, as well as two pairs of contractile vacuole pores (CVPs), each located approximately midway between the oral meridians. The two sets of cortical structures are arranged in twofold rotational symmetry, with corresponding structures opposite each other, conforming to the topology shown in Figure 4.7b. Either one or two macronuclei may be present in the interior of doublet cells. The proportion of doublets that possess single macronuclei increases with successive fissions after their formation, with the reduction in number of macronuclei accomplished by missegregation during cell division (Berger and Morton, 1980). When only a single macronucleus is present, it is larger than usual (Sonneborn, 1963), and its DNA content is 50 to 100 percent greater than that of the macronucleus of a singlet cell, corresponding to the roughly doubled protein content of doublets (Morton and Berger, 1978).

In an elegant and exhaustive study, Sonneborn (1963) proved that the determination of the inherited difference between doublets and singlets resided in the cortical region. When doublets conjugated with singlets, the doublet exconjugant gave rise to a doublet clone, whereas the singlet exconjugant gave rise to a singlet clone, despite the fact that the two exconjugant clones were shown to be genically identical. The outcome was also the same when the nuclear events of conjugation were accompanied by a massive exchange of internal cytoplasm between partners, or when exchange of old macronuclei was followed by "macronuclear regeneration" (a process in which development of new macronuclei from micronuclei is aborted and the preexisting macronucleus is revived). These experiments rule out control by organelles or symbionts that are located within the internal cytoplasm or by states of macronuclear differentiation. Finally, a genically controlled unequal division of the macronucleus was used to start singlets and doublets off with tiny macronuclei, which then grew to the sizes characteristic for singlets and doublets, respectively, demonstrating that the enlarged macronucleus of doublets is controlled by the dual cell organization rather than vice versa. Therefore, "As the only obvious remaining untested part of the cell, the cortex seems to be or contain the genetic basis of the difference between singlets and doublets" (Sonneborn, 1963, pp. 179–180).

What is this "genetic basis"? This expression will suggest to most readers some cortical repository of genetic information in the form of DNA or conceivably RNA. We saw earlier, however (Section 2.3.1), that basal bodies do not contain DNA, and although they might contain RNA, the informational role of this RNA, if any, is unknown. For Sonneborn, the genetic basis in this case lies in a structural organization rather than in the molecular code that is the usual basis of genetic transmission. We now need to inquire what this organization might be.

In *Paramecium,* the basis for inheritance of the doublet condition lies largely in the OA and the structural organization immediately around it. *Paramecium,* unlike all of the other members of our ciliate "cast," always forms new OAs close to old ones (see Section 3.2.2). This development starts from a permanent anarchic field located at the right edge of the OA, between the undulating membrane (UM) and the adjacent cell-surface ciliary rows (see Fig. 3.4). Tartar (1954a) showed that nucleated cell fragments that lacked this region could not regenerate it. Subsequently, Hanson (1962) and Hanson and Ungerlieder (1973) demonstrated, by ultraviolet microbeam irradiation of parts of one of the two oral systems of doublets, that a "primordium forming area" localized at the right-posterior edge of the buccal cavity is essential for formation of new OAs. If one such area was sufficiently damaged, the capacity to form new oral structures on that side of the doublet cell was lost. In certain cases, the primordium forming area could continue to perpetuate itself for a few cell generations in the absence of a nearby buccal cavity, but eventually it lost its capacity to form new oral structures (Hanson and Ungerlieder, 1973). These results suggest that something at or near the posterior-right margin of the buccal cavity is necessary for the formation of new oral structures. Although Hanson and Ungerlieder did not stain their cells after irradiation, a comparison of their findings with descriptive studies of oral development in *Paramecium* (see Section 3.2.2) suggests that the primordium forming area probably corresponds to the permanent anarchic field (Fig. 4.8, AF) that lies on the right edge of the OA, and that the region responsible for maintaining this area probably includes all or part of the UM.

Positive evidence for an association of the hereditary capacity to maintain a doublet with a localized region in or near the OA came from rare cases of "cortical picking" observed during conjugation between singlets and doublets. In one exceptional case in which a singlet that had conjugated with a doublet produced a doublet clone, a piece of cortical cytoplasm at a site very near the OA had been transferred from the doublet partner to the singlet partner (Sonneborn, 1963). In another case, conjugation between two singlets yielded a self-propagating doublet plus an astomatous singlet that almost certainly had donated its OA to its partner (Ng, personal communication). The fact that the *only* cases of conversions of *P. tetraurelia* singlets into doublets without total cell fusion involved such "natural grafts" strongly suggests the involvement of a critical local region in generating a new oral region.

Although critical for inheritance of dual oral systems, the possession of two primordium forming areas is not the sole determinant of the inherited doublet condition. A doublet that recently has lost one primordium forming area can propagate only one OA (the unaffected one), but nonetheless is not a true singlet; the configuration of anterior and posterior sutures and of curved ciliary fields around the sutures can persist for several fissions after the associated oral structures have been lost (Sonneborn, 1963). Thus a spatial configuration of ciliary rows is to some degree self-perpetuating. Such a spatial configuration is not an inherited property of any *individual* ciliary row, as inverted ciliary rows that shift from the dorsal surface to the circumoral region assume the curvature of the neighboring, normally

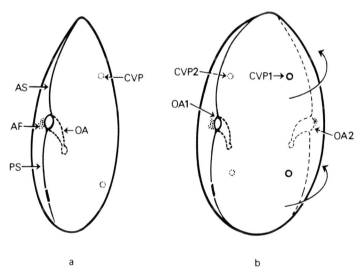

a   b

Figure 4.8 Outline sketches of a *Paramecium* singlet (**a**) and homopolar doublet (**b**) showing the oral apparatus [OA] with the permanent anarchic field [AF] at its right margin, the anterior suture [AS] and posterior suture [PS], and the contractile vacuole pores [CVP]. The structures on the other side of the cell are drawn in *dashed lines*.

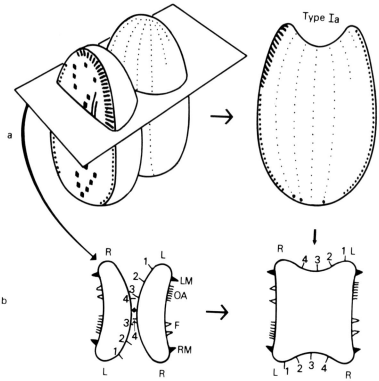

Figure 4.9 Back-to-back (type IA) homopolar doublets in *Oxytricha/Stylonychia*. (**a**) Perspective views of the presumed original back-to-back configuration of the cells to be joined *(left)* and a side view of the cells after joining *(right)*. (**b**) Schematic cross sections of the anterior region of the cells cut along the horizontal plane that is shown in (**a**), illustrated both before *(left)* and after *(right)* mid-dorsal fusion. The cross sections are viewed from the anterior end of the cell. The structures shown symbolically in the cross sections are the oral apparatus [OA], frontal cirri [F], left marginal [LM] and right marginal [RM] cirral rows, and dorsal ciliary rows 1, 2, 3, and 4 (rows 5 and 6 are restricted to the region anterior to the plane of section). "L" and "R" indicate the left and right margins of the cell. The perspective drawings are modified from Figures 1 and 3 of Grimes (1973d), with permission.

oriented, ciliary rows (Beisson and Sonneborn, 1965). Thus, even in *Paramecium,* the most "mosaic" member of our cast, larger spatial configurations are superimposed on the ciliary rows and can even be inherited.

4.3.3 Inheritance of Large-Scale Organization in *Oxytricha/Stylonychia* Doublets

Four topological classes of doublets

There are four topologically distinct types of homopolar doublets in *Oxytricha/ Stylonychia*. The most common type, here called type I, obeys the normal-neighbor principle of doublet formation illustrated in Figure 4.7. These doublets have two normally shaped (flat or convex) surfaces facing in opposite directions, separated by concave surfaces that appear as longitudinal clefts. Type I doublets are found as two geometrically different (although topologically identical) subtypes according to the relationship of the ciliary patterns to the differently shaped surfaces. In type Ia (back to back) doublets, the two sets of ventral structures are located on oppositely directed flat surfaces, whereas the dorsal structures are in the clefts; this subtype was presumably generated by dorsal-to-dorsal fusion of two singlets (Fig. 4.9). In type Ib doublets, the two sets of dorsal structures are located on oppositely directed convex surfaces with the ventral structures in the clefts (Grimes, 1973d). Type I doublets can perpetuate the doublet condition indefinitely with appropriate selection (Dawson, 1920), and the two subtypes are readily interconvertible (Grimes, 1973d). This conversion involves a large change in the shape of surfaces but no change in the organization of the cortical structures situated on these surfaces (Fig. 4.10a). The existence of these two subtypes provides an excellent illustration of the lack of an invariant relationship between shape and pattern in ciliates.

Type II (tandem side-by-side) doublets are generated by fusing the left side of one singlet to the right side of another (Fig. 4.10b). Therefore, in type II doublets, unlike type I doublets, the points of fusion of the two cells are *not* equivalent, and the topology of Figure 4.7 is *not* obeyed. The result of such side-by-side fusion is a flat homopolar doublet with two tandem sets of ventral structures on one surface and two tandem sets of dorsal structures on the other. The right and left borders of

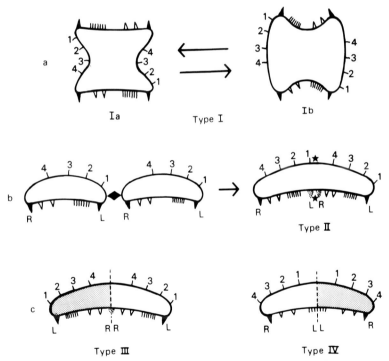

Figure 4.10   The different types of doublets in oxytrichids, illustrated as schematic cross-sections. (a) Interconversion of type Ia and Ib doublets. (b) Side-by-side fusion of two singlets (*left*) to create a type II doublet (*right*), with juxtaposed right and left pattern borders in the center [*]. (c) The two types of mirror-image doublets: (*left*) A "LRRL" mirror-image doublet (type III) with the plane of mirror symmetry (*dashed vertical line*) at the structural right margin; (*right*) a "RLLR" doublet (type IV) with the plane of mirror symmetry at the structural left margin. Regions with reversed asymmetry are shaded.

these ciliary patterns are juxtaposed in the center of the respective surfaces (Fig. 4.10b, *). Type II doublets, therefore, are *not* characterized by normal neighbors at all points. Such doublets are unstable and either revert to singlets (Hammersmith and Grimes, 1981) or regulate to mirror-image forms (Jerka-Dziadosz, 1983; Tuffrau and Totwen-Nowakowska, 1988).

Mirror-image side-by-side doublets of types III and IV (Fig. 4.10c) are like type II doublets in that they are flat and possess two sets of ventral structures on one surface and two sets of dorsal surfaces on the other. The sets of structures, however, are not arranged in tandem but instead are global mirror-images, with the mirror planes (Fig. 4.10c, vertical dashed lines) near the center of the doublets (these doublets are still considered "homopolar" because their anteroposterior axes are aligned). Type III and type IV doublets, although built on the same principle, are not interconvertible because in type III doublets the right edge of the structural pattern is central and the left edge peripheral (LRRL), whereas in type IV doublets the reverse is the case (RLLR). The inheritance of type III doublets was first reported by Tchang, Shi, and Pang (1964), and was later confirmed in four other laboratories (Grimes et al., 1980; Jerka-Dziadosz, 1983; Yano, 1987; Tuffrau and Totwen-Nowakowska, 1988). Type IV doublets form oral structures in the center of the ventral surface. Such doublets often have incomplete oral structures and can not feed (Grimes and L'Hernault, 1979; Shi and He, personal communication); they can reorganize true-to-type at least twice (Grimes et al., 1981) and sometimes are able to divide (Shi and He, personal communication).

Mirror-image doublets (types III and IV) cannot arise from simple fusions of normal cells. Instead, the regions with reversal of large-scale asymmetry (shaded in Fig. 4.10c) appear to originate as a consequence of the respecification of an axis after juxtaposition of parts that are not normal neighbors. Consideration of the nature of such respecification is postponed to Chapters 9 and 10.

Inheritance of the doublet condition

No one has repeated the exhaustive genetic analysis of the difference between the doublet and singlet condition for any ciliate other than *P. tetraurelia*. The modes of origin and reversion of all ciliate doublets, however, indicate a nongenic basis for the doublet condition, and it is no surprise that type I doublets of *Oxytricha fallax* inherit their doublet condition clonally after conjugation with singlets (Grimes, 1973d). As in *Paramecium* doublets, there is no close relationship between nuclear and cytoplasmic duality; type I doublets of *O. fallax* or *Stylonychia mytilus* may have either one or two of the bilobed nuclei that are characteristic of oxytrichids (Grimes, 1973d; Tchang and Pang, 1979).

The structural basis for the maintenance of duality in oxytrichids is, however, rather different from that of *Paramecium*. In the oxytrichids, there is no necessary relationship between any preexisting ciliary structure and the capacity to form a new oral primordium. This has been demonstrated in two ways in *Oxytricha fallax,* which normally initiates its oral primordium near a specific transverse cirrus (Grimes, 1972). First, excysting cells could reform a normal ciliature after total dedifferentiation of the preexisting ciliature during encystment (Fig. 4.11a) (Hashi-

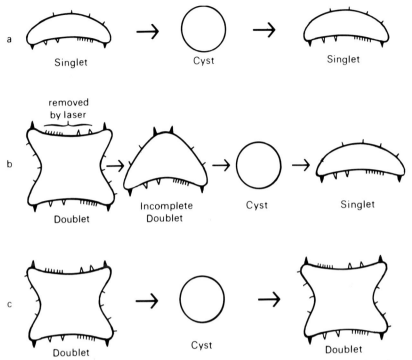

Figure 4.11 Schematic cross sections showing the results obtained after passing (a) singlets, (b) incomplete doublets, and (c) complete doublets through cysts. The conventions are the same as in Figure 4.10.

moto, 1962; Grimes, 1973a,b); second, *both* halves of equatorially transected vegetative cells could form complete sets of ciliary structures either after being allowed to regenerate normally (Fig. 4.12a) (Hashimoto, 1961) or after being taken through cysts and then allowed to excyst (Fig. 4.12b) (Hashimoto, 1964). Thus, the left-most transverse cirrus is an optional and dispensable landmark for the initiation of the oral primordium. In oxytrichids, formation of new ciliature does not show the rigid dependence on preexisting ciliature that is characteristic of *Paramecium*.

What, then, is the structural basis of the difference between singlets and doublets? The experiments that address this issue involve taking complete and incomplete doublets through the cyst stage (Fig. 4.11). As mentioned previously, ordinary singlets encyst, lose all traces of their ciliature, and then excyst as singlets (Fig. 4.11a). The completeness of the loss of preexisting ciliature is underscored by results of experiments with incomplete doublets. Grimes created these incomplete doublets by destroying the ciliature of one ventral surface of a type I doublet with laser microirradiation, leaving only the marginal cirral rows intact (Fig. 4.11b, left). This operation resulted in a cell in which two extra marginal rows (Fig. 4.11b) were propagated for several vegetative cell generations on the middle of the dorsal surface of a "humped" cell (Grimes, 1976). When such incomplete doublets with extra marginal rows on the dorsal surface were taken through cysts, the cells that emerged on excystment were normal singlets that lacked the extra marginal rows (Fig. 4.11b) (Grimes and Hammersmith, 1980). The fact that the capacity to form extra marginal rows was lost in the cyst strongly suggests that the local structural information that specifies formation of new ciliary structures truly disappears in the cyst stage.

Despite the total disappearance of preexisting ciliary structures in cysts, cells which encyst as type I doublets also excyst as type I doublets (Fig. 4.11c) (Grimes, 1973d). This perpetuation of a dual organization through the cyst stage is independent both of the size of the cyst (Grimes, 1973d) and of the number of macronuclei [many doublets that persist as such through cysts have only a single macronucleus (Grimes, personal communication)]. The demonstration of the perpetuation of large-scale organization through the cyst stage has also been extended to heteropolar doublets, in which the total loss of preexisting ciliary organization in the cyst was confirmed by serial-section electron microscopy (Hammersmith, 1976b), and, most interestingly, even to type III mirror-image doublets (Grimes, 1982, 1989). Both of these more unusual types of doublets commonly emerge from cysts with the same geometry that they had when they encysted.

What must persist in cysts is some as yet ultrastructurally unidentified system within the cell cortex (Grimes, 1982) that can retain information concerning the nature, number, and large-scale asymmetry of the organelle sets that are to be formed upon excystment. This has the self-organizing and regulating characteristic of morphogenetic fields. The self-organizing property can best be appreciated by thinking about an apparent paradox: individual ectopic marginal cirral rows can be propagated vegetatively but *not* through cysts (Fig. 4.11b), yet the capacity to form extra marginal cirral rows as part of a second normal *set* of ventral ciliature *can* be perpetuated through cysts (Fig. 4.11c). One way of resolving this apparent paradox is to assume that a self-propagating local system for maintaining marginal cirral rows is nested within a more global spatial organization that can substitute for the local system when the latter is absent. In incomplete doublets, the substratum for the global system that underlies the large-scale ventral cirral organization of one of the two components (the upper one in Fig. 4.11b) has been destroyed, so that only the local system is left. Conversely, in encysting doublets, the ciliary and microtubular structures that normally are capable of self-propagation have disappeared, but the large-scale organization perseveres.

The regulative property of the underlying organization that specifies the number of complete sets of structures was demonstrated by Grimes (1973d) when he showed that an equatorially transected type I doublet could form two doublets, both with complete ciliatures, upon excystment (Fig. 4.12d), just as they could during ordinary regeneration (Fig. 4.12c). The capacity of parts of the dual systems to regulate to form dual wholes is thus preserved through the cyst stage.

The basis of the distinction between singular and dual systems in cysts is unknown. Hammersmith (1976b) has pointed out that a pattern of grooves in the cyst wall [first described by Hashimoto (1964) as fibers] is associated with the number and orientation of the sets of cortical structures in excysted cells. He suggested that large-scale organization might be preserved in the form of differentiations within the cell surface membranes of the encysted cell. If so, these differentiations must themselves be capable of regulation.

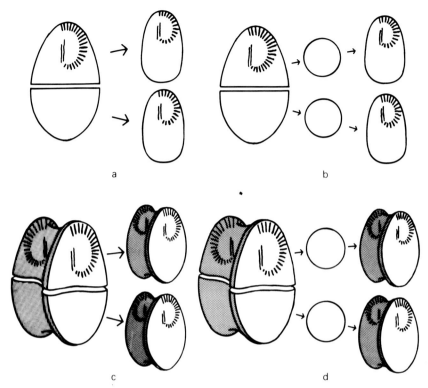

Figure 4.12 Effects of equatorial transection on the organization of singlets (**a**, **b**) and type I doublets (**c**, **d**) after direct regeneration (**a**, **c**) or passage through a cyst (**b**, **d**). The doublet is shown in a highly schematic manner, with only oral structures indicated. The shaded structures in (**c**) and (**d**) are on the opposite side of the cell.

4.4 INHERITANCE OF REVERSALS OF LARGE-SCALE ASYMMETRY

Virtually all of the cases of cortical inheritance that we have considered thus far can be thought of as variations of multiplicity or orientation of structural patterns that are internally unchanged. Even an inverted ciliary row is abnormal only in its spatial relation to the other rows. Hence, one could argue that true differences in intracellular patterns are not heritable by these nongenic means. We have, however, already noted one situation that transcends this limitation, namely the heritable mirror-image (type III) doublets of oxytrichids (Section 4.3.3 and Fig. 4.10). Because a globally reversed cell-unit cannot be superimposed on a normal one, the two must be regarded as truly different: global reversals affect not just the multiplicity but also the internal organization of the underlying morphogenetic systems. Reversals of large-scale asymmetry are considered in detail later in this book (Chapters 8 through 10). The present chapter, however, would not be complete without a summary statement that reversals of the asymmetry of arrangement of structures have repeatedly been encountered in ciliates, that they are heritable, and that in the majority of these cases the inheritance is nongenic. Justification for these factual statements is presented in Chapters 8 and 9, and the theoretical implications are explored in Chapter 10.

4.5 CONCLUSIONS

The information presented in this chapter can be summarized in the form of two major conclusions. First, differences in preexisting structural organization can be inherited during the vegetative growth and division of ciliates. Second, structural organization of many different kinds can be inherited, ranging in scale from microtubule bands to the as yet unidentified structural substratum of large-scale regulative fields.

Beause different structural systems can be inherited separately, they could be propagated by qualitatively different mechanisms. The ease of detection of inheritance of structural systems in ciliates is a consequence of ciliate topology rather than of any particular mechanism of propagation; *any* structural system using *any* mechanism of propagation can be inherited in the ciliate clonal-cylinder if it can grow longitudinally and maintain its lateral boundaries while it is growing.

These considerations provide the reason why I now prefer not to use any single term, such as cytotaxis [proposed by Sonneborn in 1964 and used since by other investigators including myself (Aufderheide et al., 1980; Grimes, 1982; Frankel, 1984; Nanney, 1985; Hjelm, 1986)], for the inheritance of structural patterns in ciliates or in other organisms. A technical-sounding term, even if carefully qualified in its usage, leads one to expect a unique mechanism or at least a unique class of mechanisms. Because this conclusion cannot be taken for granted, I prefer to use a phrase, such as structural inheritance, that emphasizes the likelihood that we are dealing with information at a supramolecular level but that has as few mechanistic connotations as possible. Even this phrase must be taken in a very broad sense, because although a regulative field certainly has a structural basis, it may also have a dynamic organization that departs from the "jigsaw" paradigm of structural assembly (see Harrison, 1982).

The demonstration that types of organization more complex than sequences of base pairs in nucleic acids can be inherited is very important and deserves reiteration in this increasingly DNA-centered era. Although ciliates are especially suited for demonstrating such structural inheritance, they are not unique. At the intracellular level, structural inheritance has been observed in mammalian cells for the form of the cytoskeleton (Albrecht-Buehler, 1977), the shapes of cells (Solomon, 1981), the numbers of cell centers (Shay et al., 1978), and in bacteria for the maintenance of the polarity of the Fe_3O_4-containing organelles (magnetosomes) responsible for orientation of certain bacteria relative to magnetic fields (Blakemore and R. Frankel, 1981; Hedges, 1985). The general importance of preexisting structure in determining new structural organization in cells other than ciliates has been emphasized for intracytoplasmic microtubule-organizing centers (Brinkley et al., 1981; McIntosh, 1983; Kirschner and Mitchison, 1986), for peripheral attachment sites of microtubules (Swan and Solomon, 1984), and for the organization of membrane domains (Poyton, 1983).

There are fewer direct parallels at the organismal level. Ciliates propagate themselves longitudinally and, therefore, can perpetuate aspects of their underlying organization directly, whereas most multicellular organisms produce eggs within a tiny region of the parental body and probably do not transmit the parental organization by direct structural means. In the unusual cases in which multicellular organisms do reproduce like ciliates, however, they also can inherit differences in parental organization like ciliates do. The classic case is that of the turbellarian flatworm *Stenostomum*, which can multiply by elongation followed by tandem subdivision. Long ago, Sonneborn (1930) generated homopolar doublet worms, and found that when these worms reproduced they inherited their doublet configuration true-to-type. When ciliates and worms grow longitudinally, the substratum of the underlying fields presumably grows along with the organisms.

Apart from providing the most convincing arguments for the insufficiency of the "central dogma" (Hershey, 1970; Nanney, 1985), structural inheritance in the ciliate cortex is valuable as an analytic tool. The capacity for a pattern to perpetuate itself offers opportunities for investigation of aspects of pattern formation that would be harder to analyze if the patterns were established de novo in every generation. One such aspect is the identification of distinct levels of control of intracellular pattern, a local level associated with preexisting ciliary organization and a more global level that is dissociable from ciliary organization. A second aspect is the regulative adjustment that can occur at the global level. The capacity of such global systems to be propagated offers opportunities for study of dynamic regulation along the length of an indefinitely growing clonal cylinder rather than in the spatially and temporally fixed confines of the developing embryo. Both of these aspects are taken up in later chapters of this book.

5

Segmental Subdivision

5.1 INTRODUCTION: CILIATE DIVISION AS SEGMENTATION

This chapter considers the cell cycle and cell division in relation to cortical patterning. This subject is complementary to the topic of cortical inheritance dealt with in the previous chapter. The very same cylindrical ciliate topology (see Fig. 2.2) that promotes continuity of specific longitudes makes the direct inheritance of particular latitudes impossible. In every cell generation, a cell equator is transformed into a new posterior and a new anterior end, and the positional value of every other cell latitude (except for the original poles) is changed. This sequence of changes ordinarily is described as an aspect of the cell division cycle. Here, however, we view it as a morphogenetic transformation akin to segmentation.

This presentation is divided into two parts, concerned with the conditions and the nature respectively of the segmental subdivision of the ciliate clonal cylinder. The first part covers the initiation of cell division. Its central theme is that preparation for cell division in ciliates is closely coordinated with cortical development. This section first reviews the evidence for an extremely loose relationship between macronuclear DNA synthesis and cell division, then shows how cell division and cortical development are physiologically connected, and finally links this to the microsurgical demonstration that initiation of division depends on the removal of a pervasive state of inhibition of cortical development.

The second part deals with the nature of the subdivision. This section centers on the fission zone. It begins with an experimental demonstration that the ciliate fission zone, like the insect segment border, is more accurately regarded as a boundary than as a structure. The remainder of this second part is devoted to an analysis of how that boundary develops.

5.2 CONDITIONS FOR SEGMENTAL SUBDIVISION

5.2.1 DNA Replication and Cell Division

For many cell biologists, the cell cycle can be reduced to an alternation of nuclear DNA replication and mitosis, with cell division as an epiphenomenon. In ciliates, this issue is complicated by the presence of two distinct kinds of nuclei, which commonly replicate at different times in response to different control mechanisms. The micronuclear cycle appears to be conventional in that replication alternates with mitosis, and these two processes remain strictly associated with each other whenever the time of mitosis is experimentally perturbed (Jerka-Dziadosz and Frankel,

1970; Torres et al., 1980; Nieto et al., 1981). However, the viability of numerous amicronucleate ciliate clones, some found in nature and others experimentally generated (Ng, 1986), indicates that the micronucleus is not essential for progression of the cell cycle. The macronucleus, on the other hand, is indispensable, but both the regulation of its S periods and the relation of these S periods to cell division are unconventional. Here I deal primarily with the latter topic, stressing the two best-studied genera, *Tetrahymena* and *Paramecium;* for recent reviews of both topics see Berger (1984, 1988).

Although the division of the *Tetrahymena* macronucleus is tightly coordinated with cytokinesis (Frankel et al., 1976), the relationship between macronuclear DNA synthesis and macronuclear/cell division is loose. In this ciliate, natural variations in macronuclear DNA content are caused by cumulative effects of unequal division and chromatin extrusion, and are regulated by varying the number of cycles of DNA replication per division cycle (Cleffmann, 1968, 1980; Doerder, 1979). When the macronuclear DNA content gets too low, two successive macronuclear S phases occur with no intervening division (Cleffmann, 1968; Worthington et al., 1976; Méténier, 1979), and when macronuclear DNA content becomes too high, two successive macronuclear divisions take place with no intervening macronuclear S phase (Doerder and DeBault, 1978; Méténier, 1979). In certain polymorphic *Tetrahymena* species, these two unusual types of cell cycles may alternate more or less regularly (Gabe and DeBault, 1973; Méténier, 1981). Such alternation can be induced in *T. pyriformis* by a sequence of synchronizing heat shocks, with two successive rounds of DNA replication without division during the synchronizing treatment followed by two successive divisions without a round of DNA replication after the end of the treatment (Jeffery et al., 1970, 1973; Andersen, 1977). In these disturbed cycles, there is no fixed phase relation between the time of macronuclear DNA synthesis and that of macronuclear or cell division (Zeuthen, 1978). Even in "normal" cycles in which single rounds of DNA replication alternate with cell division, only the first two-thirds of the DNA doubling is required for the next cell division (Andersen, 1972). This observation plus the fact that the sequence of replication of DNA molecules is not repeated from one macronuclear S period to the next (Andersen and Zeuthen, 1971) suggests that even when macronuclear DNA replication and cell division are normally "engaged" to each other, it is the attainment of a threshold DNA quantity rather than the replication of any particular DNA molecule that permits cell division to occur (Andersen, 1977). All of these lines of evidence taken together clearly indicate that the macronuclear replication cycle of *Tetrahymena* is not tightly coupled to the timing of cell division.

The same general conclusions apply to *Paramecium tetraurelia,* although some of the details are remarkably different. This ciliate regulates its macronuclear DNA content not by skipping or reiterating rounds of complete macronuclear DNA doublings, but rather by synthesizing a specified *amount* of DNA irrespective of the quantity of DNA present at the start of replication (Berger and Schmidt, 1978; Berger, 1979). The increment, however, does depend on the organizational and the nutritional status of the cell, as it is increased in doublets (Morton and Berger, 1978) and reduced in slow-growing cells maintained under conditions of nutritional limitation (Ching and Berger, 1986b). Synthesis of macronuclear DNA in the same cell cycle is necessary to permit the cell to divide, but the requirement is not for the completion of a round of DNA synthesis or even for a fixed proportion of that round. "Commitment to division" (discussed later) occurs near the middle of the macronuclear S phase under good nutritional conditions and near its end under poor ones (Ching and Berger, 1986a). Exhaustive recent work indicates that under conditions of extremely slow growth the increment may be reduced to a minimum of 30 percent of the DNA increment in well-fed cells (Berger, personal communication). Although the precise nature of the macronuclear DNA requirement for division is unknown, as in *Tetrahymena* it is unlikely to be a specific qualitative nuclear event.

The timing of the "commitment" event in *P. tetraurelia* does, however, provide a clue as to nature of the terminal events leading up to cell division. The point of commitment (the time when a mutant cell with a temperature-sensitive block in cell cycle progression becomes capable of dividing even after it is shifted to a nonpermissive temperature) is at about 73 percent of the cell cycle in rapidly growing cells (Peterson and Berger, 1976; C. Rasmussen et al., 1985), and at 96 percent of the cell cycle in cells growing much more slowly in a chemostat (Ching and Berger, 1986a). The actual clock *time* between commitment and division, however, is virtually the same under the two conditions (Ching and Berger, 1986a). The duration of the terminal events after commitment is insensitive to very large differences in the nutritional state and growth rate of the cell (Berger and Ching, 1988).

What are these terminal events? C. Rasmussen et al. (1985) noted that the point of commitment in well-fed paramecia approximately coincides with the time of onset of oral development and also of micronuclear mitosis as observed by Kaneda and Hanson (1974) under similar conditions. It is not definitely known whether this holds true in nutrient-limited paramecia as well. The constancy of the time required for oral development, however, has been documented thoroughly in *Tetrahymena*. In three separate studies, the time required for development of the oral primordium was near 100 minutes when growth rate was manipulated by different types of nutritional limitation that resulted in generation times ranging from 130 to 1030 minutes (Suhr-Jessen et al., 1977; Antipa, 1980; Nelsen et al., 1981). We can probably safely assume that this holds true for *Paramecium* as well.

The studies on *Paramecium* and *Tetrahymena* thus suggest that the appearance of the oral primordium is either the event that initiates the terminal phase of preparation for division or a reliable visible marker of some other coincident event. Once oral development is initiated, the rate of further progress toward cell division is largely unaffected by differences in rates of growth and presumably also of metabolism. This in turn suggests that the duration of the terminal phase of the ciliate cell cycle is limited by the rates of processes of assembly of supramolecular structures (which typically are exergonic) rather than of synthesis of macromolecules (which always are endergonic).

5.2.2 Commitment, Initiation, and Stabilization

Does the time of initiation of oral development mark a true "point of no return," in the sense that the developing cell cannot restore itself to its condition before the start of development? For *Paramecium,* the answer may well be "yes," although for *Tetrahymena* and *Stentor* the answer is certainly "no."

Inhibitors of macromolecular synthesis are capable of preventing or greatly delaying cell division of wild-type *P. tetraurelia* only if they are added before a "transition point" situated 1 to 1½ hours before the completion of cytokinesis (L. Rasmussen, 1966, 1967). This transition point may well be indistinguishable from the above-mentioned commitment point located at 87 ± 7 minutes before cell separation (Ching and Berger, 1986a). [Although the experiments being compared here were conducted over a decade apart in different laboratories, the foundation stock, growth medium, temperature, and average generation time (5.5 hours) were all the same, making a direct comparison possible.] Although cortical development was not examined directly in these studies, it was monitored in two other studies, one by Whitson (1964) on effects of temperature shocks and another by Wille (1966) using a drug (phenethyl alcohol) that prevents proliferation of ciliary units. Although both treatments could bring about arrest of division, neither triggered a specific resorption of the oral primordium.

The results of comparable experiments on *Tetrahymena* were altogether different. Exposure to extreme temperatures or to any of a great variety of drugs, including but not limited to known inhibitors of metabolism and of RNA and protein synthesis, brought about resorption of oral primordia with developing membranelles in *T. pyriformis* (Frankel, 1962, 1965, 1967b, 1969b; Williams, 1964; Gavin and Frankel, 1966; Nelsen, 1970; Grolière and Dupy-Blanc, 1985; Szablewski, 1985) and, less consistently, in *T. thermophila* (Gavin, 1965; Gavin and Frankel, 1966, 1969; Frankel et al., 1980b; see Frankel and Williams, 1973 for a comparative review). The resorption commonly occurred after a lag during which development continued, albeit sometimes rather abnormally (Frankel, 1967c; Grolière and Dupy-Blanc, 1985). The resorption process, once triggered, was carried to completion even after a return to optimal conditions (Frankel, 1967a). This process affected both oral development and cell division coordinately and brought the cell back to the developmental condition existing at or just before the initiation of oral development. This coordinated resorption is almost certainly closely related to the cell-age-dependent "set-back" response (Thormar, 1959) that makes synchronization of cell division by heat shocks possible in *Tetrahymena* cells (reviewed in Mitchison, 1971, chapter 10; Zeuthen and L. Rasmussen, 1972). *Paramecium tetraurelia* could not be synchronized by this method (Whitson, 1964).

Such coordinated reversal of development in *Tetrahymena* could be elicited at any time up to a "stabilization point" (Frankel, 1962), which occurs fairly late during oral development, shortly before or during the time when the fission zone is being formed. As seen first by Williams (1964), the specific stage of oral development at which stabilization takes place varies somewhat under different conditions; thus it is likely that ". . . the basis of stabilization lies outside of the (oral) primordium itself. A fundamental change may be occurring throughout the cell which indirectly stabilizes the primordium by affecting factors upon which primordium development depends" (Williams, 1964, p. 571).

Even the stabilization point does not represent an absolute point of no return. Exposure to high pressure (Simpson and Williams, 1970), as well as prolonged

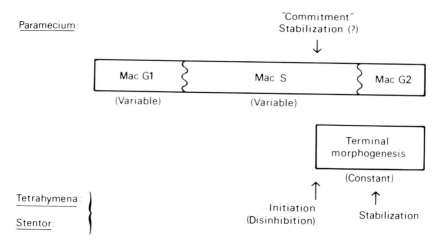

Figure 5.1   A summary scheme of events in the cell cycle of ciliates, simplified so as to emphasize the terminal events. The G1-S-G2 bar refers specifically to the macronuclear [Mac] replication cycle, whereas the terminal morphogenesis bar refers to oral development, accompanied by micronuclear and macronuclear division. The term "commitment" refers to the time at which cell division becomes independent of further macronuclear DNA synthesis; "initiation" indicates the time when oral development begins; "disinhibition" indicates the time when a stentor cell passes from a state of inhibition of division to a state of activation; and "stabilization" designates the stage at which the cell becomes no longer susceptible to a cancellation of cell division. For further explanation, see the text.

immersion in colchicine (Nelsen, personal communication), can bring about resorption of oral primordia at very late stages of development (Moore, 1972) and of anterior oral structures undergoing remodeling during division (see Section 2.3.2), *without* preventing cell division. The pressure effects may damage microtubules and other structures (Moore, 1972), and thus may prevent the execution of processes that already have been irreversibly triggered in a physiological sense.

The relationships of commitment, initiation, and stabilization in *Paramecium, Tetrahymena,* and *Stentor* (Section 5.2.3) are summarized in Figure 5.1.

5.2.3 Initiation Involves a Removal of Inhibition

To obtain further insight into the "factors upon which primordium development depend," we switch to a different organism and a different experimental approach. The organism is *Stentor coeruleus,* and the approach is that of microsurgical grafting.

Tartar (1958a, 1961, Chapter VIII) elegantly demonstrated that nondeveloping stentors could inhibit their own oral development. He first induced regeneration in a stentor by removing its membranelle band. He then excised a pie-shaped wedge that contained the oral primordium (Fig. 5.2e) from the regenerating cell, and implanted this piece into the dorsal (aboral) region of another cell that was *not* developing (Fig. 5.2a). The developing oral structures within the implanted patch were then promptly resorbed (Fig. 5.2b), whereas the patch itself, which included

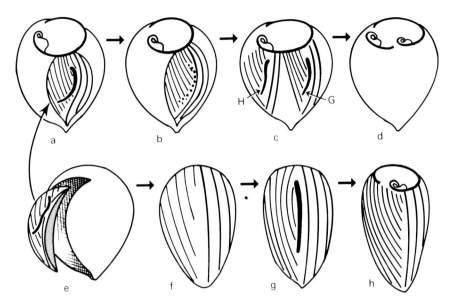

Figure 5.2 Induced resorption of the oral primordium from a regenerating stentor (**e**) after it has been grafted into a nondeveloping host stentor (**a**). After the graft, the grafted oral primordium was resorbed (**b**). This was followed (**c**) by the later synchronous formation of new oral primordia in both the grafted patch [G] and the host [H], leading to the formation of a doublet (**d**). The regenerating cell from which the oral primordium was removed healed (**f**), with a rapid formation of an oral primordium at the line of heal (**g**), followed by the reconstitution of a slender singlet (**d**). Redrawn from Figure 34 of Tartar (1961), with permission.

the "contrast zone" (see Section 3.3 and Fig. 3.8), was maintained. Many hours later, new oral primordia were initiated concurrently at the contrast zones of *both* the host (H) and the grafted implant (G) (Fig. 5.2c), and the cell reorganized to form a doublet (Fig. 5.2d) that then passed on its doublet condition to successive division products, providing us with yet another example of cortical inheritance. Similar results were obtained with wedges from cells preparing for division rather than for regeneration (Tartar, 1958b).

Was the resorption of the oral primordium of the implant perhaps due to an effect of injury caused by the operative procedure? Tartar carried out two control experiments to demonstrate that this was not so. First, if a wedge carrying an oral primordium was transplanted into another stentor that itself was undergoing oral-primordium development, then the oral primordium of the implant was *not* resorbed, but rather continued to develop, with the two oral primordia typically becoming synchronized in their stages of development (Tartar, 1966a). Second, if the wedge (carrying a few nodes of the macronucleus) simply was kept by itself without implantation into any host cell, the oral primordium also continued its development. Thus, Tartar showed that a condition specific to nondeveloping cells actively promotes resorption of developing oral primordia and coordinately prevents cells from preparing for division. Tartar labeled this condition a "state of inhibition," and contrasted it to the "state of activation" of developing cells (Tartar, 1961, Chapter VIII).

Because macronuclei could be transplanted from nondeveloping to developing cells without affecting the continuation of development, the agents establishing the states of activation and inhibition do not reside in the macronucleus (Tartar, 1961, p. 144; de Terra, 1964). This observation is consistent with numerous other experimental results (reviewed by de Terra, 1970, 1978), which indicate that macronuclear events in *Stentor* are subservient to cytoplasmic states that most likely are transmitted through the cell cortex.

What happens when cells in opposing states are joined? Tartar carried out such experiments on dividing cells and found that inhibition generally won; that is, the predividing component of a parabiotic graft complex usually took down its oral primordium and returned to the nondividing condition (Tartar, 1966c). This often happened even when the developing component of the parabiotic graft was substantially larger than the nondeveloping component. To create a maximum possibility for activation of division in a parabiotic partner, Tartar fused cells that were visibly preparing to divide with nondeveloping cells that were very large and, there-

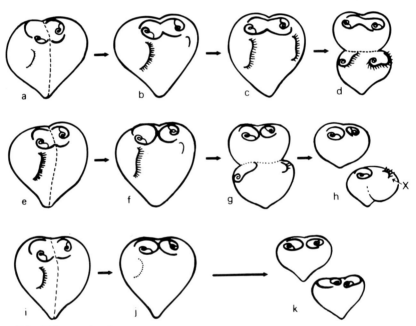

Figure 5.3 Different developmental outcomes in side-by-side parabiotic grafts of predivision stentors to large nondeveloping stentors. The three most common types of interaction from among 35 fused pairs are shown. (**a–d**, 8 cases). An oral primordium appeared in the initially nondeveloping component (**b**). The more advanced primordium then "waited" so that the two oral primordia became synchronized (**c**), and the complex divided to produce a doublet posterior daughter (**d**). (**e–h**, 4 cases). An oral primordium appeared in the initially nondeveloping component (**f**). The more advanced oral primordium completed its development while the induced primordium was arrested (**g**) and (in one of these 4 cases) the cell divided with the induced primordium represented by a vestige marked with an "X" (**h**) [in the other 3 cases fission was suppressed]. (**i–k**, 11 cases). The oral primordium of the developing cell was resorbed (**j**). The parabiotic complex produced two new oral primordia the next day and divided to produce a doublet posterior daughter (**k**). Redrawn from Figure 5 of Tartar (1966c), with permission.

fore, presumably close to initiating division themselves. Even in such cases, induced resorption of the oral primordium of the developing partner was a common result, with division as a doublet occurring only much later (Fig. 5.3i–k). In an approximately equal number of instances, however, the nondeveloping component did promptly form an oral primordium and often divided as well (Fig. 5.3a–d,e–h). In these graft complexes, an induced activation might have taken place in the nondeveloping component, although the situation is ambiguous because some of the large nondeveloping cells might have initiated division even if they had not been grafted to developing cells. In most such cases, the oral primordium of the more advanced component "waited" while the induced oral primordium "caught up" to achieve synchronization by the time of division (Fig. 5.3a–d) (see Tartar, 1966a). Sometimes, however, synchronization failed, yet the two oral systems did not simply develop independently. Rather, the more advanced one completed its oral development, whereas the less advanced one was drastically impeded, either regressing totally or forming a vestigial set of oral structures (Fig. 5.3e–h). This phenomenon, which Tartar (1966a) called *overtake*, is really a form of inhibition: the more advanced component completed its development and switched to a suppressive state before the less advanced component had lost its susceptibility to such suppression. Under such circumstances, cell division sometimes succeeded (as shown in Fig. 5.3h), but more commonly failed (not shown).

These results strongly suggest that cortical development and division are subject to control by pervasive cellular states, and that the dominant state, prevalent over most of the cell cycle, is one of inhibition. The state of inhibition is not merely unsupportive of continued development, but actively opposed to it, bringing about a return to a condition similar to that before development had begun. Activation for development can best be thought of as a disinhibition, a temporary relief from a spatially and temporally pervasive negative control. In Tartar's words, "The strong implication is that stentors are not continually building toward division as they grow but are continually inhibiting their division until a final moment when the repression is withdrawn" (Tartar, 1966c, p. 305).

The effects of exposing regenerating stentors to inhibitors of RNA and protein synthesis can easily be fit into Tartar's conceptual framework. Oral primordia were resorbed if affected at a stage before completion of membranelle formation (James, 1967; Burchill, 1968); inhibition of protein synthesis at a somewhat later stage (Tartar's stage 5) arrested development but did not induce resorption (Burchill, 1968). The stage of stabilization (near the end of Tartar's stage 4) with respect to effects of inhibitors is at or close to the stage at which resorption could no longer be induced by parabiosis with a nondeveloping stentor (Tartar, 1961, Chapter VIII). This developmental stage is structurally similar to the stage at which stabilization occurs in *Tetrahymena*.

This parallel between *Tetrahymena* and *Stentor* casts a different light on the phenomena of sensitivity to inhibitors and stabilization. If we think of the "initiation" event (see Fig. 5.1) as a shift in a dynamic balance between two pervasive and opposed cellular states, with "activation" as a provisional escape from the normal state of inhibition, then virtually any physical or metabolic disturbance could shift the balance unfavorably and thus annul the escape. Before stabilization, development that fails to go forward must instead go backward. The stabilization point is *not* the stage at which all macromolecular syntheses necessary for development are completed but rather is the point at which an induced reversion to the state of inhibition no longer is possible and the developmental program can no longer be shifted into reverse.

Although it is not my main purpose here to consider comparative cell cycles, I will point out that stimulation of cell division through disinhibition is not unique to ciliates. Fission yeast *(Schizosaccharomyces pombe)* also have a control point immediately preceding mitosis (Fantes and Nurse, 1977). The protein product of the major, positively acting control gene *(cdc2)*, whose function is essential for passage through this control point, however, is present in similar quantities throughout the fission-yeast cell cycle (Simanis and Nurse, 1986). This indicates that the *cdc2* product does not exert its effect simply by accumulation. At least one of the critical regulatory genes (*wee1*$^+$) acting at the control point exerts a negative effect (Nurse and Thuriaux, 1980) that modulates the actions of products of other cell-cycle gene(s) (Fantes, 1981, 1983; Russell and Nurse, 1987). Thus, the "trigger" to division in *S. pombe* can be thought of (at least in part) as being a relief from inhibition by the normal product of the *wee1*$^+$ gene. Additional evidence for active inhibition of mitosis in interphase cells comes from genetic analysis in the mold *Aspergillus* (Osmani et al., 1988) and from physiological studies in mammalian cells (Adlakha et al., 1983). Tartar's idea that division is stimulated by a relief from inhibition thus may have general application.

5.2.4 Formed Oral Structures Help Maintain the State of Inhibition
The inhibitory role of formed oral structures

When a nucleated wedge bearing an oral primordium was taken from a dividing stentor and maintained in isolation, the development of the fragment depended on whether or not oral structures were included. If the wedge lacked oral structures (as in the regenerating cells considered earlier), development continued. If it included a substantial portion of the oral apparatus (OA), the oral primordium was promptly resorbed and a new oral primordium was reinitiated later, now to serve for regeneration rather than division (Tartar, 1958b). These results suggest that formed oral structures might have something to do with imposition or maintenance of the state of inhibition. This idea was confirmed by a variety of other experiments, the most dramatic of which was the induction of resorption of an oral primordium in a regenerating stentor to which an intact set of oral structures was added (Fig. 5.4) (Tartar, 1958b); control grafts containing other cell parts did not induce resorption. This observation was confirmed by de Terra (1977), who also found that OAs taken from large stentors were more effective in stimulating resorption than were those from smaller stentors.

The idea that the state of inhibition is related to the presence of intact feeding structures can be applied directly to regeneration in *Stentor,* as removal of all or part of the OA serves as a powerful stimulus for regeneration of oral structures in otherwise intact cells (Tartar, 1961, Chapter VI). Subsequently, de Terra (1977) found that replacing the OA of a decapitated large stentor with the intact OA from a small stentor was sufficient to induce formation of an oral-replacement primordium by the host cell, whereas sham-operated stentors whose own OAs were

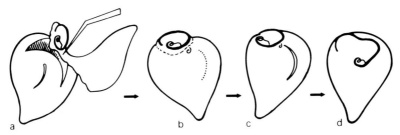

Figure 5.4 Resorption of an oral primordium of a regenerating stentor induced by grafting a set of oral structures from another stentor. A slit was made in the anterior end of an early regenerating stentor (**a**), and the apical disc taken from another stentor was implanted into it. Soon after the grafted oral structures healed into place, the oral primordium of the host was resorbed (**b**). A new oral primordium was later produced (**c**), which replaced the grafted structures (**d**). Redrawn from Figure 7 of Tartar (1958b), with permission.

removed and replaced did not form oral primordia. This suggests that reduction of the size of the OA relative to that of the remainder of the cell can trigger oral development. Other experiments, however, showed that reduction of oral size is not indispensable for stimulation of regeneration in *Stentor,* as severance of an OA into two parts that do not rejoin also could stimulate regeneration (Tartar, 1957; de Terra, 1985a). These effects of the preexisting OA appear to be mediated by structural interactions within the cell cortex, a point made originally by Tartar (1961, Chapter VIII) and buttressed by de Terra (1979; 1985a).

The oral/somatic ratio hypothesis

Both division and oral replacement are initiated while the preexisting OA is still intact and the cell has suffered no visible damage. de Terra (1969) proposed the "oral/somatic" ratio hypothesis to reconcile these facts with the idea that cortical development involves an escape from an inhibition that is *specifically* associated with an intact OA ("somatic" is the term ciliatologists use for the nonoral portions of the cell surface). She noted that although stentors grow during the interfission period, their OAs are unable to form new parts in situ (see Section 2.3.3). Thus, the oral/somatic ratio decreases steadily during the interfission interval; cells that are initiating division have an oral/somatic ratio one-half that of newly divided cells (de Terra, 1969). Therefore, de Terra (1969, 1979) proposed that a diminution of *relative* oral size in cells that exceed a certain threshold of *absolute* cell size are the jointly sufficient stimuli for initiating division in *Stentor.* She further postulated that attainment of a sufficiently low oral/somatic ratio when cell size is below the division-threshold triggers oral replacement in place of division. This interpretation was supported by the initiation of oral replacement after recapitation of a decapitated *Stentor* with an OA from a smaller *Stentor* (de Terra, 1977), as described previously.

The restrictive form of the oral/somatic ratio hypothesis as proposed by de Terra is unlikely to account for all situations of cortical development in ciliates. This is most obvious for oral replacement, which is readily observed in starving ciliates, for example in *Stylonychia* (Dembowska, 1938) and in *Tetrahymena* (Frankel, 1970; Nelsen, 1978). In those situations, cells were becoming smaller rather than larger and thus the oral/somatic ratio was increasing rather than decreasing (see Bakowska et al., 1982a). Tartar's more general view of "... reorganization as a wholly spontaneous and intrinsic response to certain disproportionalities or disarrangements of parts of the cell for the purpose of bringing them to a more normal relationship" (Tartar, 1960, pp. 103–104) may be the best generalization that we can make at the moment. Furthermore, even cell division can occur in starving ciliates; for example, anterior fragments of equatorially transected oxytrichids regenerate despite the presence of a (relatively) oversized OA and afterward often divide even if starved, indicating that growth of the "somatic" cortex relative to the OA is not always necessary for triggering cell division (Golinska and Jerka-Dziadosz, 1973).

Nonetheless, the idea of a major influence of the size of cortical domains on the normal conditions for initiation of division does have considerable experimental support. As mentioned earlier (see Section 4.3.2), cytochemical measurements by Morton and Berger (1978) showed that doublets of *Paramecium tetraurelia* have both a DNA and protein content that is nearly double that of singlets, and the experiments by Sonneborn (1963) demonstrated that nuclear size is controlled by cortical valence rather than the converse. These observations led Berger (1988) to suggest that the "size of the cell cortex" may determine the set-point for regulation of cell mass and DNA content in *Paramecium*. Nanney et al. (1975) had earlier shown that the cortical valence (singlet versus doublet) probably determines the set-point for the total number of basal bodies within all of the ciliary rows in *Tetrhymena*, a variable that may well be correlated with overall cell size. The oral/somatic ratio hypothesis would predict that the number of OAs is crucial in determining these set-points, in which case they should quickly rise when a singlet is transformed into a doublet by "cortical picking" or intentional grafting, and should fall immediately after a doublet reverts to a singlet by loss of an oral meridian. The "incomplete doublet" condition of *Paramecium tetraurelia*, in which the cell is effectively a doublet *except* for the lack of a second OA (Sonneborn, 1963) would provide an especially important test case for the oral/somatic ratio idea.

Inhibition affects expression of, but not potential for, oral development

When a cortical wedge bearing an oral primordium was transplanted from a developing to a nondeveloping stentor (see Section 5.2.3), the oral primordium regressed but the implanted wedge did not. The wedge retained its contrast-zone, which later served as an organizing center that helped to transform the complex into a doublet. Thus, in this experiment as well as in parabiotic grafts (Fig. 5.3), the preexisting oral structures have been shown to influence the *expression* of oral-primordium development, but had no effect on the *potential* for subsequent development at a site already specialized for this purpose.

This result, however, leaves open the question of whether formed oral structures can prevent the formation of *new* sites at which oral primordia could develop. The answer to this question was provided by an experiment on *Blepharisma* that was first performed by Suzuki (1957). The experiment consists of transecting the cell horizontally at a level just behind the posterior end of the OA (Fig. 5.5a), then rotating the anterior portion 180 degrees around the vertical axis, and finally setting the anterior moiety back down on the posterior half (Fig. 5.5b). The entire OA thus

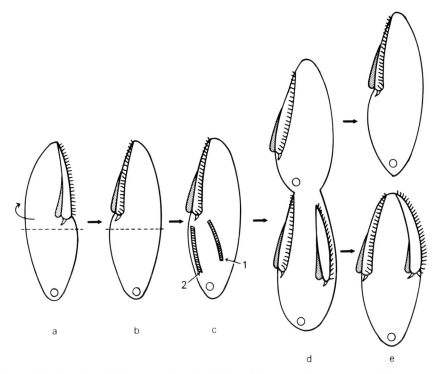

Figure 5.5 The effect of a horizontal incision followed by a 180-degree rotation of the anterior half of *Blepharisma* on the posterior half. (**a**) The cell just before the operation, showing the level of the incision. (**b**) 2.5 hours after the rotation, with no oral primordia present. (**c**) 27 hours after rotation, with two division oral primordia, one (1) at the displaced original primordium site and the other (2) at a new site posterior to the new location of the oral structures. (**d**) 30 hours, the cell in division, producing (**e**) at 31 hours, an anterior division product with one oral apparatus and a posterior division product with two. Redrawn with modifications from Figure 34B. a, b, f, g, and h of Suzuki (1957).

was shifted to a location distant from the site at which the oral primordium normally develops [this cannot be accomplished in *Stentor* because the OA encircles the anterior end of the cell]. Under these circumstances, no oral primordia formed until the normal time for the next division, about a day later. When development was finally initiated, *two* complete oral primordia formed, one at the somewhat displaced original primordium site (Fig. 5.5c,1), the other directly posterior to the new position of the OA (Fig. 5.5c,2). The cell then divided into an anterior singlet with a primordium site on the same side as the repositioned OA and a posterior doublet that propagated both old and new oral primordium sites (Fig. 5.5d,e). If the experiment was modified by excising the anterior three-quarters of the rotated portion (including most of the OA) immediately after the rotation, then new oral structures also developed on both sides, but much more promptly and in the form of regeneration/replacement rather than predivision oral primordia (Suzuki, 1957; Eberhardt, 1962). This accelerated response was presumably a consequence of removal of an inhibition exerted by the formed oral structures. The most important result, however, was that in both variants of this experiment the existing OA promoted rather than inhibited the formation of a new oral-primordium site in its vicinity.

Suzuki (1957, p. 154) summarized these results as follows: "The oral area is able to induce the formation of V-areas, and is at the same time endowed with an inhibiting function against the morphogenetic activity of V-area" ("V-area" is Suzuki's term for the oral-primordium site). The formed oral structures inhibit only the *expression* of primordium sites, while if anything promoting rather than inhibiting their *formation*. In the next chapter we consider a more plausible candidate for true positional inhibition, namely a possible lateral interaction between primordium sites.

5.3 THE NATURE OF SEGMENTAL SUBDIVISION

5.3.1 The Fission Zone as a Boundary

A central fact of ciliate life is that ciliate cortical organization is duplicated *before* the ciliate begins its actual constriction into two cells. This duplication characteristically becomes evident in two stages: first, the primordium for the oral apparatus of the posterior division product begins to develop; second, a fission zone appears. The fission zone is first seen as an equatorial ring of gaps in the ciliary rows, where the basal bodies of the rows are spaced somewhat further apart than elsewhere [*Tetrahymena:* Frankel et al., 1977; *Paramecium:* Kaneda and Hanson, 1974; *Climacostomum* (a close relative of *Stentor*): Dubochet et al., 1979]. In ciliates, such as *Blepharisma* and *Stentor,* in which longitudinal bands of pigment granules parallel the ciliary rows, the fission zone first becomes visible in living cells "as a clear band without any pigment granules" (Suzuki, 1957, p. 96; see also Tartar, 1961,

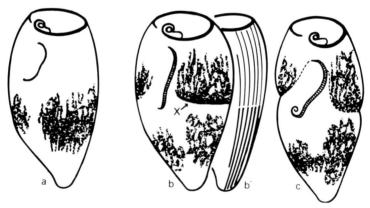

Figure 5.6 Migration and partitioning of the subcortical carbohydrate reserve *(shaded)* of *Stentor*. (a) An early predivider with the reserve beginning to migrate in an anterior direction. (b) A cell just before the appearance of the fission zone. The posterior portion of the reserve becomes more diffuse and is located more deeply in the cell, whereas the anterior portion has migrated forward and forms an even posterior border [X] where the fission zone will later appear. (b') The appearance of the pigment stripes at this stage. (c) A cell shortly after the onset of division constriction. The reserves are now partitioned into the two daughter cells. Redrawn from Figure 2 of Tartar (1959), with permission.

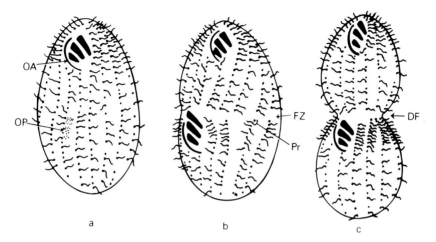

Figure 5.7 A semidiagrammatic representation of the ciliation of the ventral (oral) surface of a *Tetrahymena* cell before and during cell division. (**a**) Early oral development, showing the oral apparatus [OA] and the oral primordium [OP] in the anarchic field stage of development. All basal bodies are ciliated in approximately the anterior one-third of the cell, and about one-half are ciliated over the remainder. (**b**) Late oral development, with formation of the fission zone [FZ]. The basal bodies just posterior to the fission zone have become ciliated, with the exception of the anterior-most basal body of the apical pairs [Pr]. The impression is of a wave of ciliation spreading posteriorly from the fission zone. (**c**) A dividing cell, with a well-developed division furrow [DF]. At this time the ciliation of the closely spaced basal bodies at the anterior end of the nascent posterior division product is nearly complete, so that there appears to be a tuft of cilia at the new anterior end of each ciliary row. Freely redrawn from Figures 2, 4, and 5 of Frankel et al. (1981), using data from Table 2 of that paper.

Chapter V). Nonetheless, the fission-zone gap is *not* a complete structural break between the future fission products; the longitudinal microtubule bands (Fig. 2.6) of *Tetrahymena* and the KM fibers (Fig. 3.7) of *Stentor* both maintain their continuity across the fission zone even after cytokinesis has begun (Frankel et al., 1981; Diener et al., 1983). Thus, as is often true for insect segment borders as well (Lawrence and Green, 1975), the major ciliate pattern boundary is structurally unimpressive.

Although the fission zone itself does not amount to much, it seems to be associated with numerous other manifestations of spatial subdivision. I have already mentioned the contractile vacuole pores and basal body pairs that appear anterior and posterior, respectively, to the fission zone of *Tetrahymena* (see Section 4.2.4). Other more widely distributed features of the cell surface also become subdivided at this stage. In *Stentor*, a carbohydrate reserve takes the form of a ring situated just beneath the surface in the posterior half of the cell (Tartar, 1959); this ring becomes subdivided into two sections just before the fission zone forms, with half of the reserve moving anteriorly and coming to lie just anterior to the forming fission zone (Fig. 5.6) (Tartar, 1959). In *Tetrahymena*, all of the ciliary-row basal bodies within the anterior one-third of the cell are ciliated [except for the anterior-most nonciliated members of the apical pairs of basal bodies in rows 5 to n-1 (see Section 3.2.1)], whereas ciliated and nonciliated basal bodies roughly alternate over the remainder of the cell surface. When the fission zone develops, the basal bodies posterior to the fission zone become ciliated, creating the new anterior ciliation pattern that is completed while the cell divides (Fig. 5.7) (Frankel et al., 1981). In *Paramecium,* waves of basal-body duplication and remodeling of ciliary units proceed anteriorly and posteriorly from the fission zone (Iftode et al, 1989). Finally, in *Stylonychia,* the polarized distribution of (anterior) calcium-dependent and (posterior) potassium-dependent mechanoreceptor channels is already established at the respective ends of the daughter cells at the time when the division furrow first becomes visible (Machemer and Deitmer, 1987). Thus, the fission zone, inconspicuous if considered as a structure, is very conspicuous as an emerging pattern-border.

The concept of the fission zone as a boundary between two "nascent daughter cells" rather than a "special structure" (Tartar, 1968) was strongly reinforced in an important series of experiments carried out on *Stentor.* In the simplest experiment, Tartar (1961, Chapter V, 1968) used a glass needle to cut through the cortex completely around the equator of large nondividing cells, physically severing all of the longitudinal structural elements in cortical layer. Despite a severance *more* severe than that which occurs during cytokinesis, "the cortical structures merely heal(ed) together, often without leaving any indications of an operation" (Tartar, 1961, p. 88), and the cells did not divide. Later, de Terra grafted together equatorial regions of different nondeveloping cells, and found that pigment stripes with the same polarity tended to heal, and the adjacent ciliary rows also rejoined (de Terra, 1985b). Thus equatorial discontinuities in themselves do not lead to cell division.

Although mechanical disruption does not induce cell division in nondividing stentors, it also fails to prevent division in dividing cells; no matter how the fission zone is severed, sliced, fragmented, or transplanted, that zone or any part thereof always begins to constrict and usually completes constriction successfully, a result that was repeatedly obtained in *Stentor* (Tartar, 1961, Chapter V, 1968) and earlier

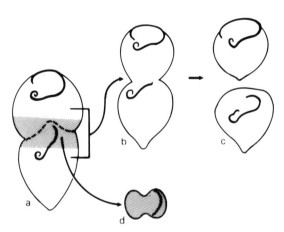

Figure 5.8 Reconstitution of the fission zone of a stentor after excision. The shaded area in (**a**) shows the excised region, which later went on to constrict partially (**d**). The remaining portions were fused together (**b**) and then divided (**c**). Similar results were obtained from this type of operation carried out at late prefission stages. Redrawn, with slight modification, from Figure 2 of Tartar, 1967a, with permission, and reproduced by permission from *Nature,* Vol. 216, pp. 695–697, copyright © 1967 Macmillan Journals Ltd.

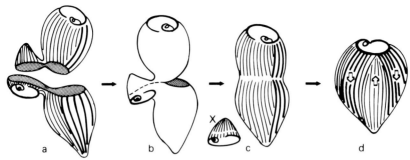

Figure 5.9 Production of tandem stentors, with subsequent interdigitation of blocks of cortical stripes. (**a**) The graft: two nondeveloping stentors were opened out, one near the tail and the other near the head. Tail and head-pieces were then pushed together (**b**), after which the smaller anucleate complex [X] was cut off and discarded, and the larger complex (**c**) was kept, with the pigment-stripes properly aligned along the horizontal plane. The complex did not divide at once; instead the cortical patterns of the two stentors interpenetrated, with wedges of posterior regions moving forward relative to the anterior regions, to generate a broad composite stentor of normal length (**d**), which later divided (not shown). Redrawn with slight modification from Figures 1 and 2A of Tartar, 1964, with permission.

in *Blepharisma* (Suzuki, 1957). A strip from a nondividing cell, however, failed to constrict even if implanted directly across the fission zone of a divider (Tartar, 1966c). These experiments show that a "prepared" surface region of a dividing cell divides autonomously, whereas *no* region of a nondividing cell can participate in constriction.

The above-described experiments indicate that the presumptive furrow region is predetermined to divide, but do not assess the competence of other regions of dividing cells to develop furrows. This problem was addressed in Tartar's most important experiment on dividers (Tartar, 1967a), in which he removed a cell-disc carrying the fission zone, and then fused the remaining anterior and posterior pieces minus the disc (Fig. 5.8a). Not surprisingly, the fission zone in the isolated disc constricted (Fig. 5.8d). More surprisingly, the remaining graft complex, which completely lacked the original equatorial ring, also divided promptly (Fig. 5.8b,c). Thus the fission zone is replaceable, provided that the material replacing it is from a dividing cell and is located between the two nascent daughter cells. In Tartar's words, "... the final 'instruction' of the division processes ... could be 'constrict between'" (Tartar, 1968, p. 33). Using different terms, the missing fission zone is restored by an intercalation process (see Section 10.4).

This intercalation of a fission zone, however, can occur only if the operation is carried out on dividing cells or cells with predivision oral primordia. An experiment on nondividing cells that is positionally equivalent to the one illustrated in Figure 5.8 is shown in Figure 5.9. In this case, the juxtaposition of the posterior region of one nondeveloping (or regenerating) cell and the anterior region of another similar cell (Fig. 5.9a–c) did *not* bring about cell division at the line of heal, despite the fact that the graft complex was now bigger than most dividing cells and was provided with a ready-made equatorial pattern-discontinuity. Neither did the graft complex merely wait for the next division and then intercalate a fission zone. Instead, it underwent a drastic shifting of cortical parts: wedges of cortex from both components interpenetrated in opposite directions (Fig. 5.9d) "... as though stripes of each set sought to reach the opposite pole" (Tartar, 1964, p. 244). Division occurred "... 2–4 days later *after* telescoping into a broad single individual, the fission line cutting indifferently across the intermingled striping" (Tartar, 1964, p. 246). The same process of interdigitation had been reported earlier by Uhlig (1960, pp. 48–51) after removal of an equatorial disc from nondeveloping or regenerating stentors, an operation that resulted in a comparable abutment of posterior on anterior cortex. This process of active interdigitation is reminiscent of the well-known sorting of metazoan cells according to differential cell affinities (Townes and Holtfreter, 1955; Steinberg, 1978).

The contrast between the response of dividing and nondividing cells to geometrically similar operations emphasizes that the entire central region of a dividing cell, and quite possibly the entire cell, is undergoing a transformation that involves much more than simply the elaboration of an oral primordium plus a constriction organelle. In Tartar's words, "Initiation of division in *Stentor* might therefore be a morphogenetic prescription or blocking out or determination of the nascent daughter cells in the cortex on either side of the equatorial line" (Tartar, 1968, p. 33). In what follows I will argue, first, that the presumptive fission zone is a by-product of an early "blocking out" of cellular territories, and second, that a subsequent more local determination event is required for formation of a visible division line along the presumptive boundary between the two separate cellular territories.

5.3.2 The Blocking Out of the Nascent Daughter Cells

Blocking out is an early event

When does the "blocking out" of the nascent daughter cells occur and how are the sizes of the blocks determined? The "when" question can be addressed microsurgically by removing anterior and posterior ends of the cell and observing the effects on the location of the fission line. As pointed out by Tartar (1954a) and Schwartz (1963), there is a great difference between the response of *Paramecium* and of other ciliates, especially *Stentor,* to such an operation. In *Paramecium,* amputation of anterior or posterior regions of nondeveloping cells resulted in the formation of unequal division products, with the division product that includes the amputated end being substantially smaller than the other division product (Calkins, 1911). Detailed studies of such amputees by Chen-Shan (1969, 1970) and Suhama (1975) indicated that "The fission plane seemed to be located in the same position it would have occupied had the cell not been cut" (Chen-Shan, 1969, p. 209). Nonetheless, substantial regulation of sizes and proportions of the fission products took place during the actual process of division. This was probably due to some regulative redistribution of zones of growth (Chen-Shan, 1969, 1970) and basal body proliferation (Suhama, 1975), although the possibility of some shifting of the fission line was not ruled out (Chen-Shan, 1969, 1970). In this ciliate, the latitude of formation of the fission zone is almost certainly predetermined, just as the site of oral development is fixed.

In contrast with the largely nonregulative response of *Paramecium,* if parts of stentors were removed before the initiation of cortical development, the products of the subsequent division were equal in size (Schwartz, 1963), indicating that the location of the presumptive fission zone is not predetermined in *Stentor* as it is in *Paramecium*. If, however, the anterior or posterior end of a stentor was removed *after* the onset of predivision cortical development, the subsequent division was markedly unequal (Tartar, 1968). Tartar, however, did observe equality of division after removal of the posterior end at one rather early stage of oral-primordium development (stage 2) but not at other earlier or later stages (Tartar, 1968). This peculiarly inconsistent response across stages led to a subsequent reinvestigation of the problem by Schulte and Schwartz (1970), who found that the degree of inequality of the division products depended on how much of a ventrally located growth zone was removed by the operation. They, therefore, ascribed Tartar's cases of regulation in size of division products to compensatory growth within the truncated posterior region rather than to a compensatory shift of the fission zone. Thus, it is probably safe to conclude that the site of the fission zone of *Stentor* becomes specified at the time when oral development is initiated.

Although the development of the oral primordium and cell division are under a joint physiological control (see Section 5.2.2), the former does not play a direct causal role in controlling the latter. Both stentors and blepharismas could divide after the oral primordium had been removed microsurgically (Tartar, 1966b; Suzuki, 1957); comparable situations are known in *Tetrahymena* (Nanney, 1967a) and *Glaucoma* (Frankel, 1961), and on the side of incomplete *Paramecium* doublets that lacks oral structures (Sonneborn, 1963). Furthermore, even though the visible fission zone gap typically first appears anterior to the oral primordium, aboral half-stentors created by longitudinal transection could intitate constriction and often divided successfully (Tartar, 1966b). A division furrow could also become established in a conical sector geometrically separate from that in which the oral primordium was located (Fig. 5.10) (Tartar, 1966c). Thus, the development of the oral primordium is only a symptom, but not a direct cause, of a globally coordinated subdivision of the cell.

I will summarize the results considered in this chapter thus far by expressing Tartar's idea of "blocking out . . . of the nascent daughter cells" in the language of positional values. Initiation of the division process is an abrupt and global switch

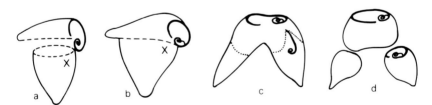

Figure 5.10 Cell division in a two-tailed stentor. The operation (**a**) involved grafting one non-developing stentor with its primordium-site excised (horizontal in diagram) onto another nondeveloping stentor whose apical disc was removed (vertical in diagram). The single primordium site of the ensuing complex (**b**) is indicated by an X. When the division was initiated in this complex (**c**), an oral primordium developed at the one primordium site, and then two fission zones appeared, one in the portion bearing the primordium, and the other in the middle of the separate "tail" that lacked an oral primordium. The fork-tail stentor then divided into three (**d**), with one of the two posterior division products lacking an oral apparatus. Redrawn from Figure 7B of Tartar (1966c), with permission.

from a single positional sequence (A-B-C-D-E-F) to a tandem dual sequence (A-B-C-D-E-F-A-B-C-D-E-F). Between the time of initiation and stabilization of division, the maintenance of this duality is dependent on continuation of a state of activation; reassertion of inhibition switches the A-B-C-D-E-F-A-B-C-D-E-F sequence back to the original A-B-C-D-E-F. The fission zone is initially not a structure in the ordinary sense but rather a by-product of cellular subdivision; its existence, maintenance, and location are expressions of the duplicated condition of the system as a whole. This view is elaborated further in Chapter 10, Section 10.4.

Spatial assessment is proportional

The biometrical analysis of the location at which cellular subdivision occurs also attests to its global nature. I have already mentioned that in *Tetrahymena, Paramecium,* and *Euplotes* the two division products obtain equal numbers of nonoral basal bodies through compensating inequalities in recruitment of preexisting basal bodies and intensity of proliferation of new ones, implying some form of global integration (see Section 3.2). A precise analysis of relative cell sizes of daughter cells, carried out on *T. pyriformis* by Hjelm (1983), indicated that the two daughter cells are nearly but not exactly equal in volume; the anterior daughter is usually slightly larger than the posterior daughter. More important, the ratio of sizes of the two daughter cells is the same irrespective of the size of the parent cell, indicating that cells must use a proportional system to assess where to place their fission lines.

Earlier studies, which concentrated on the position of the new oral apparatus of *Tetrahymena* relative to the old one, showed that when the size of the cell increased so did the distance between the old oral apparatus and the new oral primordium, irrespective of whether the distance was measured in number of micrometers or number of basal bodies (Lynn and Tucker, 1976; Lynn, 1977). Thus, the cell does not simply measure or count from a single reference point to determine the location at which it will subdivide itself. The measurement must be proportional, and therefore, uses at least two reference points.

What is the nature of the proportional system of measurement? There are indications, from two entirely different types of studies on *Tetrahymena,* that the measurement of the relative sizes of division products is sensitive to the shape of the cells. First, Hjelm (1983) observed that *T. pyriformis* cells that are forced to go through the entire cell cycle in a severely flattened condition (in which their shape is substantially modified but their capacity to grow and divide is not) have anterior division products with a larger relative volume than do cells that go through the cycle in the normal, unflattened shape. Second, in the *conical (con)* mutant of *T. thermophila,* in which cell shape is greatly modified, the oral primordium is shifted to a more posterior relative position and anterior division products are substantially larger than posterior division products (Doerder et al., 1975; Schaefer and Cleffmann, 1982). Both cases suggest that the parameter that cells measure is not relative lengths or relative volumes; it may instead be relative surface areas (Lynn, 1977; Hjelm, 1983). The biometrical analysis by Lynn is particularly ingenious. Lynn found that *within* both wild-type and *conical* clones the distance between the old oral apparatus and the oral primordium shows a linear regression on cell length, but the regression lines are different for the two clones and neither of them extrapolates back to the origin of the graph. When the distance between oral apparatus and oral primordium was compared with an estimate of cell surface area rather than cell length, however, the points for both wild-type and conical fit on the same regression line, which also extrapolated back to the origin, suggesting a precise proportionality (Lynn, 1977). Although biometrical analyses, such as this one, cannot prove the operation of any particular mechanism, they do suggest that a surface-related mechanism may operate in blocking out the presumptive division products.

The analysis of the conical mutation was based on the assumption that the inequality in size of division products was a secondary consequence of the alteration of cell shape. Are there any mutations that bring about the formation of unequal division products through a direct effect on the location of the fission zone? There are a few. In a class of mutants named *pseudomacrostome (psm)* (Frankel et al., 1984b), the original position of the oral primordium along the anteroposterior axis is variably disturbed (Frankel, 1979). Characteristically, the oral primordium is located more anteriorly than its normal position, and cells do not divide but instead undergo oral replacement to form large (hence pseudomacrostome) oral structures (Frankel et al., 1984b); sometimes, however, the oral primordium is more posterior than normal and then these cells may divide unequally to produce posterior division products somewhat smaller than anterior ones (Frankel, unpublished observations). A different newly isolated mutation, called *elongated,* is the first to bring about the consistent formation of the oral primordium more posteriorly than normal without any obvious change in cell shape (Frankel and Jenkins, unpublished observations). The way in which these mutations might affect the blocking out of daughter cells has not been analyzed in detail.

5.3.3 Final Determination of the Fission Zone

The existence of a localized aspect of cellular subdivision is implied by certain observations on division of transplanted or isolated parts of the fission zone. Tartar (1968) observed that dislocated patches of the presumptive fission zone of *Stentor* can divide when transplanted to other regions of dividing cells, or even to nondividing cells if the transplant is made after the donor has passed a stabilization point for division; this indicates that the capacity of the fission zone to constrict is spatially autonomous. Suzuki (1957) performed a detailed analysis of the division of small diagonal wedges of the presumptive fission zone of blepharismas cut at different developmental stages (Fig. 5.11), from which he concluded that the "presumptive potential related to the formation of division line" is initiated on the ventral (oral) side and then spreads to the dorsal (aboral) side (Suzuki, 1957, p. 134). These and other microsurgical results strongly suggest that the initial global blocking out must be followed by a subsequent more local determination event.

The analysis of certain temperature-sensitive division mutants in *T. thermophila* suggests that this local determination is interposed temporally between the global blocking out of presumptive daughter cells and the actual occurrence of cytokinesis. The mutants in question are alleles at the *cdaA* locus (originally called *mo1*). Cells homozygous for these mutant alleles can form oral primordia at the correct time and place even at restrictive temperatures, but are defective in the appearance of fission zone gaps within the ciliary rows (Frankel et al., 1977) and in the formation of the other structural features (CVP, basal body pairs, and ciliation

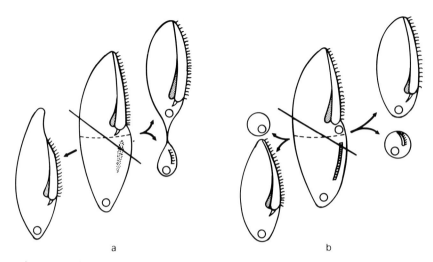

Figure 5.11 Developmental responses to comparable diagonal transections across the site of the future fission zone at two stages of predivision development in *Blepharisma*. (**a**) Transection *(solid line)* at an early stage of development (Suzuki's stage I). Constriction of the truncated presumptive fission zone *(dashed line)* was initiated and usually completed successfully on the ventral (oral) side (right), but generally was not even initiated on the dorsal (aboral) side (left). (**b**) Transection at a later stage of development (Suzuki's stage II) that is still considerably earlier than the time when the division line becomes visible. Constriction was always completed successfully on both the ventral side (left) and the dorsal side (right). Diagram (**a**) redrawn with modifications from Figure 18C, (**b**) from Figure 19B, of Suzuki (1957).

wave) normally associated with these gaps (Frankel et al., 1977, 1981). The temperature-sensitive period of a fully penetrant mutant allele of this locus *(cdaA1)* is short, beginning after the formation of the oral primordium and ending before the actual appearance of the fission zone (Frankel et al., 1980a). Thus the time of action of the *cdaA1* gene product is exactly what would be expected if that product mediated an event occurring between the global blocking out of the division products and the actual formation of the visible fission zone.

   The location of action of the presumed *cdaA* product also is largely appropriate; recently, Ohba et al. (1986) have identified a protein whose position in two-dimensional polyacrylamide gels is altered in *cdaA1* homozygotes, and which, therefore, is a strong candidate for being the *cdaA1* gene product. They have found that antibodies to this protein react strongly with the asymmetric apical crown of paired basal bodies (see Section 3.2 and Fig. 3.1), with the product becoming visible just as these pairs form immediately before the appearance of the fission zone. The strong equatorial reaction was observed in wild-type cells and also in *cdaA1* cells maintained at permissive temperatures, but the antibodies failed to react strongly with the equivalent region of *cdaA1* cells maintained at a restrictive temperature. The association of the site of strong reaction with the fission zone is nearly perfect, except for an apparent weaker reaction in those ciliary rows (2 to 4 and n-1) that participate in the formation of the fission zone but nonetheless do not form basal body pairs (Hirono et al., 1987b).

   An analysis of phenotypic mosaics has shown that *cdaA* mutants can exert their effect locally, on a row-by-row basis. When cells homozygous for a weak allele of *cdaA (cdaA2)* were maintained continuously at a restrictive temperature, or when cells homozygous for a strong allele *(cdaA1)* were exposed briefly to a restrictive temperature, some ciliary rows developed fission zone gaps, whereas others did not. Division constriction then occurred only in the region where fission zone gaps developed (Ng and Frankel, 1977). Even more important, the symptoms of cellular subdivision were always locally associated with the sites of fission zone gaps. Formation of new contractile vacuole pores and the wave of ciliary outgrowth was observed only in those ciliary rows that developed gaps, but not in neighboring ciliary rows that remained continuous (Frankel et al., 1977, 1981; Frankel, unpublished observations).

These results show that subsequent to the initial blocking out of the nascent daughter cells there is another final determinative event in which the actual formation of the fission zone and the associated visible symptoms of cellular subdivision are specified on a row-by-row basis. Again there are parallels to segmentation in insects; in morphogenetic mosaics of segmentation-deficient mutants of *Drosophila,* bristle patterns associated with separate tarsal segments differentiate only along the longitudes at which segmental boundaries form (Tokunaga and Gerhart, 1976).

5.4. CONCLUSIONS

The "segmentation model" of ciliate division can be summarized in five statements. First, cell-cycle analysis of *Tetrahymena* and *Paramecium* strongly suggests that cortical development and preparation for cell division are bound together as an integrated process that has no invariant phase relationship to the macronuclear replication cycle and probably is rate-limited by assembly rather than synthesis. Second, analysis of effects of inhibitors on *Tetrahymena* and *Stentor* shows that oral development and progress toward cell division are under a common physiological control such that they can be jointly reversed up to the time of a general stabilization. Third, microsurgical experiments on *Stentor* make the closely related point that these same processes are initiated by a suspension of a pervasive state of morphogenetic inhibition. Fourth, other microsurgical studies on *Stentor* and *Blepharisma* demonstrate that cells activated for division are tandemly partitioned into two nascent cell-units. Finally, a combination of microsurgical analysis, morphometry, and study of mutants converge to indicate that an initial global "blocking out" of presumptive daughter cells is followed by a subsequent more local determination of the site of fission. This determination precedes the visible appearance of the fission zone and probably spreads from the oral to the aboral side of the cell surface.

The above five conclusions probably apply in full to regulative ciliates such as *Stentor, Blepharisma,* and *Tetrahymena.* It is likely, however, that only the first of the five statements holds in an unmodified form for *Paramecium.* In this ciliate, the spatial course of division may be largely predetermined and development may be stabilized from the start. Both of these features are in keeping with the partially mosaic character of this ciliate. Nonetheless, in *Paramecium* as well as in the more regulative ciliates, cell division and the preceding cell subdivision may be considered largely as phenomena of cortical patterning.

6

Gradients and Intracellular Diversification

6.1 INTRODUCTION

This chapter is devoted to an experimental analysis of the issues of pattern regulation raised in Chapters 4 and 5. In Chapter 4 we saw that the heritable difference between the singlet and homopolar-doublet configuration sometimes is based on regulative systems reminiscent of classical embryonic fields, whereas in Chapter 5 we found that cell division in ciliates is preceded by a reorganization of the antero-posterior axis. Here we begin to consider the nature of the large-scale organization around the circumference and along the length of the ciliate clonal cylinder.

This chapter is devoted largely to a reexamination of the concept of control by interacting gradient systems. This idea was introduced into ciliate biology by Uhlig (1959, 1960). It was later taken up by many ciliate workers, including Sonneborn (1963), Tartar (1964), Frankel (1974), and Jerka-Dziadosz (1974), but challenged by others, notably Grimes (1976) and Martin et al. (1983). As pointed out in Chapter 1, however, the idea of a gradient is very flexible; it has been used in the context of two very different families of models of the organization of developing systems, the sequential models of Child and of Huxley and DeBeer (see Section 1.4.1) on the one hand, and the threshold–positional information models (see Section 1.4.2) on the other. These two different ways of thinking about gradients have both had their ciliatological adherents, with advocacy of the threshold model by Frankel (1974) and of a particular kind of sequential model by Sonneborn (1974, 1975a).

In this chapter, we restrict ourselves to the question of *where* the initial primordia of cortical structures (especially the oral primordium) form, without taking a close look at the internal organization of these structures. The internal organization of oral structures is the subject of Chapter 7, which in turn provides a prelude to an analysis of reversals of asymmetry (Chapters 8 and 9) and a presentation of a newer model for organization around the ciliate circumference (Chapter 10).

6.2 SPECIFICATION OF THE ORAL MERIDIAN

6.2.1 The Relation of New Structures to Old

To begin this analysis, I first review the spatial relationships of the sites at which new oral structures develop to preexisting cortical structures. The site of develop-

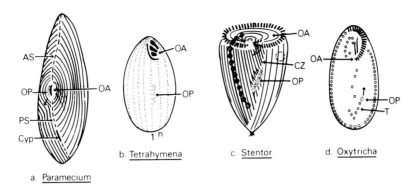

a. Paramecium

Figure 6.1   Diagrams illustrating the sites of oral promordia in (**a**) *Paramecium,* (**b**) *Tetrahymena,* (**c**) *Stentor,* and (**d**) *Oxytricha.* The oral apparatus [OA] and the oral primordium [OP] are labeled in all of these diagrams. Important special features are the anterior suture [AS] and posterior suture [PS] as well as the cytoproct [Cyp] in *Paramecium,* the two postoral ciliary rows, respectively numbered 1 (right postoral) and n (left postoral), in *Tetrahymena,* the contrast zone [CZ] in *Stentor,* and the transverse cirri [T] and the path of anterior migration (*arrow*) of the OP in *Oxytricha.*

ment of new oral primordia in four of the principal members of our ciliate cast is shown in Figure 6.1 (for further details, see Chapter 3). Inspection of this figure instantly shows that the relationship of new structures to old is not the same in all ciliates. In *Paramecium,* the new oral primordium (OP) develops in such close spatial association with the old oral apparatus (OA) (Fig. 6.1a) that the OA can meaningfully be thought of as self-reproducing (see Section 3.2.2). In the other three ciliates, the OP normally develops at different sites: the midregion of the right postoral ciliary row in *Tetrahymena* (Fig. 6.1b), a zone of breaks in many ciliary rows at the contrast zone in *Stentor* (Fig. 6.1c), and the anterior edge of the leftmost transverse cirrus in some species of *Oxytricha/Stylonychia* (Fig. 6.1d).

The stringency of the relationship between old and new structures differs drastically among these ciliate genera. This difference can be appreciated both from comparative and experimental analyses. Applying the former first, the spatial relationship between old and new structures is the same in all species of *Paramecium,* whereas it is variable within the classical *Oxytricha* and *Stylonychia* genera, with the OP originating near the left transverse cirrus in some species and in the midst of the naked ventral surface in others (Foissner and Adam, 1983; Wirnsberger et al., 1985). The most far-reaching comparative analysis, however, was carried out on *Tetrahymena,* through Nanney's ingenious exploitation of a discovery that certain clones commonly form oral primordia along ciliary rows other than the right postoral ciliary row, in one case with a clear directional bias to the cell's left. Nanney (1967a) pointed out that if this phenomenon, which he calls "cortical slippage," occurs over many cell generations, *every* ciliary row will eventually be a stomatogenic row. This conclusion is based on the fact, illustrated in Figure 6.2, that the stomatogenic ciliary row always becomes the right-postoral row (no. 1) of the posterior daughter cell irrespective of whether or not it had been the right-postoral row in the parent cell. Therefore, if the OP forms along row *n* rather than row 1, row *n* will become row 1 in the posterior daughter cell (although the anterior daughter will retain the organization of the parent cell). Thus, if there is a directional bias in

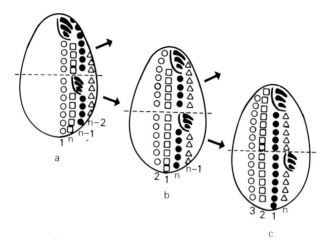

Figure 6.2 "Cortical slippage" in *Tetrahymena*. Diagram (**a**) illustrates an initial cell, and (**b**) and (**c**) show two successive generations of posterior division products, all at the stage of fission-zone formation shortly before the initiation of cytokinesis. Four individual ciliary rows are distinguished from each other by the use of different symbols to represent their basal bodies. The oral primordium is shown developing along the *left* postoral row [n] instead of the usual right postoral row [1]. As a consequence, in the posterior division products each individual ciliary row is shifted one position to the right in its relation to the oral apparatus. For example, going from the first generation (**a**) to the second (**b**), row n, the stomatogenic row, becomes row 1, row 1 becomes row 2, row n-1 becomes row n, and row n-2 becomes row n-1. The process is repeated going from the second generation (**b**) to the third (**c**). If this is extrapolated, every individual row will take its turn at occupying the postoral positions that are allocated to oral development.

cortical slippage, each ciliary row will inevitably take its turn as the stomatogenic row. Therefore, the stomatogenic ciliary row must be unique by virtue of its *location* rather than its *nature;* "... the behavior of cortical regions is not a consequence of their genetic constitution but of their position in morphogenetic fields" (Nanney, 1967a, p. 166).

The difference between *Paramecium* and the other members of our cast is reinforced by experimental analysis. In *Paramecium,* oral development has been shown to be absolutely dependent on its normal structural site, as oral development cannot proceed if that site is removed or even if it is severely damaged (see Section 4.3.2). By contrast, the normal site of oral development is dispensable in *Stentor* and in *Oxytricha/Stylonychia.* In the former, new oral primordia can be produced after the entire normal contrast zone is removed (see Fig. 5.2e–h), whereas in the latter, oral primordia can be formed after dedifferentiation of all ciliary structures in cysts (see Fig. 4.11) and also after removal of the part of the cell that includes the transverse cirri (see Fig. 4.12). Thus, although in *Paramecium* some feature of the normal structural site of oral development acts as a strict local determinant of OP development, in the other ciliates that have been analyzed suitably no *particular* cortical structure is essential for oral development.

6.2.2 Conditional Totipotency

Is the cell surface of ciliates other than *Paramecium* totipotent for oral development? The example of cortical slippage in *Tetrahymena* would appear to answer the question in the affirmative, as it showed that any ciliary row can function as a site of an OP. This experiment, however, demonstrates only the equivalence of *structures,* not of *regions.* For example, this demonstration still allows (indeed encourages) us to imagine that the old OA endows the region posterior to it with the potential for oral development, as was demonstrated for *Blepharisma* (see Fig. 5.5). If that is so, one would predict that if one could remove a lateral slice of the *Tetrahymena* or *Blepharisma* cell, it might not be capable of oral development because it has lost the suitable spatial relationships. Unfortunately, in these ciliates such a lateral piece would also lack a macronucleus, and development would certainly fail for reasons unrelated to the absence of a suitable cortical site.

Such an analysis has been performed successfully on a different ciliate, *Urostyla grandis,* by Jerka-Dziadosz (1964, 1974). *Urostyla* is a close relative of the oxytrichids that has a macronucleus subdivided into numerous small pieces scattered throughout the cell (observations of Raabe summarized in Grell, 1973, pp. 110–111). Therefore, virtually any fragment created by slicing across this dorsoventrally flattened cell has at least one macronuclear segment [each of which, if *Urostyla* is like its close taxonomic relatives, must carry numerous macronuclear genome-equivalents (see Klobutcher and Prescott, 1986)]. Thus, this ciliate is uniquely suited for testing cortical totipotency by simple transection experiments. In an early study, Jerka-Dziadosz (1964) found that fragments containing any part of the mid-ventral region of the cell (shaded in Fig. 6.3a) were able to regenerate new oral structures and associated cirral primordia, whereas nucleated fragments not includ-

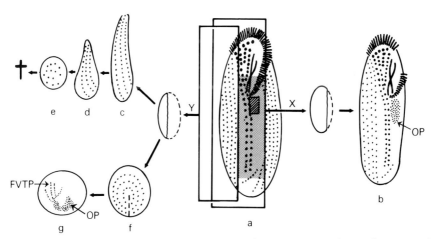

Figure 6.3 Developmental sequences in two types of fragments of *Urostyla grandis.* Diagram (**a**) shows the ventral surface of *U. grandis;* the organization area is shaded and the site at which the oral primordium normally begins to develop is cross-hatched. Diagram (**b**) depicts the site of OP formation in a wide right fragment, i.e., one whose left edge has been removed [X]. The remaining diagrams show development in a narrow right fragment [Y]: the (**c**), (**d**), (**e**) sequence indicates the most common result, with no folding of the fragment and no regeneration, whereas the (**f**) to (**g**) sequence depicts the less frequent outcome, with the fragment folding back on itself (**f**) and subsequently forming an oral primordium [OP] and frontal-ventral-transverse cirral primordia [FVTP] (**g**). Diagram (**a**) is drawn from information in Jerka-Dziadosz (1964) and Jerka-Dziadosz (1972); diagrams (**b**) to (**g**) from information in Jerka-Dziadosz (1974).

ing that region were unable to regenerate and eventually died. For example, although a "wide" right fragment that includes all or part of the postoral zone (Fig. 6.3,X) regenerated a complete cortical pattern (Fig. 6.3b), a "narrow" right fragment (Fig. 6.3,Y) failed to regenerate (Fig. 6.3c,d) and eventually rounded up and died (Fig. 6.3e). In the subsequent study (Jerka-Dziadosz, 1974), the majority of the thin lateral fragments behaved as they had in the earlier study. A few fragments, however, rounded up in a different way than before: they folded back on themselves, creating a juxtaposition of anterior and posterior ends of cirral rows (Fig. 6.3f). These folded fragments were able to form well-developed ciliary primordia (Fig. 6.3g), and could eventually regenerate as normal urostylas. Thus, although the "organization area" almost certainly defines a preferred region for oral development in *Urostyla* (Section 6.3.2), it is not irreplaceable. A nucleated fragment can reconstitute this region if certain spatial relations are achieved. The nature of these relations has not been resolved critically in *Urostyla*. For such resolution, we need to turn to the prince of ciliate microsurgery, *Stentor coeruleus*.

6.3 CIRCUMFERENTIAL ORGANIZATION

6.3.1 The Contrast Zone and a Circular Gradient in *Stentor*
The surface region is essential

Before considering the relational aspects of cortical development in more detail, it is worth asking whether the surface layer is even necessary for development of new cortical structures. Some microsurgical experiments conducted by Tartar on *Stentor* strongly suggest that it is. Tartar was able to slice off all of the surface layer of stentors. Because of the remarkable capacity that this organism has to heal by a nearly instant gelling of exposed surface, such flayed stentors did not lyse; they remained alive for a few days but nonetheless failed to regenerate any cortical structures (Tartar, 1956c; 1966d). If, however, small patches of this surface layer (distinguishable by pigment striping) were left on the surface of an otherwise denuded fragment, the patches stretched to cover the cell, which could then regenerate (Tartar, 1961, p. 108). Conversely, regeneration could also occur after virtually all of the endoplasm was removed from the cell, leaving the surface layer and the attached macronucleus, much like a toothpaste tube with most of the toothpaste pushed out (Tartar, 1961, p. 108). Thus this surface layer, which includes the structured cortex and may also include a stationary subcortical layer, is likely to be necessary and may also be sufficient for development of major cortical structures, as long as the fragment also contains a portion of the macronucleus (Schwartz, 1935; Tartar, 1961, Chapter XVI).

Demonstration of the importance of the contrast zone

Using some spectacular intracellular surgery, Tartar (1956a,b,c) demonstrated that a relational feature, the zone of contrast in width of pigment stripes, is critical in specifying the position at which an oral primordium (OP) appears. In one particularly revealing experiment, Tartar grafted a strip of broad stripes from the dorsal (aboral) surface of one stentor into the region just to the right of the contrast zone

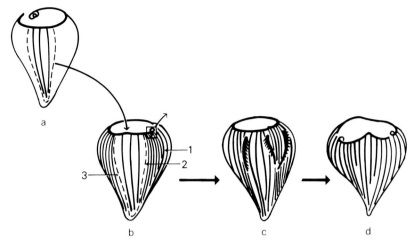

Figure 6.4 Consequences of implantation of a sector of wide pigment stripes taken from one *Stentor* (**a**) into a region of narrow stripes near the contrast zone of another (**b**). This implant created two new sites of stripe-contrast [2, 3] in addition to the original site [1]. Shortly after insertion of the implant, the mouth parts of the host cell were removed (**b**) to stimulate regeneration. Oral primordia often formed at all three sites of stripe-contrast (**c**), producing a composite OA (**d**). Subsequently, however, this system regulated to a normal singlet with a single contrast zone (not shown). (**b**), (**c**), and (**d**) are redrawn from Figure 4 of Tartar (1956c), with permission.

of another stentor (Fig. 6.4a,b), thus splitting a region of narrow stripes into two parts. As a result, three sites of "stripe contrast" were formed (Fig. 6.4b), the normal one [1] and two additional sites to either side of the implant [2,3]. Although the morphogenetic response was variable (see Fig. 11 in Tartar, 1956a), oral primordia always were formed at one or both of the supernumerary contrast zones, sometimes at all three available zones (Fig. 6.4c). This demonstration is particularly compelling because one of the extra zones [3] is fairly distant from the original contrast zone, whereas another [2] is of reversed asymmetry. Such results strongly suggest that a relational feature, the visible aspect of which is a juxtaposition of broad and narrow pigment stripes, determines the site of OP development.

Tartar's observations were not made at sufficiently high resolution to determine whether the wide-stripe area induces formation of oral primordia in the adjacent narrow-stripe region, or whether both regions collaborate in formation of the OP (Tartar, 1956c, p. 99). Uhlig (1960), using phase-contrast microscopy, observed that oral primordia form in the fine-stripe zone, not always in direct apposition to the broad stripes. More recent detailed cytological studies (using protargol staining and scanning electron microscopy) on *Stentor coeruleus* (Paulin and Bussey, 1971; Pelvat and de Haller, 1979) and on the closely related *Climacostomum virens* (Dubochet et al., 1979; Fahrni, 1985) have largely confirmed Uhlig's conclusion. Thus, in a formal sense, we can consider the formation of an OP in *Stentor* to be a result of induction.

Is a contrast zone necessary for formation of an OP in *Stentor?* When stentors were sliced into two lengthwise, so as to separate ventral (oral) and dorsal (aboral) halves, with both containing portions of the macronucleus, the dorsal half-stentors folded on themselves edge-to-edge along a vertical plane, forming a weak contrast zone at which an OP usually developed eventually, although often with substantial delay (Tartar, 1956a, 1961, p. 118). Occasionally, such aboral halves did fail to form oral primordia (Tartar, 1956a), and in narrower pieces such failure was more common (Tartar, 1966d). Development of an OP also failed to occur when the folding of an aboral half was deliberately prevented (Tartar, 1961, p. 189 and Fig. 50b). The common element in all of these cases of failure was insufficient initial contrast in stripe-widths. This implies that a threshold level of contrast might be necessary for OP development, an idea that I shall develop in more detail below.

Dynamics of the contrast zone

When development of oral primordia did eventually occur in experimental situations in which the initial degree of contrast in widths of pigment stripes was relatively slight, the original degree of contrast often was restored. The restoration occurred by a successive branching of stripes (and hence also of the ciliary rows that lie between them) within the region of not-so-narrow stripes that abutted on the wider stripes. Although Tartar (1956a, p. 95) reported that this recovery of the initial contrast mostly occurred *after* formation of the OP, Uhlig documented an extensive enhancement of contrast by branching of pigment-stripes *before* OP formation, both during normal development (Uhlig, 1960, p. 13) and in experimental situations (Uhlig, 1960, pp. 22, 27, 82 and Figs. 8, 11, 24). A requirement for such contrast-enhancement can explain the delays in OP formation commonly (although not invariably) observed at initially weak contrast zones.

The phenomenon of contrast-enhancement before OP formation was probably a major motivation for Uhlig's formulation of Tartar's stripe-contrast principle as

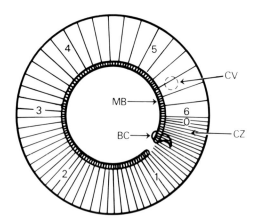

Figure 6.5 Uhlig's "circular gradient" model of circumferential organization of the *Stentor* cell surface. The cell is seen from above, with the membranellar band [MB] shown in a contracted state, so that the organization of pigment stripes *(spaces)* and ciliary rows *(lines)* on the body surface posterior to it can also be seen. Note the juxtaposition of extreme gradient values [0 and 6] at the contrast zone [CZ] posterior to the buccal cavity [BC]. The contractile vacuole [CV] is also shown. Redrawn with modifications from Figure 6 of Frankel (1984), with the symbols representing gradient-levels altered as follows: $+++ = 0$, $++ = 1$, $+ = 2$, $0 = 3$, $- = 4$, $-- = 5$, and $--- = 6$.

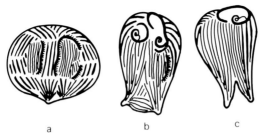

a   b   c

Figure 6.6 Regulation of a "minced" stentor. The operation consists of excising the OA and hold-fast and then generating extreme disorganization in the cortex of the remainder by "quartering" (successively rotating the anterior half on the posterior and then the left half on the right) and finally "mincing" (randomly slicing into the cortex). (**a**) shows an early stage of recovery, in which two holdfasts began to form (bottom) and multiple oral primordia appeared at abnormal loci of stripe contrast. (**b**) depicts the ensuing two-tailed abnormal doublet now reorganizing with a single oral primordium. (**c**) shows the resulting bipedal singlet, "which later fused the tails, adjusted excess striping, and became wholly normal (Tartar, 1960, p. 199)." Redrawn from Figure 10 of Tartar (1960), with permission.

a circular gradient, shown in anterior view in Figure 6.5. For Uhlig, the contrast zone is a visible manifestation of an invisible global relationship, in which OP development depends not on stripe-contrast alone, but on a rebuilding of the ends of the gradient symbolized by a clockwise sequence 0-1-2-3-4-5-6 (corresponding to Uhlig's + + +/ + +/ + /0 / − / − − / − − −) in Figure 6.5.

Although Uhlig's documentation for dynamic contrast-enhancement before OP formation is strong, it is itself not sufficient to demonstrate the requirement for a complete structural gradient for development of oral primordia. An activated cell was sometimes able to form a new OP very soon after removal of the contrast zone, without preceding stripe subdivision (see Fig. 5.2 e–g); however, this OP and the resulting OA were smaller than normal, and the cell underwent a subsequent oral replacement (not shown) (Tartar, 1958a). More dramatically, oral primordia could be formed by stentors in which the surface region had been "minced" into a crazy-quilt of misaligned patches (Fig. 6.6a) (Tartar, 1960). Realignment into a nearly normal form and pattern did eventually occur (Fig. 6.6c), but oral primordia could develop well before this realignment was completed (Fig. 6.6b). Thus, as Tartar (1960) pointed out, intact circumferential gradients, at least as manifested in the order of cortical striping, are not essential for OP development.

Is there a gradient of inhibition?

A demonstration of spatially graded inhibitory or competitive interactions around the cell circumference would argue strongly for a dynamic circular gradient. Results of certain transplantation experiments have been interpreted as indicative of lateral inhibition of OP sites, an interpretation that was first advanced tentatively by Tartar (1956a) and later promoted more confidently by Uhlig (1960) and Frankel (1974). Some of the principal experiments pertinent to the claim for inhibition will be presented here. We will see that the experimental case for the *existence* of lateral inhibition is fairly convincing but of limited scope, whereas the evidence for a spatial *gradient* of inhibition is very weak.

One classic example of putative spatially graded inhibition is the complementary pair of "segment-exchange" experiments carried out by Uhlig (1960) and illustrated in Figure 6.7. In the first type of experiment (Fig. 6.7a–e), a segment of the cell bearing the contrast zone (shaded in Fig. 6.7a) was removed and replaced by an aboral segment from the same cell (Fig. 6.7b,c). The outcome (reading clockwise around the cell perimeter in Fig. 6.7d) was the formation of one zone of weak contrast (2/4) at the place where the excision was healed and two zones of moderate contrast (5/2 and 4/1) on both sides of the inserted wedge. Use of numbers allows us to see clearly that although the 2/4 apposition had less than half of the contrast (difference in value) of the normal 6/0 site, the 5/2 and 4/1 contrast zones had one-half of the normal contrast (3 units of difference instead of 6). Uhlig observed that no contrast-enhancement or OP formation took place at the weak 2/4 site, whereas a clear contrast-enhancement by subdivision of fine stripes occurred at the half-strength 5/2 and 4/1 sites. Oral primordia subsequently were formed in the latter two regions (Fig. 6.7e) (Uhlig, 1960).

The second type of segment-exchange experiment was the reciprocal of the first. An aboral segment was removed (shaded in Fig. 6.7f) and replaced by a transposed oral segment (Fig. 6.7g,h). The contrast created by the healing of the excision (5/1) was greater than in the previous experiment, but the contrast on both sides of the sites of insertion was the same as before: 2/5 and 1/4 (Fig. 6.7i). In this case, however, oral primordia formed at the original 6/0 site and at the 5/1 site, but usually not at either the 2/5 or 1/4 sites (Fig. 6.7j).

Why did development succeed at 5/2 and 4/1 sites in the first experiment and

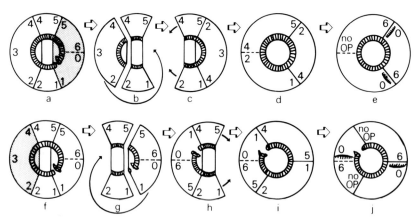

Figure 6.7 Uhlig's "segment-exchange" experiments in *Stentor*, depicted using the pictorial conventions of Figure 6.5 (**a–e**). An oral segment of a stentor (*shaded*) was removed (**a**) and replaced with an aboral segment (**b, c**). After healing of the gap left by removal of the aboral segment (**d**), contrast-enhancement occurred at the 5/2 and 4/1 zones, followed by formation of oral primordia at both of these zones (**e**) in all of three experimental cases and in their progeny. (**f–j**). The converse experiment, in which an aboral segment (*shaded*) was removed and replaced by an oral segment (**f, g**), and regeneration was stimulated. In this case healing of the gap (**h, i**) was followed by formation of oral primordia at the original [6/0] contrast zone and at the 5/1 contrast formed at the site of healing; in three out of four cases no oral primordium [OP] formed at the 1/4 and 2/5 zones (**j**). Redrawn with modifications from Figures 8 and 9 of Frankel (1974).

fail at the apparently equivalent 2/5 and 1/4 sites in the second? One major difference between the two operations was the presence of a normal (6/0) contrast zone near the half-strength zones in the second experiment and not in the first. Therefore, Uhlig concluded that the lack of activation of the 2/5 and 1/4 sites in the second experiment was due to an inhibitory influence ("Hemmwirkung") originating from the normal anlagen zone and extending over a limited distance; the reason for postulating a limited extension of inhibition was the fact that the 5/1 zone opposite to the normal contrast zone was successfully activated.

This interpretation is not entirely compelling. Another way of looking at this matter is to assume that a half-strength contrast zone (in the numbering system, a difference of 3 out of a maximum possible of 6) is near the threshold for successful contrast-enhancement and eventual formation of an OP. If the contrast is weaker than this (a difference of less than 3), no OP will form, and if it is stronger (a difference of greater than 3) an OP will develop irrespective of any possible lateral inhibitory interactions. This simple postulate accounts for the failure of OP formation at the subthreshold 2/4 zone (Fig. 6.7e) and its success at the suprathreshold 5/1 zone (Fig. 6.7j). The 5/2 and 4/1 sites of Figure 6.7a to e and the 2/5 and 1/4 sites of Figure 6.7f to j are all near the half-strength threshold, where small perturbations may be expected to drive the system one way or another. Perhaps the relevant perturbations in this case are the absence of competition from a nearby 6/0 site in the first experiment, its presence in the second. This, however, is not the only possibility; in the first experiment, the 5/2 and 4/1 sites are of normal asymmetry (i.e., the same handedness as the original 6/0 site), whereas in the second experiment the 2/5 and 1/4 sites are of reversed asymmetry relative to the original 6/0 site. Although it is clear that OPs of reversed asymmetry can readily form in stentors and many other ciliates (see Fig. 6.4, also Chapter 8), perhaps when the system is poised between enhancement and loss of contrast, there is a bias in the direction of enhancement when the asymmetry of the site is normal and a bias in the direction of loss when the asymmetry of the site is reversed.

In the light of this reformulation, it is of interest to return to the wide-stripe implant illustrated in Figure 6.4. The experiment does not help us much in evaluating whether contrast-enhancement is dependent on normal asymmetry, as OPs initially developed at the normally oriented side of the implant (marked 3 in Fig. 6.4b) in 8 out of 9 operated stentors and at the reversed side (marked 2 in Fig. 6.4b) in 7 out of the 9 operated cells (Tartar, 1956a); however, in this experimental series the initial degree of contrast was almost certainly more than half of the normal contrast. What the experiment does show clearly is that if the initial stripe contrast is fairly large, developing oral primordia do not interfere with each other, even over short distances (a consolidation to a condition of singleness did occur during episodes of oral replacement subsequent to the initial regeneration, but these subsequent adjustments can be explained in an altogether different manner, as we shall see in Chapter 10).

The best evidence for lateral inhibition or competition among primordium sites comes from some grafting experiments by Tartar (1956a) on aboral half-stentors (Fig. 6.8). I have already mentioned that if a *single* aboral half-stentor was allowed to fold back on itself (Fig. 6.8b), it generally produced an OP at the line of heal, but

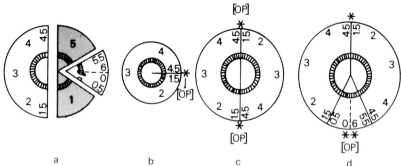

Figure 6.8 Tartar's experiments involving combinations of aboral halves of stentors with each other and with a wedge that includes the contrast zone, depicted following the pictorial conventions of Figure 6.5. The portions of the stentors used in these operations are unshaded in (**a**). In (**b**–**d**), normal 6/0 contrast zones are labeled by **, while half-strength 4.5/1.5 contrast zones are indicated by *. (**b**) represents a single folded aboral half. An OP was usually formed at the single half-strength contrast zone. In (**c**), two aboral halves were fused together. An OP initially developed at only one of the two half-strength contrast zones, although two out of 13 such complexes eventually formed oral primordia at both contrast zones and became doublets. In (**d**), two aboral halves were fused, and an oral wedge was inserted into one of the two junctures. An OP formed at the normal 6/0 site, but not at the half-strength 4.5/1.5 site, and the complexes remained singlets in all of 6 cases. Drawn from information in Tartar (1956a) and (1966d).

occasionally failed to do so. Presumably, in those cases the near-threshold contrast was resolved in the wrong direction, creating one of the very few morphogenetic dilemmas from which a stentor cannot extricate itself (see Tartar, 1966d). If two aboral half-stentors were grafted (literally) back-to-back (Fig. 6.8c), OP formation usually took place at only one of the two potential sites to produce a singlet, occasionally at both to create a doublet (Tartar, 1956a). Perhaps in this case the success of development at one site was causally related to failure of the other, but the case for competition in this experiment is not entirely convincing. It is more so in the experiment in which a wedge including the contrast zone was inserted into one of the back-to-back junctures (Fig. 6.8d). Then an OP always formed at the implanted 6/0 site and failed to appear at the opposite 4.5/1.5 site. In this case, then, a half-strength contrast zone that usually would produce an OP on its own failed to do so if a full-strength contrast zone was active on the opposite side of the cell. This is consistent with possible lateral inhibition. The fact, however, that stronger OP sites are *not* inhibited even when close to the normal 6/0 juncture (Fig. 6.4), whereas half-strength sites are inhibited regardless of their placement, suggests that if inhibition does occur it affects only OP sites that are near the threshold for effective enhancement, and it shows no apparent relationship to distance. These and other experiments involving doublets (Tartar, 1956a) are consistent with a possible generalized competition for a diffusible substrate, but offer no support for a *gradient* of inhibition.

A structural incompatibility gradient

There is one other manifestation of a circular gradient in *Stentor* that is operationally different from any that we have yet considered. Recall from the previous chapter (see Section 5.3.1) that if nondividing stentors were transected and re-fused at the *same* anteroposterior levels, the disrupted stripes and ciliary rows rejoined, whereas if different anteroposterior levels were fused, the outcome was interpenetration rather than rejoining (see Fig. 5.9). Here we will add that rejoining after fusion at the same anteroposterior levels occurred *only* if the fused halves were aligned along the same longitudes as well as the same latitudes. If the anterior half was instead rotated relative to the posterior half, the pigment stripes failed to rejoin, and the posterior stripes gradually extended anteriorly, whereas the anterior moiety was gradually resorbed (Tartar, 1956a). Uhlig (1960) carried out an extensive set of rotations at different anteroposterior levels and found incompatibility in all cases. If the level of rotation was equatorial (as in Tartar's experiments) or apical, the more basal component replaced the more apical (as Tartar also had observed). If the rotation was "deep-basal" (about three fourths of the way down the cell), then the complementary result was achieved, with extension of the apical component and resorption of the basal component. With "basal" rotations (about three fifths of the way from apex to base), the cell's response was the same as that resulting from anteroposterior incompatibilities (Section 5.3.1): blocks of cortex interpolated themselves in both directions, creating complex disturbances of oral development.

This circumferential incompatibility is clearly not just a result of a purely geometrical inability of ciliary rows to make contact; Tartar (1956a, p. 101) pointed out ". . . that when fine-(stripes) met wide-stripes the arrangement was often that of two of the former abutting each wide stripe in neat regularity, yet without fusing." These findings are reminiscent of observations first made by Locke (1960) on insect segments: when epidermal patches taken from one segment were grafted into another, the cuticular patterns observed after the subsequent molt were harmonious if the graft and surrounding host were at the same anteroposterior levels within their respective segments, but displayed characteristic disturbances near the graft-edges if the anteroposterior levels of graft and host were mismatched. Although this "incompatibility gradient" was interpreted by Stumpf (1966) as reflecting interactions of different levels of a diffusible substance, subsequent work by Nübler-Jung (1977, 1979) suggested that the gradient was really a consequence of a gradation of structural differences on the surfaces of cells at different segmental levels.

Before summarizing the diverse results pertinent to the "circular-gradient" in *Stentor,* I should briefly point out what the reader may already have inferred: the circumferential incompatibility is not as great in *Blepharisma* as it is in *Stentor:* in Suzuki's 180-degree rotation (see Section 5.2.4 and Fig. 5.5), the transposed anterior and posterior components appeared to persist and communciate developmentally in a manner not observed in *Stentor;* however, in *Blepharisma,* there was a tendency toward derotation (see Fig. 5.5c), which is suggestive of some degree of positional incompatibility. The same phenomenon was also observed in rotated patches of insect cuticle (Nübler-Jung, 1974).

A provisional summary

In my view, the experimental results obtained by Tartar and Uhlig provide no decisive evidence either for a dependence of oral development on global circumferential order, or for a gradation of inhibition over distance. The existence of a circumferential gradation in the spacing of ciliary rows and of an apparent incompatibility of different points along that gradation, however, do indicate that there is some large-scale structural order. Finally, the phenomenon of contrast-enhancement before the development of oral primordia at new sites suggests that this structural order is not static but dynamic. This subject is taken up again in Chapter 10, after a consideration of reversals of asymmetry.

6.3.2 Lateral Stability in *Urostyla*

A structurally based positional gradation might be expected to undergo regulative shifts more slowly than one based on diffusion. One way of investigating such regulation is to find out whether primordia undergo a shift in location when the geometry of the system is altered. The flexibility of the surface of *Stentor* makes this organism relatively unsuitable for asking this question. The ideal ciliate for such a study would be one in which development is regulative, the cell surface is physically rigid, the geometry is simple, transections are easy to accomplish, and a portion of the macronucleus is present in every conceivable fragment. *Urostyla grandis* (closely related to oxytrichids) fulfils all of these requirements, and the issue of regulative shifts of sites of formation of oral primordia was systematically investigated in this organism by Jerka-Dziadosz (1974). Her basic result is easily summarized: although the sites of oral and other ciliary primordia could readily be shifted along the anteroposterior axis, they were resistant to dislocation along the lateral (right–left) axis. This result [also obtained by Grimes and Adler (1978) in *Stylonychia*] is illustrated in Figure 6.3b. If the left edge was removed from the cell, the OP nonetheless formed at its usual site, which is near the new healed edge of the laterally truncated cell. Thus the longitude along which the OP normally forms has a fair degree of inertia, even though the OP can form along another longitude if the favored longitude is missing (see Section 6.2.2. and Fig. 6.3f,g).

When attempting to account for her results, Jerka-Dziadosz (1974) adapted Lawrence's "sand model," which had been devised to account for presumed rapid positional regulation within insect segments (Lawrence, 1966). This model, presupposing rapid shifts in the peak of the "sand hill" (representing the oral primordium site), accounts nicely for lability along the anteroposterior axis (see Section 6.4.2). To accommodate the results along the mediolateral axis, however, Jerka-Dziadosz was forced to wet the sand so that it would not flow rapidly when a lateral flank of the sand ridge was suddenly removed. In my view this is equivalent to asserting that the basis of any mediolateral gradation is unlikely to be the unrestricted diffusion of a small molecule.

6.3.3 The Positioning of Contractile Vacuole Pores

Thus far we have considered only the positioning of a single structure, the oral apparatus. Assessment of any putative gradient would be greatly facilitated by data on the *relative* placement of two or more structures. I have already reviewed some studies of this kind pertinent to the location of the new OA along the anteroposterior axis of cells preparing for division (see Section 5.3.2); here I will describe the placement of the contractile vacuole pore (CVP) along the circumferential axis.

CVP-cytogeometry in Tetrahymena

New CVPs in *Tetrahymena* develop just anterior to the fission zone in a specific geometric relationship to particular basal bodies (see Section 4.2.4 and Fig. 4.6). CVPs, however, are restricted to certain ciliary rows located to the right of the oral meridian (see Section 3.2.1 and Fig. 3.2). The question then is, which ciliary rows? Nanney's crucial discovery was that the cell's choice of which ciliary rows to use for CVP-production depends on a *relative* distance from the oral meridian: the greater the total number of ciliary rows, the greater also is the number of ciliary row-intervals between the oral meridian and the CVPs (these row-intervals are relatively uniform and thus probably provide a valid measure of circumferential distance). The rough proportionality between the position at which the CVPs form and the cellular circumference was described by Nanney (1966a) using the metaphor of a constant "central angle" (Fig. 6.9a,b). However, Nanney also pointed out that this central-angle description cannot be taken literally; it breaks down as soon as one considers homopolar doublets that possess two oral meridians with a CVP set to the right of each. In such doublets, the central angle is reduced to one-half of what it is in singlets (Fig. 6.9c). Nanney concluded that "the cell in some way measures a fraction of the distance between one stomatogenic meridian and the next (or possibly between other cortical features) and regulates the field size in relation to that distance" (Nanney, 1966a, p. 316).

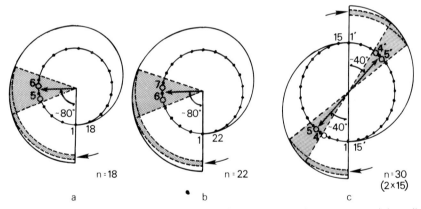

Figure 6.9 Schematic diagrams representing the geometry of CVP formation around the cell circumference of *Tetrahymena*. The central circles studded with large dots represent cross sections of *Tetrahymena* cells seen from the anterior, with dots indicating ciliary rows. (a) and (b) depict singlet cells with 18 and 22 ciliary rows, respectively, whereas (c) symbolizes a homopolar doublet cell with oral apparatuses directly opposite one another and a total of 30 ciliary rows, 15 in each semicell. The numbering of ciliary rows in (a) and (b) follows the conventional scheme, in which ciliary rows are enumerated clockwise starting with the right postoral ciliary row. The doublet in (c), with two right postoral ciliary rows, therefore, has two tandem sets of numbers, here indicated as 1 to 15 and 1' to 15', respectively. Nanney's central-angle formulation is shown inside the cells, whereas a reformulation in terms of thresholds within circumferential gradients is shown outside. The shaded region indicates the width of the area within which CVPs are formed and the corresponding range of values of the postulated gradient that determines formation of CVPs. Redrawn from Figure 7 of Frankel (1984).

The idea of a rough proportionality between the position of the CVPs and the total number of ciliary rows has been confirmed repeatedly in several members of the *"T. pyriformis"* sibling-species swarm (Nanney, 1967b, 1971a; Nanney et al., 1980; Frankel, 1972) and for the closely related species *Glaucoma scintillans* (Klug, 1968; Cho, 1971). Furthermore, it has been shown that CVPs can form along the expected CVP rows at atypical locations along the anteroposterior axis (Frankel, 1979 and unpublished), indicating that the circumferential measurement specifying the longitude at which CVPs develop is probably independent of the system that specifies the latitude of CVPs (see Frankel, 1979, pp. 240–241). This permits us to reformulate Nanney's fundamental insight in terms of Uhlig's circular gradient, provided that we also assume that *Tetrahymena* has the equivalent of a contrast zone despite uniform spacing of its ciliary rows. First assume that a "cliff" of a specified height (corresponding to the 6/0 contrast-meridian in *Stentor*) defines the oral meridian (Fig. 6.9). Then extend a circular gradient from one oral meridian to the next, i.e., from the top to the base of the cliff, similar to the 0-1-2-3-4-5-6 gradation in Figure 6.5, except that the lowest number happens to correspond to the highest point. CVPs would be formed at a particular range of *levels* of the gradient, corresponding to the shaded areas in Figure 6.9. This formulation synthesizes Nanney's fractional distance between stomatogenic meridians and Uhlig's circular gradient into a single unified threshold model (see Section 1.4.2), and tempts one to imagine that CVPs might be formed by an intracellular interpretation of a specific level of positional information.

The location of the contractile vacuole of Stentor

In addition to the extensive data accumulated for *Tetrahymena,* there is clear evidence of relational positioning of the contractile vacuole (CV) for *Stentor* as well; in this ciliate a single CV is located in the zone of widely spaced stripes a short distance to the left of the line of contrast with the fine stripes (Fig. 6.5) (Tartar, 1961, pp. 40–41). This spatial relationship is probably determinative, as whenever the contrast zone is reconstructed with a reversal of the normal asymmetry the CV is still under the wide stripes, but now to the *right* of the line of stripe-contrast (Uhlig, 1960). In *Stentor,* however, it is totally unknown whether the distance from the contrast zone to the CV is measured in constant or relative terms. Parallel conclusions apply to the macronucleus of *Stentor,* which is invariably situated directly underneath the cell surface on the side of the narrow-stripe region that is opposite to the CV, i.e., the macronucleus is located to the right of the narrow-stripe zone in normal stentors, to the left of this zone in reversed stentors (Tartar, 1960; de Terra, 1981).

Reference points for CVP-positioning in Chilodonella

I will now introduce a ciliate that is not a member of the cast presented in Chapter 3; it has been the subject of detailed biometrical analyses of CVP locations. This organism, *Chilodonella,* is a dorsoventrally flattened cell with CVPs scattered over the ventral surface much as cirri are in oxytrichids, but with the difference that the CVPs of *Chilodonella* form at their final locations. The investigations on *Chilodonella* are difficult to summarize briefly, in part because three types of organisms with different numbers of CVPs on the ventral surface have been used: *Chilodo-*

nella cucullulus stocks B, S, and K with 3 to 5 CVPs on the ventral surface (Kaczanowska and Kowalska, 1969; Kowalska and Kaczanowska, 1970; Kaczanowska, 1971), *C. cucullulus* stock X with 8 or more CVPs (Kaczanowska, 1974, 1975), and *C. steini* with about 15 CVPs (Kaczanowska, 1981; Kaczanowska et al., 1982). The conclusions drawn from these studies partially reinforce and partially qualify the main conclusions of Nanney's analysis of *Tetrahymena*. In general, new CVPs tend to develop along longitudinal bands on the ventral cell surface that are positioned with reference to the oral meridian and cell margins (Kowalska and Kaczanowska, 1970; Kaczanowska, 1981). In the forms with low and intermediate numbers of CVPs *(C. cucullulus),* the new CVPs within these bands also form at certain preferred distances from an inductive center located at or near the OA (Kaczanowska, 1974). The quantitative values of the distances between the bands in stocks B and K (Kowalska and Kaczanowska, 1970) as well as of the radii in stock X (Kaczanowska, 1975) are closer to being absolute rather than relative. In *C. steini*, an ingenious global analysis suggested that the CVP distribution was relationally controlled by an overriding "scaling factor," with the oral apparatus serving as a crucial reference point (Kaczanowska et al., 1982).

Provisional summary

The analysis of positioning of CVPs in several ciliates has provided further strong support for Tartar's concept of an "intimate relatedness of pattern components" over the ciliate surface. In one of these systems *(Tetrahymena)* the evidence strongly favors proportional positioning, whereas in other systems the evidence relevant to this issue is somewhat ambiguous *(Chilodonella)* or lacking *(Stentor).* It is, however, clear that in both of the latter ciliates there must be some form of directionally oriented measurement from one or more reference points, implying a large-scale spatial order.

6.4 ANTEROPOSTERIOR ORGANIZATION

6.4.1 The Basal–Apical Gradient in *Stentor*

In a thorough experimental analysis of *Stentor*, Uhlig (1960) provided the best evidence for the existence of an anteroposterior gradient in any ciliate (especially if this is taken together with the results of his follower, Eberhardt (1962), on the worm-shaped heterotrich *Spirostomum*). Here I give a brief précis of Uhlig's three general conclusions, all of which were documented in several types of experiments. First, even a strong contrast zone is not sufficient to stimulate oral primordium (OP) development; such development is more likely to be suppressed the closer one gets to the tail pole. Second, the posterior portion of the OP tends to bend away from the contrast zone and become more horizontal (see Fig. 3.8g). This tendency is more pronounced the further posteriorly the OP is located, and probably is an expression of the negative posterior influence on OP development. Finally, during late stages of oral development, the posterior region of the cell promotes the formation of a buccal cavity and cytostome at the nearest (usually most posterior) end of that primordium.

The posterior inhibition of formation of oral primordia was subsequently confirmed by Tartar (1964) when he observed that tandem doublets of the kind shown in Figure 5.9 could form oral primordia before interdigitating in their anterior but not posterior components. "Inhibition of the posterior component . . . appears to be the strongest yet demonstrated" (Tartar, 1964, p. 249), and was attributed by Tartar to an extension of Uhlig's basal influence over the entire tandem system. Such inhibition, however, was not observed in every tandem-grafted complex, nor is it consistent with de Terra's later observation that oral development occurs preferentially in the posterior component of stentors with an equatorial discontinuity (de Terra, 1985b). The strongest evidence for the reality of the basal-apical gradient pertains not to inhibition of formation of an OP, but rather to induction of a buccal cavity within an already formed OP. We will postpone this subject to the next chapter (see Section 7.4.2) where it will be taken up as an aspect of the development of the OP.

6.4.2 Anteroposterior Lability in Oxytrichids

In oxytrichids and their close relatives, the latitudes at which ciliary primordia develop appear to regulate more readily than do the longitudes (Jerka-Dziadosz

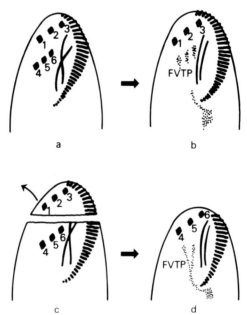

Figure 6.10 Development of the left-most three FVT streaks in the anterior portion of the ventral surface of *Paraurostyla weissei*. (**a–b**). Development in an intact cell. (**a**) The starting configuration, with frontal cirri numbered. (**b**) The three frontal-ventral-transverse primordia [FVTP] formed from disaggregation of frontal cirri 4, 5, and 6. (**c–d**). Development after amputation of the anterior tip, including frontal cirri 1, 2, and 3. (**c**) shows the operation, and (**d**) illustrates the ensuing appearance of frontal-ventral-transverse primordia [FVTP] posterior to the remaining intact cirri 4, 5, and 6. In this case, taken from Figure 28 of Jerka-Dziadosz and Frankel (1969), only two FVTPs were formed, but in other similar experiments 3 or 4 FVTPs appeared. Redrawn with modifications from Figure 8 in Frankel (1984).

and Janus, 1972; Jerka-Dziadosz, 1974). The most interesting type of "test case" for such regulative shifts arises when a particular new structure normally forms from or near a *specific* preexisting structure. One can then ask, is that particular preexisting structure still used after a microsurgical operation brings about a shift in the relative location of that structure along the anteroposterior axis of the cell? In other words, is it a *structure* or a *location* that really counts in determining the site of origin of such a ciliary primordium?

The clearest case in which location proved to be critical was in the formation of the three left-most frontal–ventral–transverse (F-V-T) streaks of *Paraurostyla weissei* (Jerka-Dziadosz and Frankel, 1969). During normal development, the three most anterior frontal cirri remain quiescent, whereas the second tier of frontal cirri disaggregate to form the ciliary streaks that later develop into new F-V-T cirri (Fig. 6.10a,b). But when the anterior tip of the cell was ablated, the cirri that would otherwise have formed these streaks were newly located near the anterior end of the cell, and then were not used. Instead, the ciliary streaks developed more posteriorly, at a distance from the anterior tip that was similar to the distance between the normal F-V-T primordium and the tip of the unoperated cell (Fig. 6.10c,d).

One case in which relative location appeared not to be all-important has also been reported, in the oxytrichid *Laurentiella acuminata.* Martin et al. (1983) stated that when the posterior tip of this organism was removed, the OP still formed adjacent to its normal "guidepost," the anterior-most transverse cirrus (see Fig. 6.1d), despite the fact that in such a fragment the *relative* position of this guidepost was now more posterior than normal. In *Laurentiella,* however, like its close relative *Oxytricha/Stylonychia,* the OP undergoes a rapid anterior migration immediately after its initial formation, and thus the final rather than initial position of the OP might be subject to regulative positioning.

6.4.3 Proportional Positioning of the Cytoproct in *Paramecium*

Of all of the members of our ciliate cast, *Paramecium* would appear to be the least promising for demonstration of gradients. Its incapacity to regulate its oral primordium site after removal or damage (see Section 4.3.2) and its relatively slow reorganization of form after transection (see Section 5.3.2) suggest that it is the most "mosaic" of all of the experimentally investigated ciliates (Tartar, 1954a; Schwartz, 1963). Nonetheless, this organism is known to be capable of regulating the placement of its cytoproct along the anteroposterior axis and sometimes of forming ectopic cytoprocts at unusual sites around the cell circumference. I will describe the anteroposterior regulation of cytoproct position here, and postpone the circumferential lability for discussion in another context (see Section 10.3.3).

Recall that the cytoproct, or cell-anus, is a narrow slit located in the posterior suture (see Section 2.2.5 and Fig. 3.3a). Chen-Shan (1969) removed the posterior third of the cell, and found that, in keeping with the relatively leisurely pace of recovery of normal body proportions characteristic of truncated paramecia, it took three generations for the postoral regions of successive posterior division products to attain their normal length. Within the abbreviated posterior sutures of the first and second posterior division products of transected cells, the cytoprocts were both shorter and closer to the oral apparatus than they were in uncut or third generation

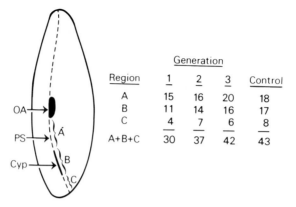

Region	Generation			Control
	1	2	3	
A	15	16	20	18
B	11	14	16	17
C	4	7	6	8
A+B+C	30	37	42	43

Figure 6.11 Regulation of the location of the cytoproct within the posterior suture of *Paramecium tetraurelia* in the first three successive posterior division products (cell generations) after excision of the postoral region. The diagram *(left)* indicates the location of the cytoproct [Cyp] within the posterior suture [PS] relative to the external opening of the oral apparatus [OA], and shows the three postoral regions [A, B, C] whose mean length in micrometers is given in the table to the right. The controls are uncut cells measured at the same stage. Data from Table 4 of Chen-Shan, 1969.

(normal length) cells (Figure 6.11) (Chen-Shan, 1969). Thus, both the position and the length of the cytoproct vary more or less coordinately with the length of the posterior suture.

Dubielecka and Kaczanowska (1984) pursued this matter further by carrying out a correlational analysis of the position and the length of the cytoproct in doublets of *Paramecium tetraurelia* that were regulating to singlets and were unusually variable in cell length. They observed that both the length and the location of the cytoproct were strongly correlated with the length of the cell. There is thus a strong indication of something akin to a gradient determining both the length and the location of the cytoproct along the anteroposterior axis. This analysis also strongly suggested that this gradient involves the entire length of the cell, and not just the segment occupied by the postoral suture.

6.4.4 Serial Intraclonal Dimorphism

Two cases are known in which two alternative patterns are consistently observed in equal numbers in a clonal ciliate culture. In both of these, the numerical equality at the population level is due to a difference between the two products of cell division. In *Euplotes vannus,* anterior division products form three right caudal cirri, whereas posterior division products produce two (Hufnagel and Torch, 1967); in *Chilodonella cucullulus* (stock B), the anterior division products make three CVPs and the posterior ones generate four (Kaczanowska and Kowalska, 1969). In both of these cases, the relevant structures are formed anew in *both* division products, each at its definitive location.

An even more dramatic example of this type of intraclonal differentiation is found in *Folliculina* (Fig. 6.12a), a sessile marine heterotrich closely related to *Stentor* (Mulisch and Hausmann, 1984) that lives in a house (lorica) that it constructs itself. In this ciliate, the two division products are very different (Fig. 6.12c): the anterior division product is a motile swarmer with a nonfeeding OA that swims off and eventually settles down and builds a new house, whereas the posterior division product forms an expanded anterior end and continues to occupy the old house (Fauré-Fremiet, 1932; Uhlig, 1963b; Mulisch and Hausmann, 1983). The OA of the anterior division product is much smaller than that of the posterior division product (Fig. 6.12b). In *Folliculina,* unlike other heterotrichs, before division the old anterior OA is resorbed and the new small anterior OA is built from a separate small OP that develops at the same time as the much larger OP of the posterior division product (Uhlig, 1963a).

Only one example of such intraclonal dimorphism has been subjected to an experimental analysis. Ng (1976a), working with a mutant strain of *Paramecium tetraurelia,* observed some consistent differences between anterior and posterior division products in the expression and locations of extra cytoprocts. As a result of some ingenious experimental tests that ruled out other possibilities such as differential nuclear expression or differential growth, Ng suggested that the difference between the cytoproct patterns of the two daughter cells might depend on translation of different positional information at different points of the parent cell surface (Ng, 1976b, p. 175).

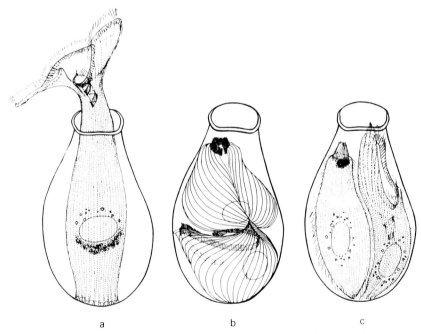

a   b   c

Figure 6.12 Unequal division in *Folliculina ampulla.* (**a**) A cell shortly before the onset of cell division. The large membranelle band is extended over two "wings" at the anterior end of the cell. (**b**) The dividing cell. Note the very small oral structures of the anterior daughter, and the large ones of the posterior daughter. Both sets of oral structures developed from new oral primordia, after resorption of the entire parental OA. (**c**) The two division products, still within the old house. The anterior daughter (viewer's left) will soon swim away, while the posterior daughter (viewer's right) will extend its OA and appear similar to the cell in (**a**). From Figures 1, 6, and 8 of Fauré-Fremiet (1932).

These examples suggest that the representation of ciliate division simply as a reorganization of an A-B-C-D-E-F sequence into an A-B-C-D-E-F-A-B-C-D-E-F sequence (see Section 5.3.2) is insufficient. As Ng (1976b) suggested, the *different* patterns in the two daughter cells must somehow be predetermined within the parent cell. This might be written as A-B-C-D-E-F transforming into A′-B′-C′-D′-E′-F′-A″-B″-C″-D″-E″-F″. This notation emphasizes that (for example) the two new E locations in the two daughter cells might be subtly different, and that they are *not* specified by the single old E location of the parent cell; E′ comes from a point near C, E″ from a point near F. The different parental locations somehow specify the variants of E (E′ vs. E″) in the immediate progeny. It is probably no coincidence that such persistent intraclonal dimorphism always involves structures that are formed anew in *both* division products, or else are formed anew in one and remodeled extensively in the other.

This type of dimorphism, in which the two patterns are generated continuously within a single culture but neither is inherited by a clonal cell line, contrasts sharply with differences such as those between singlets and doublets (see Section 4.3) in which each pattern can be inherited clonally. The difference between these two types of cortical variants is rooted in the geometry of the clonal cylinder. In terms of this geometry, the former type of dimorphism is *serial* and the latter *parallel.* The sharp differences between the two reflect the fundamental distinction between ciliate latitudes and longitudes.

6.5 CONCLUSIONS

We can draw four major conclusions concerning the location of the oral primordium (OP), and probably of other major ciliary primordia, on the ciliate surface. The first conclusion is that, except in *Paramecium* (and possibly other ciliates not yet investigated), the potential for forming an OP is likely to be unrestricted, in the sense that any part of the cell cortex may acquire the capacity to generate an OP. The second conclusion qualifies the first; realization of this potential depends in large part on the attainment of appropriate spatial relations. The third conclusion is that pattern regulation, when it occurs, is typically positive: a cell fragment usually either fails to develop new structures but retains the structures that already are present, or else it develops a complete normal set of cortical structures.

The fourth conclusion is that regulation generally occurs more easily and quickly along the anteroposterior axis than along the mediolateral (or circumferential) axis; latitudes shift more readily than longitudes. A corollary to this conclusion, most clearly demonstrated in *Stentor,* is that a coherent gradient may well be required for normal cortical development along the anteroposterior axis, but probably is not essential around the cell circumference. In my view, this last conclusion and its corollary are closely connected with the cylindrical topology of ciliate clonal growth, which entails periodic transformation of the positional values of latitudes (Chapter 5) at the same time that it can maintain indefinite stability of the values of longitudes (Chapter 4). Thus it is not surprising that even *Paramecium* can display considerable flexibility in positioning of at least some of its structures along the anteroposterior axis.

Parts of this chapter have attempted to go beyond these four relatively straightforward conclusions to some deeper formulation of the nature of the positional systems involved, especially with reference to the more stable circumferential axis. It is possible to imagine that the oral meridian is the site of a self-maintaining pattern discontinuity and that the longitudes of other structures are specified in relation to this discontinuity. Figure 6.9 illustrates one version of this idea. This type of thinking, whether expressed in terms of organizers and gradients in the style of Huxley and De Beer, or updated in the language of positional information, has dominated speculations concerning pattern formation in ciliates for a quarter-century (e.g., Tartar, 1956c; Uhlig, 1960; Sonneborn, 1963, 1975a; Frankel, 1974, 1982). Recent studies of situations involving reversals of large-scale asymmetry, however, have suggested that a more comprehensive and more predictive hypothesis of pattern formation in ciliates might be based on the premise of continuity rather than discontinuity. I will return to these more general issues in Chapter 10.

7

The Formation of an Organelle Complex: The Oral Apparatus

7.1 INTRODUCTION: ASSEMBLY AND "META-ASSEMBLY"

In Chapters 5 and 6, we have considered the positioning of cortical structures, such as the oral apparatus, but not their internal organization. This chapter attempts to fill this gap by analyzing the formation of the oral apparatus, which is at the same time the most complex and the most intensively investigated organelle system in the ciliate cortex. To do this, we must first take a close look at what is meant by the concept of assembly.

According to one contemporary dictionary, the term "assembly" means "the putting together of manufactured parts to make a completed product" (Morris, 1976). For cell biologists, the "manufactured parts" are generally macromolecules, the "completed product" is a structure, such as a virus particle or a ribosome, and the "putting together" occurs through close spatial interactions of the macromolecules either with each other or with partially assembled structures. Sometimes the component parts themselves seem to be sufficient for assembly of the structure, in which case the process often is called "self assembly" (Crick and Watson, 1956). This commonly used expression, however, hardly conveys the extraordinary complexity and variety of well-understood assembly processes. For example, in the production of bacteriophages, complex parts (heads, tails, and tail fibers) are assembled separately and then become linked together in an invariant order (Wood, 1980). Construction of the parts involves specific sequences of additions of specific molecules to partially assembled structures (Kikuchi and King, 1975). The assembly sequence requires participation of specific classes of molecules that do not end up in the final structure (Wood, 1980). These include ordinary enzymes involved in proteolytic cleavages (Laemmli, 1970), assembly-catalysts required for the efficient fitting together of structural components (Wood and Henninger, 1969), molecular scaffolds that specify shape (King et al., 1978; Hendrix, 1985), and length-determining protein yardsticks (Hendrix, 1985; Katsura, 1987). A different type of complexity is found in the assembly of microtubules, where the structure is composed of similar tubulin subunits, but successful assembly of the structure depends on the existence of nucleating sites (Tucker, 1979; McIntosh, 1979) and associated proteins (Horio and Hotani, 1986) that affect transitions between states of tubulin with radically different proclivities for disassembly (Kirschner and Mitchison, 1986).

Despite this great variety and complexity, all of these putting together processes follow the principle of a jigsaw puzzle in one key sense: the parts are fit together by specific, short-range interactions. This principle does not, of course, require that all of the components be close together at the start, especially when a fairly abundant molecule is being added onto a growing structure, as in microtubule formation. The principal idea is that the actual assembly steps occur through a direct and short-range interaction of the specific parts that are involved in the process.

The idea of short-range assembly can take us a long way; it is in principle sufficient, for example, to account for the assembly of the mitotic spindle (Kirschner and Mitchison, 1986), for the formation of complex cytopharyngeal structures in the nassulid ciliates (Pearson and Tucker, 1977), and in large part for the perpetuation of longitudinal microtubule bands and even ciliary rows in ciliates such as *Paramecium* and *Tetrahymena* (see Section 4.2). Indeed, Sonneborn (1974, 1975a) attempted to account for much of the development of *Paramecium* in terms of a specific assembly model which he called "sequential nearest-neighbor interactions."

Can a suitably expanded assembly model of the jigsaw type provide a complete understanding of the formation of intracellular structures? One arena for considering this question is the ciliate oral apparatus. This is indubitably an intracellular structure, formed anew in each cell generation, yet its complexity is far beyond that of an elementary cellular structure such as a microtubule. Can formation of such a structure be accounted for solely by short-range assembly mechanisms?

Before moving on to the evidence, I should briefly introduce an alternative idea, stated in its most general form by Williams and Honts (1987). They noted that in a ciliate oral apparatus the constituent subunits are themselves large and complex structures (ciliary units), which are initially not in direct contact. These structures "assemble" (into membranelles and the undulating membrane) in a manner suggesting that intersubunit interactions are subject to some long-range coordination or guidance. The term that they give for such long-range process(es) is "meta-assembly." This term is deliberately noncommittal as to specific mechanisms, but *is* committal to the idea that some pervasive influence(s) beyond short-range intersubunit interactions are at work when a membranelle or an oral apparatus is formed.

I hope in this chapter to show that assembly mechanisms as commonly defined and understood are necessary but not sufficient for understanding what is known about the formation of the ciliate oral apparatus. This will lead us to define those processes that add the "meta-" to the assembly.

7.2 THE ORGANIZATION OF CILIARY UNITS IN ORAL MEMBRANELLES

7.2.1 Steps in the Organization of Membranelles

The oral apparatus (OA) is a complex structure made up of highly organized sets of ciliary units and associated fibrillar systems (see Section 2.2.3), as well as specialized membrane domains. Given the complexity of this structure, we need to consider it part by part. I begin with the membranelles, then bring in the undulating

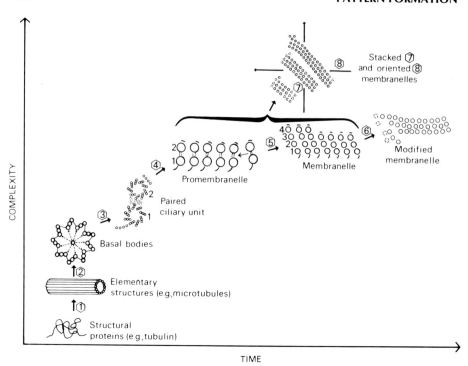

Figure 7.1 Steps in the formation of membranelles. For further explanation, see the text.

membrane (see Fig. 2.5), and finally consider the morphogenesis of the buccal cavity and cytostome.

I start by reviewing the steps in the formation of an oral membranelle (Fig. 7.1). These are shown partially as a pathway in the ordinary sense, and partially as a conceptual breakdown of the process of membranelle formation into its constituent levels of organization; thus some of the "steps" proceed simultaneously. The first step beyond macromolecular synthesis is the assembly of certain proteins, such as tubulin, into elementary structures such as microtubules. The second step (concurrent with the first) is the coordinate assembly of microtubules and other elementary structures into a basal body. The third step is the formation of paired ciliary units (see Fig. 2.4b), either by two preassembled basal bodies joining or by an old basal body nucleating formation of a new one. In the fourth step, paired ciliary units line up side-by-side to form two-rowed promembranelles. The fifth step is a second round of basal body proliferation that adds extra rows of basal bodies anterior to the two primeval rows. In ciliates that have many membranelles (the heterotrichs and oxytrichids of our cast) this step is the final one in the formation of an individual membranelle. In ciliates with only three membranelles, particularly *Tetrahymena* and its relatives, there is an additional, sixth step, a displacement of basal bodies at the right ends of the membranelles to create unique arrangements that are specific and different for each membranelle.

The steps considered thus far relate only to the construction of each individual membranelle. Steps 7 and 8 in Figure 7.1 are not further steps in a pathway of formation of individual membranelles, but rather aspects of membranelle formation that can be described only by reference to spatial organization external to the individual membranelle. *Stacking* (step 7) refers to the organization of membranelles with respect to each other, *orientation* (step 8) to the configuration of membranelles with respect to other structures in the cell. Stacking and orientation proceed concurrently while the membranelles are being formed (steps 4 to 6). Even though all of these processes occur simultaneously, the fact that they can be dissociated in various ways indicates that they are separate processes.

Understanding an assembly process requires analysis either by reconstitution or by perturbation, or both. For the early assembly steps, little information is available for ciliates beyond the descriptive: there has been some reconstitution of parts of basal bodies (Gavin, 1984) and of fibrillar elements of the OA (Honts and Williams, personal communication); the only perturbation analysis is the occasional incomplete basal body assembly in *sm19* mutations of *Paramecium tetraurelia* (Ruiz et al., 1987). Both perturbation and reconstitution analyses are more advanced in the unicellular alga *Chlamydomonas* (Goodenough and St. Clair, 1975; Huang et al., 1982; Wright et al., 1983) than they are in any ciliate. In *Chlamydomonas*, however, the organization of ciliary (flagellar) units is at the level of a paired unit, and steps 4 and beyond effectively do not exist. Ciliates make their special contributions to the understanding of organelle assembly at these higher levels of intracellular organization.

In the sections below, I first justify considering the development of membranelles within the oral primordium as a process independent of neighboring ciliary structures. Then I summarize experimental evidence relevant to each of the steps in membranelle formation from step 4 onward.

7.2.2 The Oral Ciliature is Organized Independently of the Nonoral Ciliature

Oral primordia generally form adjacent to other ciliary structures. In *Tetrahymena,* for example, the oral primordium (OP) forms next to a ciliary row (see Fig. 3.2). If the early "anarchic field" primoridum consists of lateral extensions of this ciliary row, the field might retain features of the organization of that row. The anarchic field, however, does at some stage appear truly anarchic, in the sense that new basal bodies are formed in varying orientations with reference to old ones and then are arranged more or less at random (see Section 2.3.2). These observations led Grain and Bohatier (1977) to the conclusion that the orientation of the new basal bodies in an OP does not depend on the orientation of the "parental" basal bodies of the stomatogenic ciliary row.

Grain and Bohatier's conclusion was subsequently confirmed in three different ways. First, Ng found that a clone of *T. thermophila,* which fortuitously had a ciliary row inversion along the stomatogenic ciliary row, nonetheless formed membranelles and an undulating membrane with completely *normal* orientation and asymmetry, despite the fact that the oral field within which these structures formed had appeared on the right side of a ciliary row instead of the usual left side (Ng, personal communication). Second, in a *disorganized (dis)* mutant, the ciliary rows were severely disrupted, yet the internal organization of the OA was normal or nearly so (Frankel, 1979 and unpublished). Finally, in the logical converse of the above two situations, the internal organization of the developing OA was severely affected by reversals of large-scale cortical asymmetry despite the completely normal internal geometry of the neighboring ciliary rows (see Section 7.2.6). These reciprocal dissociations strongly confirm Grain and Bohatier's original conclusion: the organization that emerges as membranelles develop within the OP is new.

7.2.3 Assembly of Promembranelles Can Be Dissociated from Stacking and Orientation

The integration of paired ciliary units into promembranelles might be thought of as simple assembly on a magnified scale. Promembranelles are formed by sequential addition of basal-body pairs, generally proceeding from cell's right to left. Considered abstractly, this process might be no different in principle from the growth of a microtubule by addition of tubulin dimers.

At present, we do not have the information to prove or disprove this view. If it is correct, however, we should be able to observe this assembly process occurring normally even when virtually all other aspects of membranelle development have gone awry. Such observations were first made in *Tetrahymena pyriformis* grown for many hours at a temperature (33° C) just below the upper physiological limit for oral development and cell division in this species. Under these conditions, basal-body pairs began to assemble into short promembranelle fragments (often no larger than basal-body quadruplets) that often appeared to be oriented at random; some of these fragments then extended further to form promembranelles that later matured into an array of membranelles that had little or no overall order (Fig. 7.2b). Every OA then was different from every other, with a range from nearly nor-

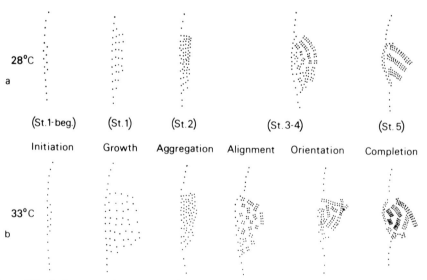

Figure 7.2 A diagrammatic summary of successive stages of oral development in *Tetrahymena pyriformis* as observed at (**a**) 28 °C and (**b**) 33 °C. The dots indicate basal bodies, and the labeled stages are the standard ones used in describing *Tetrahymena* development (Frankel and Williams, 1973). Redrawn from Figure 13 of Frankel (1964).

mal to a nearly random number and orientation of membranelles (Frankel, 1964). Because the promembranelles, nonetheless, always appeared *internally* as typical double files of basal bodies, I then concluded that the alignment of basal bodies into double files is an event dissociable from the stacking and the orientation of membranelles that normally occurs at the same time (Fig. 7.2a).

In this early study, observations were made entirely on silver-stained cells prepared for light microscopy. Subsequently, similar observations were made on oral development in *membranellar pattern (mp)* mutants of *Tetrahymena thermophila* (Kaczanowski, 1976; Frankel, 1983; Burg and Frankel, unpublished observations), and in a stable high-temperature-sensitive variant (a presumed macronuclear mutant) of *T. thermophila* that is incapable of phagocytosis under restrictive conditions (Orias and Pollock, 1975). This latter putative mutant, called NP1, is of special importance because at restrictive temperatures it expresses a deficiency in overall spatial order of newly formed OAs that is even more extreme than that observed in wild-type *T. pyriformis* at 33°C; nonetheless, promembranelles still were formed as locally ordered double files of basal bodies (Frankel and Williams, unpublished observations). Ultrastructural and immunochemical observations on NP1 cells indicated that the basal bodies and all accessory structures found in the OA appeared normal (Williams and Honts, 1987).

The robustness of the process of promembranelle formation under conditions inimical to orderly oral development is consistent with (although it does not compel) the view that the formation of promembranelles (step 4 in Fig. 7.1) is dependent only on the intrinsic properties of the building blocks, i.e., basal body pairs. One might even think of promembranelle formation as a self-assembly process operating on a large scale. In promembranelle formation, however, the subunits are vastly larger and more complex than are those involved in the assembly of structures such as microtubules, and the relevant distances are in the micrometer rather than the nanometer range. Thus, although the idea of promembranelle formation by magnified self-assembly is abstractly appealing, it is difficult to imagine the actual mechanisms by which it might come about.

7.2.4 Secondary Proliferation of Membranelle Basal Bodies Is Subject to Regulation

Although membranelles are always initiated as two-rowed promembranelles (Bakowska et al., 1982a), the number of rows of basal bodies within completed membranelles varies greatly in different ciliates. In a few cases (in heterotrichs) this number remains at the original two (Peck et al., 1975; Mulisch and Hausmann, 1984). The most common number is three or four, but in certain ciliates, notably in *Glaucoma ferox,* it may be as high as 15 (de Puytorac et al., 1973). During conversion of a promembranelle into a mature membranelle, the third row is always added adjacent to the second row of the promembranelle, the fourth row is added next to the third (Fig. 7.1), the fifth row (if present) next to the fourth, and so on. Because the orientation of any additional ciliary row is rigidly specified by its relation to the initial two rows, the process by which these additional rows are formed appears to be an excellent example of local assembly with strict short-range dependence on preexisting structural organization.

Nonetheless, the addition of rows of basal bodies in membranelles raises a question similar to one that we have already encountered for nonoral ciliary rows. There, we saw that we could predict with certainty the *orientation* of all new ciliary units formed within a row, but we could *not* predict *which* units would become paired and which would remain single without making reference to cellular features extrinsic to the rows (see Section 4.2.4). An analogous distinction might apply to secondary basal-body proliferation within membranelles. Why is the short fourth row of basal bodies in membranelles of *Tetrahymena* formed only at the right end of the membranelle (as shown in Fig. 7.1) and not at the left end, or along the entire length of the membranelle? It is unlikely that the other basal bodies of the third row are inherently incompetent to nucleate fourth-row basal bodies. Ectopic fourth-row basal bodies, and sometimes complete fourth rows, have been observed in a variety of mutants affecting oral development (Frankel et al., 1984b; Frankel, unpublished observations) and in wild-type cells with reversals of global asymmetry that force abnormal membranelle organization (Nelson et al., 1989b). Fourth and sometimes even fifth and sixth rows have been observed as one aspect of the general disorganization of membranelles in the NP1 mutant (Frankel and Williams, unpublished observations; see also Fig. 3C in Williams and Honts, 1987). The general association between disturbance of patterning and relaxation of regulation of secondary proliferation suggests that the control of this proliferation is not entirely intrinsic to each membranelle.

The degree and manner in which secondary proliferation of basal bodies is regulated has been investigated in two organisms, *Paraurostyla weissei,* which has many membranelles (Bakowska and Jerka-Dziadosz, 1980), and *Tetrahymena thermophila,* which normally has only three membranelles (Bakowska et al., 1982a). Both of these studies were biometrical analyses of number and internal organization of membranelles in cells of different sizes. Although some of the details are fairly intricate, the main results are rather simple, and most of them can be summarized with reference to the electron micrographs of two typical membranelles from the same (ventral) region of the membranelle band of *P. weissei,* the first from a normal-size cell (Fig. 7.3a) and the second from a much smaller cell (Fig. 7.3b). Five features are apparent: (1) membranelles of miniature cells are *shorter* than those of normal cells. Thus membranelles, unlike basal bodies, do not have a constant size. (2) However, membranelles do have a constant "thickness"; *all* membranelles have four rows of basal bodies, even those of truly tiny cells in which the number of membranelles is one-third that of normal cells and their length less than half of normal (Bakowska and Jerka-Dziadosz, 1980). (3) The length of row 1 is always equal to that of row 2. This is to be expected for rows that arise from the side-by-side assembly of basal-body pairs (Fig. 7.1, step 4). (4) In *P. weissei,* row 3 is *always* shorter than row 2. The row 3 to row 2 ratio is *roughly* constant, although when certain specific comparisons are made, significant departures from strict proportionality do appear (Bakowska and Jerka-Dziadosz, 1980). (5) Row 4 of *P. wiessei* generally has 3 basal bodies, although in the tiniest cells it has only two.

These observations indicate that secondary proliferation of basal bodies in membranelles of *P. weissei* is obligatory and strongly suggest that it is spatially regulated. The most critical evidence for a true coordinated regulation of membranelle

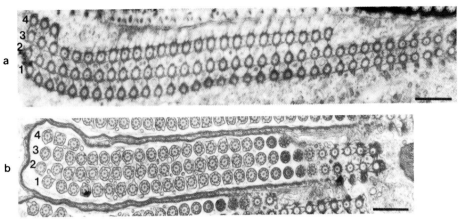

Figure 7.3 Transmission electron micrographs of cross sections of membranelles of (a) normal and (b) miniature *Paraurostyla weissei*, located in the same relative position within the ventral region of the membranelle band. The rows of basal bodies are numbered from posterior to anterior. The scale bars on the right represent 1 μm. Photographs kindly supplied by Dr. Maria Jerka-Dziadosz. Panel (a), from Fig. 11A of Bakowska and Jerka-Dziadosz (1980), used with permission.

organization, however, comes from analysis of the membranelle *set* as a whole, a subject to which we now turn.

7.2.5 The Membranelle-Set is Patterned as a Unified Whole

In *Paraurostyla*, the membranelle band has two sections, the frontal and ventral, in which membranelles differ in size, spacing, and details of location of the additional rows of basal bodies (the pattern described in the previous section applies to ventral membranelles) and in other features as well (Bakowska and Jerka-Dziadosz, 1978). Cells with a widely differing number of membranelles manifest a near-exact proprotionality between the number of frontal and ventral membranelles, with close to 28 percent of the membranelles of the frontal type, 72 percent of the ventral type (Bakowska and Jerka-Dziadosz, 1980). This proportionality indicates that individual membranelles must "know" where they are in relation to the whole set, and adjust their pattern of secondary proliferation (and other processes) accordingly.

In *Tetrahymena*, there are far fewer membranelles than in *Paraurostyla*, but the internal organization of each membranelle is more complex. The basal bodies at the right ends of the first and second membranelles and all of the basal bodies of the third membranelle are arranged in a seemingly irregular manner (Fig. 7.4) (McCoy, 1974; Bakowska et al., 1982a). These apparent irregularities, however, are the same in all OAs of wild-type cells grown under normal conditions (Williams and Bakowska, 1982; Bakowska et al., 1982a). Thus, the individual pattern of each membranelle is an easily recognizable "signature" of that membranelle. This signature arises from coordinated displacements of basal bodies that take place during the final stages of oral development, *after* secondary proliferation of basal bodies is completed. We, thus, can work out the path of displacement of each basal body

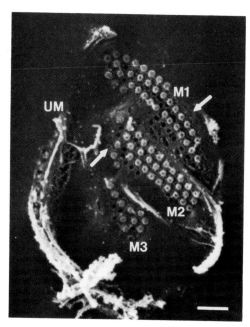

Figure 7.4 A scanning electron micrograph of an isolated oral apparatus from a lysed wild-type (inbred strain B) cell of *Tetrahymena thermophila*. Ciliated basal bodies appear as dense rings, nonciliated basal bodies are barely visible as much fainter rings *(arrows)*. Membranelles 1 [M1], 2 [M2], and 3 [M3], as well as the undulating membrane [UM] are labeled. Most of the basal bodies of M1 and M2 form regular parallel columns, whereas some basal bodies at the right (viewer's left) end of M1 and M2, as well as all of the basal bodies of M3, are arranged in a less regular manner. The scale bar represents 1 μm.

(Fig. 7.5). When we do, we find that the displacements within the three membranelles are related: the more posterior the membranelle, the greater is the displacement of basal bodies within that membranelle. In the third membranelle, the combination of extreme displacement with resorption of many of its basal bodies results in an apparent total loss of the original "row and column" arrangement (Bakowska et al., 1982a,b).

Knowing the normal signatures of all three membranelles allows one to ask how these signatures are specified. Is there a separate, unique design for each membranelle? One way to answer this question is to find out whether variation in the number of membranelles affects their design. The number of membranelles is rather highly canalized in tetrahymenas, so that even mutations that greatly alter the length of membranelles typically do not change their number (Frankel, 1983; Frankel et al., 1984b). However, there are a few exceptions: the membranelle-pattern mutations *mpC* and *mpD* tend to increase the number of membranelles (Frankel et al., 1984c), whereas the *mpG* mutation tends to reduce that number (Frankel and Jenkins, unpublished observations). All three of these mutations are variably expressed, so that some cells have OAs with the normal three membranelles, whereas others have OAs with more *(mpC, mpD)* or fewer *(mpG)* than three membranelles. In these mutants (especially *mpD* and *mpG*), those OAs that have three membranelles are generally normally organized; unusual features were observed

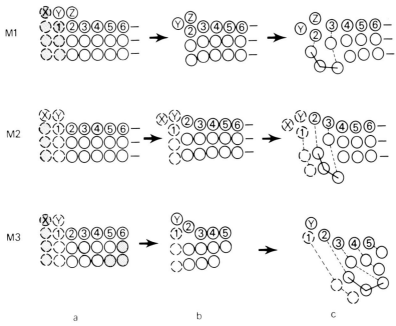

Figure 7.5 A diagram illustrating the modifications occurring during the final stages in the development of membranelles 1 [M1], 2 [M2], and 3 [M3] of *Tetrahymena thermophila*. Circles *bounded by solid lines* indicate ciliated basal bodies, and *circles bounded by dashed lines* represent nonciliated basal bodies. Basal bodies that are destined for subsequent resorption are *stippled*. The numbers inside basal bodies enumerate the basal body columns, and the letters [X, Y, Z] indicate the individual basal bodies of the short fourth row. Column (**a**) shows the membranelles just before the onset of modification. Column (**b**) illustrates an intermediate stage of modification. Resorption of basal bodies at the right ends of M1 and M2, and both ends of M3, has been completed. Two *ciliated* fourth-row basal bodies [Y, Z] persist in M1, two *nonciliated* fourth-row basal bodies [X, Y] persist in M2, and one *ciliated* fourth-row basal body [Y] persists in M3. Column (**c**) shows the membranelles after completion of basal-body displacement. The *dashed lines* show the paths of displacement of the basal bodies of each column. Three corresponding basal bodies that are arranged in a prominent "hook" configuration are connected by *solid lines*.

primarily in OAs with an unusual number of membranelles. This internal control allows one to be reasonably confident that unusual features of membranelles are directly related to their unusual position within an atypical membranelle complement and are not separate effects of the mutations on membranelle organization.

Tracings of arrangements of cilia of typical membranelle-sets from OAs with two, three, four, and five membranelles are shown in Figure 7.6. The crucial observation is that an altered number of membranelles is associated with intermediate organizations that are not found in OAs with three membranelles. In OAs with four membranelles, the third membranelle, although resembling the second membranelle in most ways, shows a greater displacement of the basal bodies at its right end, and exhibits a specific feature (ciliation of the "Y" basal body, filled in and marked with arrows in Fig. 7.6c), that is a normal characteristic of the third membranelle (Figs. 7.5 and 7.6b). In OAs with five membranelles, the third membranelle, which is now in the center of the membranelle set much as the second mem-

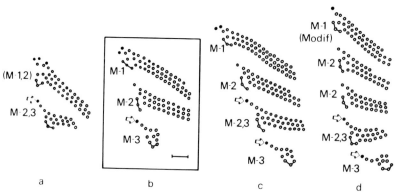

Figure 7.6 Tracings of scanning electron micrographs of membranelle sets from isolated oral apparatuses of *Tetrahymena thermophila,* all drawn at the same scale (scale bar = 1 μm). Only ciliated basal bodies are shown. The *closed circles* indicate ciliated fourth-row basal bodies, with *arrows* pointing to the ciliated Y basal body that is characteristic of membranelle 3 in normal OAs (see Fig. 7.5c). *Solid lines* connect the basal bodies that make up the "hook" configuration, as shown in Figure 7.5. The membranelles are enumerated not by their sequential locations but by their internal patterns: M-1, M-2, and M-3 label the patterns that are found in *normal* membranelles 1, 2, and 3, respectively (panel **b**, boxed); M-1,2 and M-2,3 indicate intermediate patterns observed in OAs that have a total number of membranelles different from the normal three. A two-membranelle OA from an *mpG* cell is shown in panel (**a**), and three-, four-, and five-membranelle OAs from *mpD* cells in panels (**b**), (**c**), and (**d**). Panels (**b**), (**c**), and (**d**) are tracings from Figures (8), (9), and (10), respectively of Frankel et al. (1984c).

branelle is in normal three-membranelle OAs, also has the normal membranelle-2 (M-2) pattern. The fourth membranelle now is a clear intermediate between the M-2 and membranelle-3 (M-3) patterns, hence is called an M-2, 3 pattern (Fig. 7.6d).

These intermediates between the patterns normally observed in the second and third membranelles indicate that membranelle patterns cannot be specified by unique individual blueprints. Instead, the existence and nature of the observed intermediate patterns strongly suggest that the organization of each membranelle depends in a systematic way on the relative position of that membranelle within the set as a whole. This in turn suggests that the patterns of the developing membranelles are integrated in the sense of being governed by a coordinated set of controlling forces. The territory in which these forces operate thus obeys the classical definition of an embryonic field (see Section 1.4.1).

The results and conclusions just summarized were published at a time (Frankel et al., 1984c) when the only mutants available expressed a number of membranelles *greater* than the normal. The idea of an integrated oral field would lead us to predict that if only two membranelles were formed, the first one should express a pattern intermediate between those of normal first and second membranelles, the second a pattern intermediate between those of the normal second and third membranelles. In an initial study, the patterns of small two-membranelle OAs produced by tiny severely starved wild-type cells agreed only partially with these predictions (Bakowska et al., 1982a). More recently, we have isolated a new mutation *(mpG)* in which about one-half of the OAs of well-fed cells have two membranelles. In these, the first membranelle is almost invariably abnormal, with an appearance suggestive of an intermediate between the usual patterns of the first and second membranelles. The second membranelle is of a clear-cut intermediate M-2, 3 pattern, coinciding in organization with the range of intermediates already encountered in four and five membranelle OAs (note the resemblances of the second membranelle in Fig. 7.6a with the third membranelle in Fig. 7.6c and the fourth in Fig. 7.6d) (Frankel, unpublished observations). Observations on OAs of *mpG* mutants, therefore, strongly support the conclusion of a coordinated specification of membranelle displacement patterns.

The mechanism of this coordinated specification of membranelle patterns is unknown. There are indications, however, that this coordination may be effected at least in part by mechanical rather than chemical means. The displacement process is temporally coincident with the formation of the buccal cavity (Bakowska et al., 1982b), a time at which there is also extensive formation of oral fibers (Williams et al., 1986). Furthermore, unlike all earlier steps of membranelle formation, the displacement of basal bodies is largely reversible; before cell division the membranelles of the anterior OA temporarily return to a more compact arrangement of basal bodies resembling that depicted in column b of Figure 7.5 (Bakowska et al., 1982b). This transient reversal of basal-body displacement coincides with the temporary disappearance of the buccal cavity and loss of oral fibers (Williams et al., 1986). However, basal-body displacement does occur, albeit abnormally, in OAs that lack both an undulating membrane and a buccal cavity (Frankel et al., 1984a). Thus, displacement due to a simple mechanical stretching during formation of the buccal cavity is likely to be only a part of the story.

The general idea of spatially coordinated specification of unique features of repeated structures is by no means a new one. This idea was emphasized by Bateson (1892, 1894) with reference to variation in serially duplicated parts such as vertebrate teeth and digits. Bateson found that when a space normally occupied by three molars is instead occupied by four, several or all members of the series are altered so that they do not correspond precisely to members of the typical series. He, therefore, concluded that "... the whole Series of Multiple Parts is bound together in one common whole" (Bateson, 1892, p. 111). In such a modified series, the "general configuration of the series" is conserved but not the structure of the individual elements that make up that series. Bateson also extended this principle to cases of polydactly in vertebrate limbs, although in those cases the evidence was somewhat more ambiguous (see Frankel et al., 1984c, pp. 92–93, for a brief review of these studies). In the near-century since Bateson expressed these views, they have been not so much refuted as ignored, despite their obvious relevance to problems such as the way in which morphogens might control repetitive patterns (e.g., Slack, 1987). My own view is that this clear ciliate example of Bateson's insight reinforces the idea of a deep and possibly general integration within morphogenetic fields, such as has been emphasized by few authors after Bateson (see, however, Goodwin and Trainor, 1983).

7.2.6 The Orientation of Membranelles Is Extrinsically Controlled

The prominent internal asymmetries within membranelles allow assessment of abnormal orientations of membranelles that result from reversals of large-scale cellular asymmetry and polarity. In *Tetrahymena*, as membranelles assemble they are

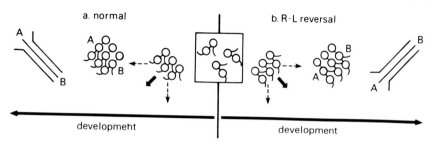

Figure 7.7 Diagrams indicating orientations of a developing membranelle of *Tetrahymena* cells of (**a**) normal and (**b**) reversed asymmetry. The initial state with unoriented basal-body pairs in an anarchic field is shown in the center, and subsequent stages of membranelle formation are shown to either side: a promembranelle closest to the center, a membranelle after completion of basal-body proliferation next, and a fully-modified membranelle (using lines to represent rows of basal bodies) farthest out. In the innermost diagrams, the vector-components representing asymmetry and polarity are shown by *dashed arrows,* and the resultant vector by a *solid arrow.* In the other diagrams, the two dissimilar ends of the membranelle are indicated by the letters A and B. For further explanation, see the text.

oriented with a specific tilt, normally from the cell's anterior-right to posterior-left. In cells or portions of cells in which large-scale right–left asymmetry is reversed (as judged by independent criteria), the direction of this tilt is often reversed, suggesting a mirror-image of the normal OA (Jerka-Dziadosz and Frankel, 1979). If one notes the location of the modified ends of the membranelles in the "reversed" OAs, however, one discovers that what at first sight appears to be a mirror-reversal is really a rotational permutation of the normal organization (Frankel et al., 1984a; Nelsen et al., 1989b; see Section 8.3.3).

The key to understanding this situation is to consider that ciliary structures of unchanged internal asymmetry are responding to a reversal of a superimposed large-scale cortical asymmetry. The way in which this might work is shown in Figure 7.7. There, the cellular asymmetry and polarity are drawn as the X (circumferential) and Y (anteroposterior) components of a vector, which in the normal cell (Fig. 7.7a) points to the cell's posterior right, and in the cell with a reversal of global asymmetry (Fig. 7.7b) is directed to the cell's posterior left. We can then make two critical postulates. The first is that when promembranelles form, the paired ciliary units become oriented in conformity to the cell's polarity/asymmetry vector, like a compass needle is oriented in conformity to the polarity vector of the earth's magnetic field; the result is that whenever there is a reversal of large-scale cellular asymmetry, the promembranelles necessarily become mirror-images of the normal in their direction of tilt, but rotational permutations in their internal organization. Furthermore, a necessary consequence of the fact that ciliary units are both polar and asymmetrical (and that all basal-body pairs within a promembranelle have the same polarity and asymmetry) is that every membranelle *must* have two dissimilar ends, which I have called the "A" and the "B" ends (Fig. 7.7). This fact provides the basis for the second postulate, namely that the displacement process can only work on one structural end of a membranelle (the "A" end in our convention). This modification will, therefore, make the inherent structural polarity of each membranelle readily visible in suitably stained light-microscopic preparations.

This analysis yields two complementary conclusions: first, that the orientation of membranelles *does* respond to cues extrinsic to the developing membranelles, and second, that the response is constrained by the inherent asymmetry of the structural units that make up the membranelles.

7.2.7 Summary

The reader should be convinced by now that formation of a membranelle set is a complex process. To try to bring order into what might appear to be conceptual chaos, I will attempt to classify the processes discussed previously according to two dichotomously arranged classifying variables: first, whether the process is *intrinsic* to the developing oral apparatus (OA), i.e., controlled entirely within the developing OA, or *extrinsic,* i.e., controlled at least in part by structures or forces that extend beyond the developing OA, and second, whether the process requires only *short-range* interactions, or involves influences working over a *long-range.* Because our current understanding is very incomplete, any such summarization must be in part speculative; the classification used in Table 7.1 has been biased in favor of intrinsic and short-range controls; other options are chosen only if there is clear evidence in favor of rejecting the conservative hypothesis of assembly processes working over molecular distances within the OA (the numbers in parentheses refer to the numbered steps in Fig. 7.1).

The processes that have been placed in the short range category are those for which there is no compelling evidence that assembly mediated by short-range interactions is insufficient to account for the process. This classification assumes that oral development has already been initiated at a particular site in the cell cortex. A question mark is placed next to promembranelle formation because basal-body pairs are not really close to one another in the anarchic field, and it is hard to imagine how they could become mutually aligned without some force(s) acting over distances that are long relative to molecular dimensions.

All of the presumed short-range interactions are intrinsic. The fact that the category of "extrinsic short-range" processes is empty is a contingent fact rather than a logical necessity, as it *could* have been true that the geometry of the developing OA would have been influenced by the preexisting organization of nonoral ciliary rows.

Three processes are placed in the "long-range" category. Two of these, secondary proliferation and membranelle modification, are long-range because they are

Table 7.1 A Classification of Processes Involved in Development of Membranelles

	Intrinsic	Extrinsic
Short-range	Basal-body formation (2) Pair-assembly(3) Promembranelle formation (?) (4) Secondary proliferation(5)	None
Long-range	Secondary proliferation (5) Membranelle modification (6)	Spatial orientation (8)

The numerals in parenthesis refer to the steps in membranelle development illustrated in Figure 7.1

proportionately regulated within the membranelle set as a whole (see Section 7.2.5) but probably are intrinsic as the spatial regulation is not known to involve anything outside of the membranelle band. Secondary proliferation of basal bodies is entered twice in Table 7.1, because although the nucleation of new basal bodies during secondary proliferation occurs over short intracellular distances, the decision as to *which* preexisting basal bodies will nucleate new ones is subject to long-range influences. Finally, regulation of the spatial orientation of membranelles is both long-range and extrinsic, as reversals in orientation of developing membranelles are generally coordinated with reversals in the asymmetry of cell-surface configurations that extend far beyond the borders of the oral system itself.

The general conclusions of this section are twofold. First, one can speak of the "assembly of an oral apparatus" only in the loosest sense, even if one restricts oneself to the membranelles (it will become looser yet when we add the undulating membrane and the cytostome). Second, the kinds of processes designated as "meta-assembly" by Williams and Honts (1987) are quite diverse phenomenologically, extending over different scales and possibly involving different mechanisms.

7.3 RELATIONSHIP BETWEEN MEMBRANELLES AND THE UNDULATING MEMBRANE

Thus far, I have dealt with only one aspect of oral development, namely the arrangement of basal bodies within membranelles. However, as we have seen throughout, the OAs of the ciliates in our cast all have a second ciliary component, the undulating membrane (UM), located on the right side of the OA (see Figs. 2.5, 2.10, 3.2, 3.4, 3.8, 3.9, 7.4). In heterotrichs *(Stentor, Blepharisma),* oxytrichids, and *Paramecium* the anarchic field becomes separated early during oral development into two components, one of them destined to form the membranelles and the other the UM. In *Tetrahymena,* on the other hand, the UM and membranelles appear to arise from one contiguous anarchic field. The analytic evidence, however, strongly suggests that two distinct oral subsystems exist in this ciliate as well.

There are three main lines of evidence suggesting a different regulation for the organization of the UM and of the membranelles in *Tetrahymena thermophila.* The first line of evidence is biometrical: although there is a strong correlation between the length (measured in number of ciliary units) of different membranelles in both wild-type and mutant cells, the relationship between length of membranelles and that of the UM is weak (Bakowska et al., 1982a; Frankel et al., 1984b). This analysis suggests that ". . . much of the variation in UM length is generated by causes unrelated to membranelle length" (Frankel et al., 1984b, p. 56).

The second type of evidence is provided by mutations. Although mutations of *Tetrahymena thermophila* that generate extreme oral disorganization, such as the NP1 putative mutation considered earlier (see Section 7.2.3), affect both membranelles and UM, many mutations affect one system or the other, but not both. The *membranellar pattern (mp)* mutations affect the organization of membranelles but generally have little if any effect on the UM (Kaczanowski, 1975, 1976; Frankel, 1983; Frankel et al., 1984b), whereas the converse is true for the *misaligned undu-*

lating membrane (mum) mutation. This mutation brings about the formation of multiple parallel UM-fragments in place of the normal single extended UM, while having only minimal effects on the membranelles (Fig. 7.8) (Lansing et al., 1985). A detailed comparison of oral development in wild-type and *mum* cells suggests that the multiplicity of UMs observed in the latter arises as a consequence of a temporal distrubance; in *mum* cells, basal body proliferation in the territory of the developing UM continues beyond the stage at which a single "pro-UM" normally forms (Lansing et al., 1985). The *mum* mutation, therefore, reveals the existence of a developmentally distinct UM-subfield, which is not visible anatomically in oral primordia of normal tetrahymenas.

The third line of evidence for independence of the UM is the imperfect coordination of responses of the two systems to reversals of global asymmetry (Frankel et al., 1984a; Nelsen et al., 1989b). In cells that have undergone such reversals, the location and even the internal organization of the UM often is not in close register with that of the membranelles.

Taking the three lines of evidence together, it becomes clear that the oral apparatus is a composite, developmentally as well as structurally. Its composite nature will become even more obvious when we add to our inventory of oral structures the nonciliary portion of the OA, as represented by the buccal cavity and cytostome.

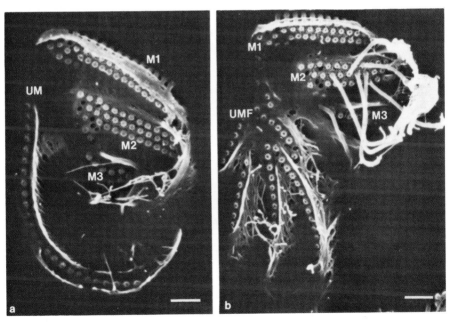

Figure 7.8 Isolated OAs from (**a**) wild-type and (**b**) *mum* cells. The three membranelles [M1, M2, M3] are labeled in both wild-type and *mum* cells. Note that the basal-body patterns at the right ends of the membranelles are identical in the two OAs, but the undulating membranes are very different, with a single structure [UM] in the wild-type cell and several shorter UM-fragments [UMF] in the *mum* cell. The scale bars represent 1 μm. From Figures 1 and 2 in Lansing et al. (1985), with permission.

7.4 DETERMINATION OF THE BUCCAL CAVITY AND CYTOSTOME

The formation of the buccal cavity and cytostome occurs late in oral development. These processes are morphogenetic in the strict sense of genesis of new form. The most obvious question to ask about these processes is *how* they take place, in a mechanistic sense. Little is known about this, although the oral filaments can hardly escape being involved. A second question is a traditional embryological one, namely *when* are these events determined to occur? A third question is whether these morphogenetic processes are specified extrinsically or intrinsically, that is whether or not they are dependent on relations that extend outside the developing oral system itself. The second and third questions have answers, and the answer to the third question is not the same in all ciliates. We will see that specification of the site of the buccal cavity and cytostome is extrinsic in *Stentor* and its relatives, whereas it is intrinsic in *Tetrahymena* and *Glaucoma*. This distinction is of major importance in understanding the fate of cells of reversed asymmetry, the topic of the next two chapters.

7.4.1 Determination of Parts

One of Tartar's simplest experiments was to bisect a regenerating stentor horizontally across the middle of the oral primordium (OP). The outcome of this experiment differed according to the stage of development at which it was performed. If performed before the differentiation of membranelles, both half-primordia continued to develop, and they produced structurally complete and more or less full-size OAs, each with a complete set of "mouth parts" (the buccal cavity plus cytostome as distinct from the oral apparatus as a whole) (Fig. 7.9a–c). If the transection was performed after membranelles had differentiated, the anterior half produced a membranelle band with no mouth parts (or very defective ones), and the posterior half generated complete mouth parts but formed only a very short membranelle band distal to these mouth parts (Fig. 7.9d–e) (Tartar, 1957). Positive regulation was thus observed in the first case, no regulation in the second. The transition between the two, or determination of parts, coincided with the time at which membranelles are being formed within the OP.

These results are easy to comprehend with respect to the ciliary structures of the OP. As long as an anarchic field is still present, primary proliferation of basal bodies can occur. Then, reduction in size of the field could serve as a signal for continued basal-body proliferation within the field, so that more membranelles could be made. After the membranelles have actually been formed at the expense of this anarchic field and no "uncommitted" basal bodies remain, such compensatory proliferation no longer is possible.

This simple explanation does not, however, account for the determination of the location of the mouth parts. Membranelle differentiation in itself gives us no basis for understanding when, where, or how particular regions of a developing OP could become determined to participate in a specific later morphogenetic process. In particular, the experiment leaves completely open the question of whether the

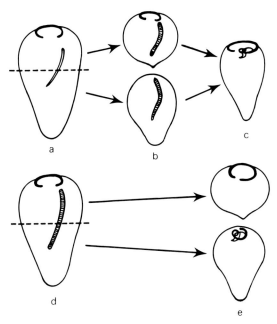

Figure 7.9 The development of halves of oral primordia in regenerating stentors equatorially transected at a stage (**a–c**) before and (**d–e**) after differentiation of membranelles. In all cases, new oral structures are derived from parts of the transacted oral primordium. After early equatorial transection (**a**), the two half-primordia grew in length as they produced membranelles (**b**), and both developed into complete and normally proportioned oral apparatuses (**c**). After later transection (**d**), the two half-primordia did not increase significantly in length (**e**), and the anterior portion formed a ring of membranelles with no mouth parts, whereas the posterior half produced well-developed mouth parts but only a very short membranelle band. Redrawn from Figure 2 of Tartar (1957), with permission.

process leading to such determination is intrinsic to the developing oral system, or whether it involves some signal originating outside of the OP.

7.4.2 Extrinsic Versus Intrinsic Control of Formation of Mouth Parts

Extrinsic control in heterotrichs

The first indication that the OP of *Stentor* does not determine the site of its own mouth parts was obtained in experiments in which regions of fine stripes, often containing the contrast zone, were rotated 180 degrees around the *horizontal* axis, so that the resulting composite stentor carried a longitudinal patch bearing ciliary rows of a polarity opposite to that of the rest of the cell. The main motivation for this experiment was to find out whether oral primordia could form at contrast zones in which the interacting cortical regions were of opposed polarity. The answer was that they can (Tartar, 1956b). This experiment also demonstrated that such heteropolar grafts tended eventually to migrate off their hosts in the direction indicated by the arrows in Figure 7.10a, reinforcing the idea that differential affinities can be expressed at an intracellular level (see Section 5.3.2). The important

 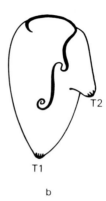

Figure 7.10 Induction of mouth parts of *Stentor* by the tail pole. (**a**) The result after excision of a segment bearing the contrast zone, followed by rotation of this segment around the horizontal axis of the cell and heteropolar reimplantation. The *dashed line* marks the perimeter of the inverted implant. The *arrows* show the intrinsic polarity of the host and the graft. The tail poles of the host and graft are indicated by T2 and T1, respectively. The oral primordium formed mouth parts at both ends. (**b**) Induction of incomplete mouth parts in the middle of an oral primordium after implantation of an extra tail pole [T2] from another stentor. (**a**) redrawn from Figure 8(B) of Tartar (1956b), (**b**) redrawn from Figure 44(C) of Tartar (1961), both with permission.

result in the present context is that when mouth parts were formed before the graft had separated from the host, they appeared at *both* ends of the OP; one normal set of mouth parts was formed at the intrinsic posterior end of the OP near the graft's tail-pole (T1), and a second defective set developed at the intrinsic *anterior* end of the OP of the rotated graft, near the tail-pole (T2) of the host. This second set would not have formed if mouth parts were determined entirely intrinsically within the developing OP.

Further experiments demonstrated that a grafted tail-pole could induce the formation of a second, incomplete set of mouth parts within an adjacent portion of a developing membranelle band (Fig. 7.10b). Such experiments, as well as others by Uhlig (1960) that also demonstrated development of incomplete mouth parts near ectopic tail-poles, suggest that something in the posterior region of a stentor might serve as an inductor of mouth part formation. This view is supported by the fact that when large numbers—up to 100—stentors were grafted together, they failed to reorganize into a typical *Stentor* form, lacked tail poles, and "produced long garlands of membranellar banding in joined loci of stripe contrast, but no mouth" (Tartar, 1966d, p. 132; for details see Tartar, 1954b). One could imagine that the failure to form mouth parts was associated with the lack of an appropriate tail-pole inductor.

Uhlig (1960) suggested that formation of mouth parts is not a consequence of a localized induction by a specific neighboring structure, but rather is a response to a particular level of a basal–apical gradient (see Section 6.4.1). Using this interpretation, the inability of large *Stentor* masses to produce mouth parts is due to their failure to reconstitute a coherent basal–apical gradient, conceivably because they are above a maximum size for setting up a diffusion-based gradient [see Frankel (1974, pp. 473–474) for application to *Stentor* of the "size-limit" suggestion originally made by Crick (1970)].

A developmental system based on a gradient should show more or less proportional regulation after disturbance. The closest approach to a biometrical analysis of such regulation was provided by Eberhardt (1962), working with a ciliate that is uniquely suited to such an investigation. This ciliate is *Spirostomum* (Fig. 7.11a), which for our purposes can be thought of as a stretched-out *Blepharisma* (see Fig. 3.8). In one of his most interesting experiments, Eberhardt removed the posterior tip of a spirostomum that was preparing to divide, in which the OP had formed its membranelles but had not yet begun to develop its mouth parts. The result that is

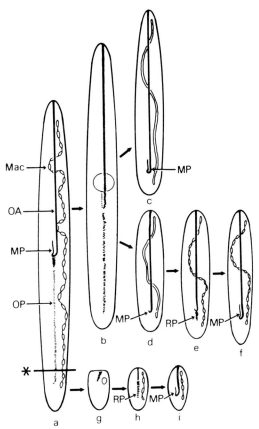

Figure 7.11 Formation of mouth parts after transection near the posterior end of the oral primordium of a dividing cell of *Spirostomum ambiguum*, carried out after membranelles had begun to form. (**a**) A spirostomum at the time of transection. The structures labeled are the macronucleus [Mac], oral apparatus [OA], mouth parts [MP], and oral primordium [OP]. The plane of transection is indicated by * (**b**) The long anterior fragment shortly after transection. This fragment divided on schedule to produce (**c**) a normal anterior division product and (**d**) a posterior division product with underdeveloped mouth parts [MP]. Later, an oral replacement primordium (RP) was produced (**e**), resulting in (**f**) a cell with normal mouth parts (MP). (**g**) The posterior fragment shortly after transection. The posterior tip of the oral primordium is located in the anterior half of this fragment. No mouth parts formed. Instead (**h**) an oral replacement primordium [RP] was produced that extended the membranelle band further posteriorly, after which (**i**) a normal set of mouth parts [MP] developed. Redrawn with modifications from Abb. 20 of Eberhardt (1962), with permission.

of greatest interest here concerns the small posterior fragment, which contained the posterior end of the original OP plus a portion of the long beaded macronucleus. If the operation had been not performed, mouth parts would have formed at the posterior end of the OP. Within the detached posterior fragment, however, mouth parts were formed *only* if the operation was done so that the posterior end of the OP extended back over one-half or more of the length of the fragment. If the OP did not extend beyond the midpoint of the posterior fragment (as in Fig. 7.11g), then no mouth parts were formed, and the fragment later underwent oral replacement (Fig. 7.11h) to produce a new OP that now stretched sufficiently far posteriorly to form mouth parts (Fig. 7.11i) (Eberhardt, 1962).

This result is inconsistent with a simple inductive influence emanating from some structure located at the posterior end, as such an influence should work over a fixed distance and not be responsive to changes made elsewhere. It is, however, exactly what would be expected on the basis of an actively regulating gradient, which might be expected to distribute itself proportionately over the available territory.

The remaining results of this same experiment are also consistent with the idea of a gradient. The long anterior portion of the transected *Spirostomum* went on to divide as it would have if the operation had not taken place. The posterior division product formed mouth parts despite the loss of its posterior end. These mouth parts were abnormally small (Fig. 7.11d) and were later replaced (Fig. 7.11e,f), probably because the operation was performed close to the time of determination of parts (the mouth parts were better developed after similar operations carried out at earlier stages) (Eberhardt, 1962). Tartar (1967b, p. 65) had also found that mouth parts could be formed after cutting off the tails of stentors bearing oral primordia. Although one could imagine that a posterior structure responsible for induction of mouth parts could regenerate very rapidly, in view of the other results of the same experiment it is more reasonable to conclude that the formation of mouth parts is not dependent on the presence of any unique "inducing" entity.

Intrinsic control in tetrahymenids

Formation of mouth parts in *Tetrahymena* and *Glaucoma* is almost certainly under intrinsic control. In these ciliates, oral primordia that complete their development in an inverted orientation nonetheless can form a normal cytostome, even when this is wedged against the anterior tip of the cell (Suhama, 1985; Nelsen et al., 1989b). Bipolar oral differentiation, as in Figure 7.10a, is never observed in these smaller ciliates; a cytostome, if it forms at all, is always produced in a more or less correct relationship to the internal geometry of the rest of the oral system no matter what the orientation of that system is in relation to the rest of the cell. Thus it appears to be determined entirely intrinsically.

This difference between tetrahymenids and heterotrichs is not surprising if one considers the different relations between membranelle band and mouth parts in the two kinds of ciliates. In tetrahymenids, all of the membranelles are located within the buccal cavity, whereas the UM is located at the rim of this cavity. The mouth parts of a tetrahymenid encompass the entire oral system, not just its posterior end as in the larger heterotrichs. Therefore, oral morphogenesis, in the strict sense of the complex maneuvers involved in hollowing out of a buccal cavity, is an operation that includes the entire oral system of tetrahymenids, while it works on a small part of that system in the multimembranellar ciliates such as *Stentor, Blepharisma, Spirostomum,* and *Oxytricha/Stylonychia.*

One reason for emphasizing this difference in control of the formation of mouth parts is that it is of immense importance in understanding both the origin and maintenance of reversals of asymmetry, as we will see in the next two chapters.

7.5 RELATIONSHIP OF ORGANELLAR SIZE TO CELL SIZE

One of the classical problems in regeneration of single cells, which engaged the interest of experimentalists, such as F. R. Lillie (1896) and T. H. Morgan (1901), was that of the lower size limit for successful regeneration. Tartar came up with a novel answer to this old question. He studied very small *Stentor* fragments, less than 1 percent of the volume of normal stentors, and observed that "Although these tiny stentors had much fewer than the usual number of membranelles, the width and length of these organelles when measured proved to be very nearly the same as in large animals, and these relatively oversized organelles caused the anterior end of the tiny animals to shake and shudder with their beating" (Tartar, 1961, p. 121). Therefore, he concluded "that a limit to the reconstitution of normal form is imposed by the fact that the units of ectoplasmic structure are each of a nearly constant size or incapable of 'miniaturization,' so that with decreasing volume there will come a point beyond which the formation of anything like a normal set of feeding organelles is impossible with such units" (Tartar, 1961, pp. 121–122).

A detailed analysis of the size of compound ciliary structures in miniaturized ciliates indicates that Tartar's conclusion is likely to be incorrect in detail but correct in principle. Where ultrastructural analysis has been carried out, it is clear that the length of each membranelle (Bakowska and Jerka-Dziadosz, 1980; Bakowska et al., 1982a) is *not* constant; small oxytrichid cells have shorter (but not narrower) membranelles than larger ones of the same species (compare Fig. 7.3a and 7.3b). Cirri also are smaller in miniaturized cells (Bakowska, 1980, 1981). Although the *size* of the individual ciliary units that make up each compound ciliary structure is almost certainly invariant, the *number* of such units is not.

The degree of diminution in the size of membranelles is not commensurate with the reduction of the size of the cell that possesses these membranelles. We have already seen that the number of ciliary rows formed by secondary proliferation within membranelles is resistant to change (see Section 7.2.4). In light-microscopic studies of starved *Tetrahymena,* the average area of the OA was reduced by about one-third in cells whose average total surface area was reduced by two-thirds (Bakowska et al., 1982a). Thus, as cells get smaller their OAs (probably produced by oral replacement) become proportionately larger, creating an impression of near constancy of oral size in all but the tiniest cells (Fig. 7.12). Thus, despite the lack of absolute constancy of size, Tartar's conclusion that the size of a regenerating fragment might be limited by the minimum size of units needed to make a functional organelle system could nonetheless be correct.

The lack of strict proportionality in sizes of organelle systems relative to the cells in which they are found, and also of different organelle systems relative to each

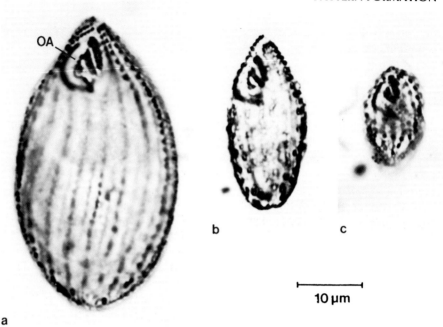

a

b   **c**

├────────────┤
10 μm

Figure 7.12 Relative size of the oral apparatus [OA] of *Tetrahymena thermophila* cells maintained under different nutritional conditions. The scale bar indicates 10 μm. (**a**) A silver-stained preparation of a cell grown in 1 percent proteose peptone medium, illustrating the typical size and proportions of well-fed cells. (**b**) and (**c**) Silver-stained cells fixed after 4 days of starvation in a buffered inorganic medium. (**b**) represents one of the larger cells of this sample, with a three-membranelle OA, and (**c**) shows one of the tiniest, with a two-membranelle OA. Magnification in (**b**) and (**c**) is the same as in (**a**). From Figures 17–19 of Bakowska et al. (1982a), with permission.

other (Bakowska, 1980), violates the simplest interpretation of the threshold model of diversification (see Section 1.4.2 and Fig. 1.6). If the limits of territories within a developing system were to depend on a reading of thresholds in a simple linear gradient, then the *relative* size of the parts should always be the same, much as the proportion of a French flag that is blue, white, and red is always the same irrespective of the size of the rectangle from which it is made (Wolpert, 1969, 1971). This "French flag" property is not strictly necessary in gradient-threshold systems (Meinhardt, 1982), but the presence of this property is a strong argument for a system operating in such a manner rather than being controlled by sequential interactions. We have already seen that in ciliates the "French flag" rule of proportionality does hold for the location of the fission zone (see Section 5.3.2) and for the position of the contractile vacuole pores around the circumference of *Tetrahymena* cells (see Section 6.3.3). The absence of such proportionality when the sizes of different compound ciliary structures are compared, or when the size of an organelle system is compared to that of the cell, is probably due to strong intrinsic components of the development of ciliary organelle systems, which by their nature fail to respond (or respond incompletely) to changes elsewhere in the cell. Thus, although positional information models might apply to certain global aspects of proliferation and positioning of cortical organelle systems, they do not apply to the control of the internal differentiation of these systems (for a more detailed version of this argument, see Frankel et al., 1984c).

7.6 CONCLUSIONS

One major and obvious conclusion is that the formation of the oral apparatus of ciliates is a complex process. The fact that a single cell has the capacity to build such a highly ordered supramolecular structure is itself amazing; it is no surprise that Lwoff (1950) viewed basal bodies of ciliates as virtual analogs of cells of multicellular organisms.

Although no aspect of the development of the oral apparatus of ciliates is thoroughly understood, the component processes have been sufficiently dissected by a combination of mutational, microsurgical, and biometrical analysis to indicate the presence of at least three hierarchically nested levels of control, based on the short-range versus long-range and intrinsic versus extrinsic dichotomies (Table 7.1). The first level is that of traditional assembly mechanisms operating over short distances. Such processes might conceivably carry the oral primordium as far as the formation of promembranelles, and in addition would constrain the sequence through which promembranelles mature to membranelles. The second level is that of longer range processes that still are intrinsic to the developing oral system. This type of control is seen most clearly in the spatial integration of membranelle organization; it corresponds to the classical embryologist's notion of a self-differentiating embryonic field, that is, one in which the integrative mechanism is contained within the field itself. The third level of control is that of pervasive influences, in the form of a basal–apical gradient and of a large-scale cellular asymmetry, that affect the arrangement and orientation of all cortical structures. The large-scale asymmetry is probably present in all ciliates, whereas the specification of mouth parts through a basal–apical gradient probably applies only to those ciliates in which the OA is a highly extended structure most of which is outside the buccal cavity.

The interplay of controls at these three levels creates certain constraints on the evolution of the organization of the oral apparatus. One of these is a constraint imposed by spatial coordination. Any change that brings about a major modification in one membranelle will probably affect the others as well. A second constraint results from the interplay of controls working at different levels of organization. For example, the invariant local asymmetry of ciliary units prevents a reversal of global asymmetry or polarity from bringing about a genuine mirror-reversal of oral organization. This may help to explain why no ciliates have true bilateral symmetry (Frankel, 1983). These constraints, however, do not give a full explanation of the surprising uniformity in detail of these structures in some groups of species and even genera with many differences in oral proteins (Williams and Bakowska, 1982; Williams, 1984). Stabilizing selection of an optimal design could account for this uniformity (Frankel, 1983). The constraint imposed by spatial coordination may, however, serve to reinforce the effect of stabilizing selection by tending to resist possibly adaptive changes in one feature that might have coordinated nonadaptive effects on another.

8

Reversals of Asymmetry

8.1 INTRODUCTION: STENTORS IN DILEMMAS

Much of the remainder of this book is focused on reversals of cortical asymmetry, considering first the nature and the propagation of established reversals (this chapter), then the way in which the reversals come into existence (Chapter 9), and finally a tentative model of intracellular positional information based largely on analyses of these reversals (Chapter 10). Reversals of large-scale asymmetry are especially illuminating because they create a spatial context in which developmental operations that normally produce functional structures instead generate defective ones. Analysis of these defects provides an opportunity for the investigator to tease apart the separate developmental strands that normally are intertwined flawlessly to bring about a normal end-result.

The cellular dilemma created by a pattern reversal is manifested strikingly in stentors with reversed contrast zones. The capacity of stentors to reorganize their cortical pattern successfully after an extraordinary variety of microsurgical deletions and disruptions—some of which have been illustrated in the preceding three chapters—makes those few situations in which this organism *cannot* regulate to a viable state especially significant. Recognizing this, Tartar gathered these exceptional cases into a short review titled "Stentors in dilemmas" (Tartar, 1966d). Normal reconstitution in the presence of a nucleus, a cortex, and a contrast zone was stymied in only two situations: (1) when large fusion masses formed oral primordia that failed to form mouth parts, presumably due to inability to generate a normal basal-apical gradient (see Section 7.4.2) and (2) when a spatially reversed contrast zone was created.

Tartar provoked stentors into constructing a reversed contrast zone by a rather bizarre operation. First, he excised the anterior and posterior ends of the cell, then he cut the cell across the middle and rotated the anterior half of the cell 180 degrees around the vertical axis, and finally he made a longitudinal cut and rotated the left half of the cell 180 degrees around the horizontal axis and rejoined it to the right half (Fig. 8.1b). Some cells responded to this drastic rearrangement by regenerating a reversed contrast zone: narrow stripes were now to the left of wide stripes (Fig. 8.1c) rather than to their right as in normal cells (Fig. 8.1a). An oral primordium (OP) appeared, as always, in the region of narrow stripes adjacent to the wide stripes (Fig. 8.1c). A buccal cavity and a cytostome developed, as expected, at the posterior end of the OP, and the entire new oral system moved anteriorly to its expected final site (Fig. 8.1d).

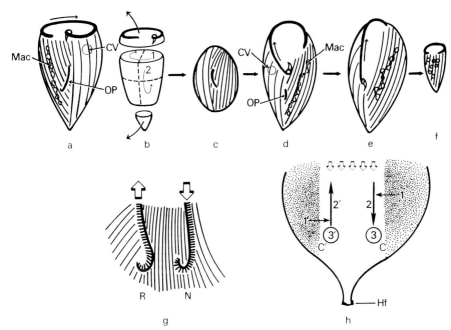

Figure 8.1 Formation of oral primordia in stentors with reversed contrast zones. (a–e) Development at a reversed contrast zone that is formed subsequent to "quartering" of a stentor. (a) A normal *Stentor* forming an oral primordium (OP) in a normal contrast zone. The macronucleus [Mac] is located underneath the narrow stripes, the contractile vacuole [CV] underneath the broad stripes. The membranelle band beats in the direction (arrow) of the oral funnel. (b) The operation (see text for details). (c) Subsequent formation of a reversed contrast zone and an oral primordium. (d) Completion of development. The locations of the macronucleus [Mac] and contractile vacuoles [CV] are reversed along with the pattern of striping. Oral membranelles beat away from the mouth *(arrow)*. A second oral primordium (OP) forms. (e) The completion of a repeated oral replacement, with membranelles again beating away from the mouth *(arrow)*. (f) A tiny incomplete stentor resulting from numerous unsuccessful rounds of oral replacement. (g) The geometry of oral primordia that are formed within a zone of narrow stripes with wide stripes on both sides [R, reverse, N, normal]. The *broad open arrows* indicate the direction of ciliary beat within the membranelle band. (h) A schematic diagram illustrating the geometry of oral primordia in adjacent normal and reversed contrast zones. The uniform polarity of the ciliary rows is indicated by the *small open downward-pointing arrows* at the top. The two contrast meridians (C and C') are shown as a border between a shaded and unshaded area. The *horizontal arrows*, 1 and 1', indicate the direction of asymmetry relative to the two contrast meridians; the *long vertical arrows*, 2 and 2', show the resulting polarity of the membranelle band; the *circles*, 3 and 3', indicate the location of the mouth parts, which always are formed at the end of the oral primordium that is closest to the holdfast [Hf]. For further explanation, see the text. (a–f) redrawn with slight modifications from Figure 13A of Tartar (1960), Figure 49 of Tartar (1961), and Figure 14 of Tartar (1966d); (g) drawn from Figure 24(f) of Uhlig (1960).

At this time, however, a flaw became apparent: the cilia of the membranelles, instead of beating posteriorly toward the mouth parts, beat anteriorly, *away* from the mouth parts. Hence the cell could not feed (the mouth parts themselves were also structurally incomplete). A new OP was, therefore, formed and replaced the defective oral structures (Fig. 8.1d,e). Unfortunately for the cell, the new structures were no less defective than those that they replaced. The cell repeated this useless performance as many as 12 times, gradually becoming smaller (Fig. 8.1f) and finally starving to death. It could not extricate itself from its geometrical dilemma.

This result confronts the experimenter with an interpretive problem. The easiest way out is to imagine that the 180-degree rotation of the left half of the cell around the horizontal axis resulted in a massive inversion of the entire fine-stripe zone, creating an inverted OP. Tartar, however, did not believe that any part of the contrast zone was physically inverted, and instead suggested that a relational factor was at work: "during its formation the direction of polarization of the membranellar band is determined by the sense of the adjacent striping, and not by the polarity of the stripes [within which the band forms]" (Tartar, 1966d, p. 129).

The correctness of Tartar's interpretation cannot readily be appreciated from the rather complex operation shown in Figure 8.1b. Uhlig (1960), however, independently obtained results similar to Tartar's from geometrically less complex experiments. The result pertinent to the issue at hand is shown in Figure 8.1g, freely copied from a photograph (Fig. 24f) in Uhlig's 1960 monograph. A wedge of narrow pigment stripes had been deflected so that it had broad stripes on both sides. There could thus be little doubt that the intrinsic polarity of the ciliary rows was the same throughout this wedge of closely spaced ciliary rows. Oral primordia appeared on both sides of the wedge of narrow stripes. The OP that was formed on the cell's left (viewer's right) side of the wedge was completely normal, whereas the OP formed at the cell's right (viewer's left) side of the wedge was structurally a mirror-image of the normal OP; this can be seen in Uhlig's photograph by the location of the "boundary stripe" (rendered as a heavy longitudinal line in Fig. 8.1g) on the fine-stripe side of both oral primordia. This observation vindicated Tartar's conclusion and showed that the asymmetry as well as the polarity of the developing membranelle band is determined by its spatial relation to the meridian of contrast ("sense of the adjacent striping") and not by the polarity of the ciliary rows within which it was formed.

Figure 8.1h pictorially summarizes the Tartar–Uhlig interpretation of asymmetry reversals in *Stentor*. The interpretation involves three logical steps, numbered in Figure 8.1h. In step 1, a directional influence emanating from (or extending through) the contrast meridian determines the asymmetry of the membranelle band. A reversed contrast meridian causes the membranelle band to have reversed asymmetry. In step 2, which can be thought of as concurrent and coordinated with step 1, the polarity of the membranelles adjusts itself to be in harmony with their asymmetry. If an imaginary observer stands at the position of the numerals 1 and 1′ facing the asymmetry arrowheads, the polarity arrowheads (2 or 2′) always point to that observer's left. Because the polarity of the membranelles depends on their asymmetry and on nothing else, at the reversed contrast zone (C′) inverted membranelles will develop within a region of normally oriented ciliary rows [recall the converse demonstration that normally oriented membranelles can develop next to an inverted ciliary row (see Section 7.2.2)]. Finally (3 and 3′) the mouth parts (circles) develop at the end of the OP nearest the tail pole of the cell, ignoring the polarity of the membranelles. The irremediably maladaptive nature of the oral apparatus (OA) that develops in the reversed contrast zone, thus, allows us to begin to decipher those rules that constrain development even in the most regulative of ciliates.

The *Stentor* reversal, illuminating as it is, leaves many open questions that have subsequently been at least partially resolved with other ciliates. Viable clones of ciliates with large-scale pattern reversals were maintained either by propagation of mirror-image doublets in which the nonfeeding reversed component is sustained by a complementary normal component, or by growth of reverse singlets. The doublet strategy has been extensively pursued in oxytrichids, and viable reverse singlets have been obtained and analyzed in *Tetrahymena* and *Glaucoma*. In both of these groups cytological observations have confirmed the essence of the Tartar–Uhlig analysis and have filled in many missing details.

8.2 MIRROR-IMAGE DOUBLETS IN OXYTRICHIDS

8.2.1 Cortical Organization

The most spectacular of all mirror-image doublets are found among the oxytrichids. Such doublets were observed first by Tchang et al. (1964) and shortly thereafter by Totwen-Nowakowska (1965). The remarkable stability of these doublets was noted from the very start. In the title of their initial report, Tchang et al. (1964) proclaimed that this "induced monster ciliate" was "transmitted through three hundred and more generations." Since then, such doublets (classified as type III doublets of Fig. 4.10) have been induced in at least four ciliate genera by researchers in six laboratories. The species that I focus on in this description is the one in which the original mirror-image doublets were obtained, *Stylonychia mytilus*. The principal sources for this description are Tchang and Pang (1981), Shi and Frankel (1989a), Grimes et al. (1980), Grimes and Goldsmith-Spoegler (1990a), and Tuffrau and Totwen-Nowakowska (1988), with additional ultrastructural information derived from *Paraurostyla weissei* mirror-image doublets analyzed by Jerka-Dziadosz (1983, 1985).

The mirror-image doublets (originally called "jumelles" by the Chinese workers) are always joined side-by-side, with the two sets of ventral ciliature on one plane and the two dorsal sets on the opposite plane. The left component is normal, the right component is mirror-reversed. The topology of the mirror-image doublets thus resembles that of one's two hands held side by side, with both palms facing in the same direction.

The ventral and dorsal surfaces of singlets and mirror-image doublets are contrasted in Figure 8.2 (photographs kindly supplied by Dr. Xinbai Shi). Recall that we use the same left–right conventions for description of these flattened cells as we would for examining a person. Thus, for the ventral surface the observer's left corresponds to the cell's right, and for the dorsal surface, the observer's left corresponds to the cell's left.

The general arrangement of ciliary structures of the left component (Fig. 8.2b,d) corresponds to that of a normal singlet cell (Fig. 8.2a,c), whereas the general arrangement of the structures of the right component forms a mirror-image of that in the normal singlet cell. This reversal is almost perfect for the oral structures, where the membranelle band and the undulating membrane together form a lopsided "V" that is normal in the left component and mirror-reversed in the right

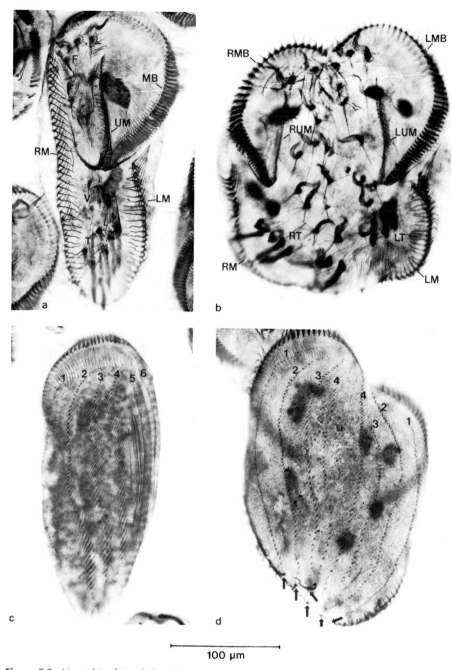

Figure 8.2 Ventral (**a**, **b**) and dorsal (**c**, **d**) surfaces of normal (**a**, **c**) and mirror-image doublet (**b, d**) cells of *Stylonychia mytilus*. The ventral structures labeled in (**a**) and (**b**) are the membranelle band [MB], undulating membrane [UM], frontal [F], ventral [V], transverse [T], right marginal [RM], and left marginal [LM] cirri. In the mirror-image doublet (**b**), the two sets of ciliary structures are marked according to location, the prefix L referring to the left component, R referring to the right component. The dorsal structures labeled in (**c**) and (**d**) are the ciliary rows 1, 2, 3, 4, 5, and 6. The caudal cirri are indicated by arrows in (**d**). Scale bar, 100 μm. Previously unpublished photographs, kindly supplied by Dr. Xinbai Shi and used with permission.

component (Fig. 8.2b). The mirror-image arrangement is less obvious for the cirri of the ventral surface, where the total number is less than double that of the normal set (Tchang and Pang, 1981) and the arrangement is much less regular than is normally observed. One can, however, usually notice a rough mirror-symmetry in the arrangement of transverse cirri: in normal cells these form an array that slopes from the cell's posterior-right to the anterior-left (Fig. 8.2a, T), and in the posterior region of mirror-image doublets there is generally a group of cirri sloping from the posterior-right to the anterior-left in the left component (Fig. 8.2b, LT), and another group of cirri sloping in the reverse direction in the right component (Fig. 8.2b, RT).

The arrangement of marginal cirri is mirror-imaged in a more subtle manner. A normal cell has a row of marginal cirri on its left and right borders, designated the left marginal cirral row (LM) and right marginal cirral row (RM), respectively (Fig. 8.2a). The cirri of these two sets are similar but not identical in organization (Grim, 1972; Shi and Frankel, 1989a). In the mirror-image doublet, there are no marginal cirri visible along the line of union of the two components, leaving only one marginal cirral row at each edge of the composite cell. The row located at the left edge of the left component (LM) has typical left marginal cirri. The row located at the right edge of the right component, labeled a right marginal cirral row (RM) because of its topographical position in the right component, is actually made up of the same type of cirri that normally are found at the *left* margin of the left component (Shi and Frankel, 1989a).

On the dorsal surface, most mirror-image doublets of *Stylonychia mytilus* possess dorsal ciliary rows 1 to 4 in the form of two arrays in rough mirror-image arrangement (Fig. 8.2d), with caudal cirri produced at the posterior ends of rows 1, 2, and 4 in both sets (Fig. 8.2d, arrows) (Grimes and Goldsmith-Spoegler, 1990a; Shi and Frankel, 1989a). The short dorsal rows 5 and 6 are absent from *both* components of most mirror-image doublets in *Stylonychia;* however, these rows are present in corresponding mirror-image doublets of *Paraurostyla weissei* (Jerka-Dziadosz, personal communication).

In summary, one can describe the *arrangement* of cell-surface structures of the two halves of these cells as being mirror-images around a plane of symmetry in the center of the doublet, with the only imperfections being the irregular arrangement of the cirri that cover the ventral surface plus the frequent absence of structures normally produced near the plane of symmetry.

The impression of a mirror image vanishes as soon as one looks closely at the internal organization of the individual ciliary elements. All of these are normally organized (Grimes et al., 1980; Jerka-Dziadosz, 1983; Shi and Frankel, 1989a). The probable reason for this is that ciliary structures on both sides of the mirror-plane are constructed of basal bodies of normal asymmetry, making formation of a ciliary unit of reversed asymmetry impossible (Jerka-Dziadosz, 1983).

As with units of ciliary rows considered earlier in this book (see Section 4.2.2), however, an internally normal ciliary structure may be oriented in various ways relative to the cellular axes. In the right component of mirror-image *Stylonychia* doublets, all cirri are normally oriented, whereas the membranelles and the undulating membranes are inverted (Fig. 8.3) (Grimes et al., 1980; Grimes and Goldsmith-Spoegler, 1990a; Shi and Frankel, 1989a) (recall that an inversion is a 180-

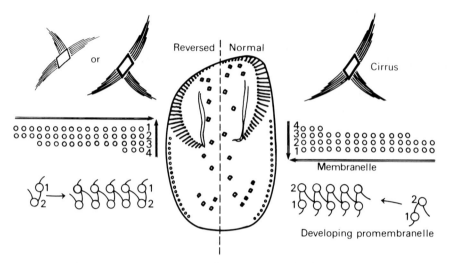

Figure 8.3 A diagrammatic representation of a mirror-image doublet of *Stylonychia mytilus* and of selected ciliary structures of such a doublet. The arrangement of structures on the ventral surface is indicated in the central panel, with the plane of bilateral symmetry separating the normal from the reversed component represented by the *vertical dashed line*. On either side, the orientations of cirri (top), mature membranelles (center), and developing promembranelles (bottom) of the normal (right in diagram) and reversed (left in diagram) portions of the doublet are shown. In the reversed component, membranelles are always inverted, whereas cirri are usually normal but occasionally (in *Paraurostyla*) inverted. For further explanation, see the text. Redrawn with modifications, in part from Grimes et al., 1980, in part from Grimes, 1989, with permission.

degree rotational permutation of a normal structure and thus is not a mirror-image of the normal structure). In *Paraurostyla* mirror-image doublets, some of the cirri at the right margin also are inverted (Jerka-Dziadosz, 1985).

Given these observations, it is not surprising that only the left component of a mirror-image doublet can feed (Tchang and Pang, 1977). The structural inversion of membranelles in the right component of mirror-image doublets of *Stylonychia* leads to a reversal in direction of ciliary beat (Shi, personal communication), just as in membranelle bands formed in reversed contrast zones in *Stentor*. In addition, the reversed OA of *Stylonychia*, again like that of *Stentor*, contains defective mouth parts at its posterior end (Tchang and Pang, 1981; Shi and Frankel, 1989a). Thus, in both *Stylonychia* and *Stentor* the OAs produced in regions of reversed large-scale asymmetry are mirror-reversed in the large-scale arrangement of ciliary elements, 180-degree rotated in the organization of individual ciliary structures, and defective in the morphogenesis of accessory buccal elements. Shi (personal communication) regarded this syndrome as a consequence of "development meeting with obstruction," the nature of which we consider later.

8.2.2 Cortical Development

The mirror-image arrangement of mature ciliary structures in mirror-image doublets is a consequence of a prior mirror-image arrangement of ciliary primordia. Two striking examples suffice to illustrate this point: First, although in normal singlet cells the oral primordium (OP) initially arises adjacent to the left-most transverse cirrus (see Section 3.4 and Fig. 3.9a), in mirror-image doublets the left OP (Fig. 8.4a, LOP) arises near the left-most transverse cirrus of the left (normal) component and the right OP (ROP) forms next to the *right*-most transverse cirrus of the right (reversed) component (Fig. 8.4a) (Grimes and Goldsmith-Spoegler, 1990a; Shi and Frankel, 1989a; Pang, personal communication). Second, the two sets of frontal-ventral-transverse (F-V-T) primordia (LFVTP, RFVTP) that are destined to form the new frontal, ventral, and transverse cirri also are arranged in a clear mirror-image configuration (Fig. 8.4b), suggesting that the obscuring of symmetry in the arrangement of mature cirri is a secondary phenomenon that occurs at a late stage of development, perhaps during migration of nascent cirri.

How do ciliary structures that are *not* mirror images of the normal ones develop within ciliary primordia that are arranged in a mirror-image reversal of the normal pattern? One way in which this might come about in the OP is illustrated in Figure 8.3, in which two-rowed promembranelles are shown assembling on both sides of the mirror-image doublet by the usual sequential addition of paired ciliary units (see Section 7.2.1). The units are added from cell's right to left on the normal, left side of the mirror-image doublet (Fig. 8.3, right), and from cell's left to right on the

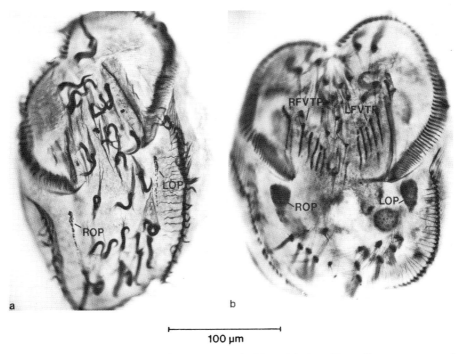

100 μm

Figure 8.4 The ventral surfaces of mirror-image doublets of *Stylonychia mytilus* at two stages of cortical development. (a) An early stage, fixed when the left oral primordium [LOP] and right oral primordium [ROP] are migrating anteriorly from sites of origin at the left-most and right-most transverse cirri respectively. (b) A later stage, in which the two oral primordia are now located immediately posterior to the two respective OAs, and the two FVT primordia (LFVTP, RFVTP) are present as ciliary streaks. Scale bar, 100 μm. Previously unpublished photographs kindly supplied by Dr. Xinbai Shi and used with permission.

reversed, right side of the mirror-image doublet (Fig. 8.3, left). Thus the sequence of addition of ciliary units to the growing promembranelle is spatially reversed on the two sides of the doublet. The ciliary units themselves, however, are of identical asymmetry on both sides. To accommodate the reversal of spatial order of addition of units on the right side of the mirror-image doublet, the ciliary units must assemble in an orientation that is a 180-degree rotational permutation of that on the non-reversed left side of the doublet. This scheme, proposed by Grimes (1989), fully reconciles the reversal of large-scale asymmetry and the normalcy of local asymmetry, and shows how one consequence of this reconciliation is that the membranelles on the reversed side *must* become inverted.

There is some difference of opinion concerning the more global details of oral development on the reversed right side of mirror-image doublets. Grimes (1989) reports that although *each* membranelle develops in an inverted manner as shown in Figure 8.3, the order of formation of membranelles is from anterior to posterior in both components of the mirror-image doublet. Because each membranelle differentiates from the cell midline to the periphery as shown in Figure 8.3, in Grimes' view the development of the OP of the right component is a gross mirror image of that of the left OP. Shi and Frankel (1989a), however, have deduced something rather different and more complicated. Shi observed that the arrangement of nascent membranelles at early stages in development of the OP of the reversed right component is not a simple mirror image of that of the OP of the normal left component, and concluded that the anteroposterior order of formation of separate membranelles is inverted, so that the earliest assembled membranelle becomes the most posterior one (Shi and Frankel, 1989a).

The mouth parts (buccal cavity and cytostome) invariably form at the cell's posterior end of the OA, presumably in response to an extrinsic signal as in *Stentor* and *Spirostomum* (see Section 7.4.2). In the normal (left) OA this corresponds to the morphological posterior ends of the membranelles and UMs; in the reversed (right) OA it corresponds to the intrinsic anterior end of the UM and of each individual membranelle. As the mouth parts have major fibrillar components, and many of these fibrillar components are known to develop in specific association with particular ciliary structures, it is not surprising that when a cellular signal directs mouth parts to form at the "wrong" ends of ciliary structures, the resulting mouth parts cannot be normal. This is a particularly clear example of how a reversal of large-scale asymmetry causes development to meet with a major obstruction.

8.2.3 Separation of Normal and Reversed Components

As mirror-image doublets typically possess two bilobed macronuclei, one in each component (Tchang and Pang, 1977), they can be cut apart along the mirror plane (Fig. 8.5a, vertical line) to produce two nucleated components, the left one of normal (Fig. 8.5c) and the right one of reversed (Fig. 8.5b) large-scale asymmetry. As first shown by Tchang and Pang (1977), after this was done the normal left component could feed, divide, and produce a clone of normal singlet cells, whereas the reversed right component could not feed and eventually starved to death. Subsequent detailed analysis of microsurgically separated components by Grimes and Goldsmith-Spoegler (1990b) and by Shi (Shi and Frankel, 1989b) indicated that the

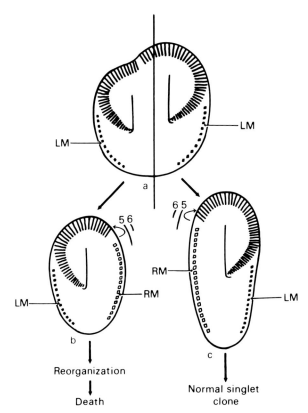

Figure 8.5   Derivation of singlets (**b**, **c**) by transection of a mirror-image doublet of *Stylonychia mytilus* (**a**) along the plane of symmetry [solid vertical line in (**a**)]. Only oral ciliature and marginal cirri are shown, with the left marginal cirri [LM] filled in and the right marginal cirri [RM] open (the cirri are here labeled according to organization rather than position; this differs from the convention used in Fig. 8.2). The site and direction of formation of the fifth and sixth dorsal ciliary rows in the derived singlets is also indicated. Redrawn with slight modifications from an unpublished drawing provided by Dr. Xinbai Shi, with permission.

reversed right component could nonetheless carry out ciliary primordium development both by regeneration and reorganization (spontaneous oral and cirral replacement) and in one case even divided three times before succumbing (Grimes and Goldsmith-Spoegler, 1990b). A surgically generated "reverse singlet" of *Stylonychia* is thus comparable to the reversed singlet stentors that try, but never succeed, in extricating themselves from their morphogenetic dilemma.

One feature of the microsurgical derivation of singlets from mirror-image doublets is of special interest. Recall that in mirror-image doublets the structures that ought to be found at the right margin of the left component and the left margin of the right component—the right marginal cirri and the short dorsal ciliary rows (rows 5 and 6) normally produced by these cirri—generally are missing from the central region of the mirror-image doublet. As soon as the fragments produced by a longitudinal transection of the mirror-image doublet underwent cortical development, however, these structures were formed (Fig. 8.5b,c) (Grimes and Goldsmith-Spoegler, 1990b; Shi and Frankel, 1989b). The left component thereby became a complete normal singlet. The reversed right component, although doomed to eventual death by starvation, nonetheless first managed to reconstitute a complete set of cortical structures; these included a complete set of F-V-T cirri and marginal cirri at the new cell margin created at what previously had been the mirror-plane of the mirror-image doublet. Not surprisingly, although the new margin was topographically a left margin, the cirri that were formed there were structurally and morphogenetically *right* marginal cirri, indicating a complete reversal of asymmetry in arrangement of all of the ciliary structures (Shi and Frankel, 1989b).

This completion of ciliary patterns by these derived singlets tells us something interesting about the original mirror-image doublets. Such doublets are structurally incomplete at and around the mirror-plane. One may think of this incompleteness as being a simple consequence of the absence of a physical cell margin at the mirror-plane, but this does not explain the irregular and incompletely developed F-V-T cirral patterns on either side of that plane. More likely, there is some genuine interaction taking place across the plane of symmetry that prevents normal development of structures that would otherwise form in that region. Such interaction could either prevent development of ciliary primordia or cause regression of marginal cirral primordia that sometimes do develop in this region (Grimes and Goldsmith-Spoegler, 1990a). We return to the possible nature of this interaction in Chapter 10.

Mirror-image doublets normally are stable as long as they retain their global symmetry. Cases have been reported, however, in which mirror-image doublets may spontaneously "revert" to singlets. One method of reversion appears to be through a spontaneous self-transection, in which a mirror-image doublet reorganizes itself to produce marginal cirral rows at the plane of symmetry (Shi and He, personal communication), after which the two components may pull apart or divide to produce two separated singlets, one normal and viable and the other reversed and nonviable (Pang et al., 1984; Shi and He, personal communication). Another sequence of return from the mirror-image doublet to the singlet condition involves resorption of the reversed right component (Pang et al., 1984); the way in which this occurs has not been described in detail.

8.3 MIRROR-IMAGE FORMS AND REVERSE SINGLETS IN TETRAHYMENIDS

8.3.1 Organization of Mirror-Image Forms

janus *cells in* Tetrahymena

The first mirror-image forms to be seen in *Tetrahymena* were a phenotypic consequence of a gene mutation called *janus* (*jan;* now *janA*). After the orginal *janA* mutation was genetically characterized (Frankel and Jenkins, 1979) and its phenotype described (Jerka-Dziadosz and Frankel, 1979; Frankel and Nelsen, 1981), two additional *janus* mutations, *janB* and *janC,* were isolated and examined (Frankel et al., 1987; Cole et al., 1988; Frankel and Jenkins, unpublished observations).

These mutations are nonallelic, yet are phenotypically similar, although *janB* differs somewhat from the others by being semidominant and temperature sensitive.

To provide an efficient general description of the large-scale cortical organization that is altered by *janus* mutations, I employ a polar representation of cortical geometry, which I introduce with wild-type cells. The polar projections are maps of the *entire* surface, as can be seen by comparing ventral views (Fig. 8.6, top row) to polar projections (Fig. 8.6, bottom row). For dividing cells, the fission zone is shown as a ring halfway between the center and the periphery (Fig. 8.6b), like the equator in a polar projection of the earth. The inclusion of both ends of the cell on the same polar map allows visualization of spatial relationships of the two major markers of circumferential geometry, the anterior oral apparatus (OA) and the posterior contractile vacuole pores (CVP).

In Figure 8.7, we compare a polar map of the cytogeometry of dividing *janus* tetrahymenas to that of wild-type homopolar doublets produced by a parallel parabiotic (Siamese twin) fusion of two normal cells. If viewed very superficially, the *janus* cells (Fig. 8.7b) appear rather like the parabiotic doublets shown in Figure 8.7a, as they both have oral structures located opposite one another. However,

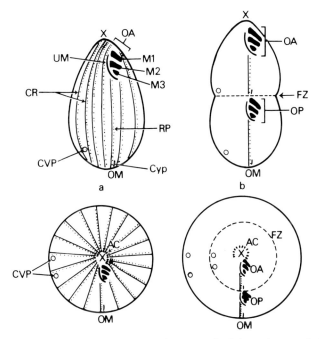

Figure 8.6 Construction of polar maps of *Tetrahymena* cells. Schematic ventral views are shown in the top row, corresponding polar projections in the bottom row. An X marks the location of the anterior end of the cell. (**a**) A nondividing normal singlet, showing all ciliary rows [CR] with adjacent longitudinal microtubule bands, the oral area [OA] with an undulating membrane [UM] and three membranelles [M1, M2, M3], the contractile vacuole pores [CVP], and the cytoproct (Cyp). The location of the right postoral ciliary row [RP] marks the oral meridian [OM]. The apical crown (AC) is evident only in the polar projection. (**b**) A normal singlet that is beginning to divide, with the fission zone [FZ] indicated as a *dashed line,* and an oral primordium [OP] just posterior to it. In this pair of diagrams, only the right postoral ciliary row, marking the OM, is shown.

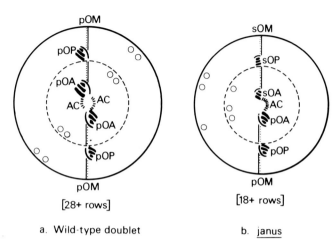

a. Wild-type doublet   b. janus

Figure 8.7 Schematic polar projections of dividing wild-type parabiotic doublet (**a**) and *janus* (**b**) cells of *Tetrahymena thermophila*. The normal (primary) oral structures and oral meridians of the parabiotic doublet (**a**) are prefixed with a "p" [pOA, pOP, and pOM]. The *janus* cell (**b**) possesses both normal primary oral structures [pOA, pOP] that define a normal primary oral meridian [pOM] and abnormal secondary oral structures [sOA, sOP] that define a secondary oral meridian [sOM]. The arrangement of CVPs shown in the map of the *janus* cell is one of a variety actually observed. The basal-body pairs that make up the apical crown (AC) are asymmetrically arranged in the parabiotic doublet, and symmetrically arranged in the *janus* cell.

there are two major differences. One is in the location of the CVP sets: in parabiotic doublets CVP sets are located to the cell's right of both OAs, in *janus* cells one CVP set is to the right of one oral meridian whereas the other CVP set is to the *left* of the opposite oral meridian. Thus, in *janus* cells both CVP sets are within the same hemisphere of the cell, in sharp distinction to parabiotic doublets that have one CVP set in each hemisphere. A second critical difference between *janus* cells and typical parabiotic doublets is that in *janus* cells the oral structures formed along the two opposite oral meridians are not equivalent. Although parabiotic doublets maintain two sets of normal OAs, both designated *primary* OAs (pOAs, Fig. 8.7a), *janus* cells produce normal or nearly normal oral structures (pOAs) along one oral meridian (the primary oral meridian or pOM), and form abnormal oral structures (secondary OAs or sOAs) along the other oral meridian (the secondary oral meridian or sOM) (Frankel et al., 1984a; Frankel and Nelsen, unpublished).

The organization of the sOA of *janus* cells is extremely variable; it may include membranelles with normal or reversed orientation, or sometimes some of each. The undulating membrane is absent, or when present is small and variably arranged relative to the membranelles. The buccal cavity is often rudimentary and sometimes absent. *janus* cells use their primary but not their secondary OAs for taking in particulate food (Jerka-Dziadosz and Frankel, 1979).

The two major manifestations of the *janus* phenotype are geometrically coordinated: the pOM is associated with a CVP set to its right and the sOM is associated with a CVP set to its left (Fig. 8.7b). Thus, although balanced parabiotic doublets exhibit perfect twofold rotational symmetry, *janus* cells manifest a clear, although imperfect, bilateral symmetry.

The local asymmetry of the individual ciliary rows is totally unaffected by the *janus* mutations (Jerka-Dziadosz and Frankel, 1979; Frankel et al., 1984a). The "fine-positioning" of all structures that develop in close spatial relationship to the ciliary rows is similarly unaffected; oral primordia develop immediately to the left of the stomatogenic ciliary rows, and CVPs form to the posterior-left of basal bodies. This normal local asymmetry is not surprising in view of the fact that ciliary units are ultrastructurally normal in *janA* cells, even within the sOA (Jerka-Dziadosz, 1981d).

Although ciliary rows themselves are geometrically normal, the spatial arrangement of specialized ciliary structures that are formed in some ciliary rows but not in others is mirror-symmetrical in the *janus* cells. This is particularly evident for the arrangement of basal-body pairs at the anterior end of the cell. In normal *Tetrahymena* cells, paired ciliary units are found at the anterior ends of ciliary row 5 to row *n*-2, and together make up an asymmetrical apical crown (see Fig. 3.1b). In *janus* cells, the apical crown (Fig. 8.7, AC) is symmetrical, because the arrangement of basal-body pairs is reversed around the sOA, i.e., although basal-body pairs appear at the anterior ends of ciliary rows close to the *left* side of the pOA, they develop close to the *right* side of the sOA (Fig. 8.7b) (Frankel et al., 1984a). Thus, as in mirror-image doublets of *Stylonychia*, *janus* tetrahymenas can express a mirror-reversed arrangement of ciliary structures that are of normal internal organization.

A mirror-image global body plan is consistent with a wide variety of possible relative locations of the primary and secondary oral meridians (Fig. 8.8). The two oral meridians may or may not be directly opposite one another. If they are not opposite, the cell-sector bearing CVPs may be either narrower (Fig. 8.8a) or broader (Fig. 8.8c) than the sector that lacks CVPs. In *janA* cells, the sector bearing the CVPs typically is slightly narrower than the opposite sector (Jerka-Dziadosz and Frankel, 1979; Frankel and Nelsen, 1981). This spatial relation is similar in *janC*, and is not affected by temperature differences in either of these mutations (Frankel, unpublished observations). *janB* cells, however, tend to form sOMs at relative

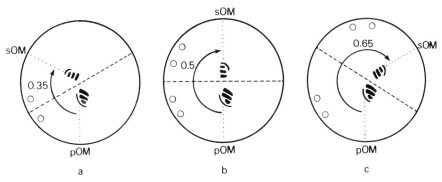

Figure 8.8 Polar projections of variations of cell-surface organization found in both *janus* mutant cells and *janus* phenocopies in *T. thermophila*. The *dashed lines* represent the planes of symmetry, and the decimal numerals indicate proportion of the cell circumference occupied by the sector between the primary and secondary oral meridians that contains the CVPs. Abbreviations are the same as in Figure 8.7.

positions greater than 0.5, with a shift to the right (clockwise in Fig. 8.8) during continued exposure to a nonpermissive temperature. Cells doubly homozygous for *janA* and another pattern mutant called *broadened cortical domains (bcd)* (Cole et al., 1988) also have arrangements approaching those shown in Figure 8.8c (Cole et al., 1988). Under all known circumstances, the average midpoint of *janus* CVP locations is close to halfway between the oral meridians (Frankel and Nelsen, 1981, 1986b; Cole et al., 1988). Thus, *janus* cells really are bilaterally symmetrical in their global organization.

The conclusion that *janus* cells have a bilaterally symmetrical organization must, however, be qualified in one crucial respect. Although OAs invariably are present along the pOM, oral development along the sOM is sporadic. There are many "incomplete" *janus* cells, with two widely spread CVP sets but no secondary oral structures. The proportion of secondary oral structures observed in *janus* cells varies from near zero to close to 100 percent, depending on the culture medium (higher in richer medium, for *janA* at least), the growth phase (highest in exponential phase, very low in stationary phase), the temperature (higher at elevated temperatures, for *janB*), the presence or absence of an enhancer gene (higher if the enhancer is present together with *janA;* also higher in the presence of *bcd*), and the number of ciliary rows (higher the greater the number of rows, at least for *janA*).

Variability in manifestation of secondary oral structures sharply distinguishes *janus* mutant cells from ordinary parabiotic doublets. In parabiotic doublets of the type shown in Figure 8.7a, both OAs are faithfully propagated until one of them is lost permanently, with (sometimes) a brief transitional period of unstable expression. In *janus* cells, the instability is permanent. One might imagine that this incomplete expression of secondary OAs is a consequence of "development meeting with obstruction," with the extreme situation being one in which an sOA cannot get started properly despite the presence of the correct signal delivered at the correct time and place. In the absence of markers that could detect the location at which the signal is delivered independently from the response to that signal, one cannot rule out the alternative possibility that the *janus* condition itself might be unstable, switching from an asymmetric large-scale organization to a mirror-image organization and back again.

Phenocopies of janus in Tetrahymena

Close imitations of the *janus* configuration have been obtained in *Tetrahymena* without prior mutagenesis or fertilization. These *janus*-like phenocopies, found in studies on the regulation of parabiotic doublets (see Section 9.2.2), displayed a gamut of patterns ranging from that depicted in Figure 8.8a to that shown in Figure 8.8c (Frankel and Nelsen, 1986a; Nelsen and Frankel, 1989). They also exhibited variable expression of secondary oral structures. The *janus*-like phenocopies differed from true *janus* cells in that they were transient and always had a number of ciliary rows that was well above the normal stability range of 18 to 21 (see Section 4.2.1).

The second difference provides the reason why I have described the *janus* mutants as mirror-image *cells* rather than mirror-image *doublets*. In oxytrichids, every part of the cell surface is marked by regionally distinctive structures, so that one can know exactly what structures are lost at the symmetry line when a mirror-image doublet is formed. In *Tetrahymena,* where ciliary rows are basically all alike, one cannot tell how much of the large-scale pattern is lost in a mirror-image cell. One and the same pattern could be interpreted either as a singlet in which the dorsal half of the cell has been lost and replaced by a ventral surface of reversed asymmetry, or as an entire *Tetrahymena* circumference that has been mirror-duplicated. In the former case, one would have a double half-cell, in the latter a double whole-cell. *janus* mutant cells commonly have numbers of ciliary rows similar to those of wild-type cells, suggesting that they might be double half-cells. The *janus*-like phenocopies typically have a number of ciliary rows comparable to that of parabiotic doublets and thus come close to being double whole-cells. In Chapter 9, I present evidence concerning the origin of these forms that supplements this anatomical argument.

Stable mirror-image doublets of Glaucoma

A stable mirror-image doublet clone of *Glaucoma scintillans* that originated in 1978 during laboratory cultivation of a strain isolated from nature has been maintained for the last decade by Suhama (1982, 1983, 1984). *Glaucoma* is a close relative of *Tetrahymena,* resembling it in most respects. Several small differences are indicated in the polar map shown in Figure 8.9a. The total number of ciliary rows is greater in *Glaucoma* (about 34), as is the number of postoral ciliary rows (4 to 6). There is usually only one CVP, located somewhat further to the right of the oral meridian than in *Tetrahymena.* Finally, the dividing micronucleus of *Glaucoma* lines up longitudinally directly underneath the cell surface, about two ciliary rows to the right of the CVP (Suhama, 1983); in *Tetrahymena,* the dividing micronucleus also aligns itself longitudinally underneath the cell surface, but at no preferred meridian (Frankel, unpublished observations).

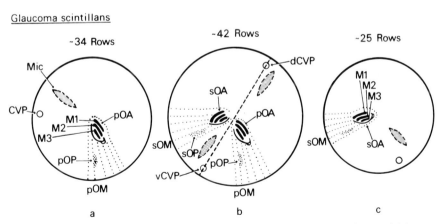

Figure 8.9 Polar projections of **(a)** normal singlets, **(b)** mirror-image doublets, and **(c)** reverse singlets in *Glaucoma scintillans*. Conventions and abbreviations are the same as in Figures 8.7 and 8.8, with the three membranelles [M1, M2, and M3] of the OA labeled in **(a)** and **(c)**. In these diagrams, all postoral ciliary rows plus one additional row on each side of the OA are shown, as well as the dividing micronucleus [Mic] at a typical position. In mirror-image doublets CVPs may be found at one of two locations, ventral [vCVP] or dorsal [dCVP]. Drawn from information in Suhama (1983, 1985).

Mirror-image *Glaucoma* doublets have a number of ciliary rows that is variable but always higher than that of singlets. They possess two oral meridians, with OAs reliably formed at both meridians in the clone maintained by selection of "perfect doublets" (Suhama, 1982). At one of them, labeled pOM, normal primary OAs (pOAs) are formed from oral primordia produced along the right postoral ciliary row. At the other, labeled sOM, abnormal secondary OAs (sOAs) are formed from oral primordia produced along the *left* postoral ciliary row. The sOM is located only about 25 percent of the cell circumference to the right of the pOM. CVPs are present approximately midway between the two oral meridians at one or both of two diametrically opposite positions: either in the center of the short "ventral" sector to the right of the pOM and the left of the sOM (Fig. 8.9b, vCVP), and/or in the middle of the long "dorsal" sector (dCVP) (Suhama, 1984). Micronuclei divide close to these CVP meridians, more frequently to their right than to their left. Micronuclei, however, may be present without CVPs, or CVPs may be present without micronuclei, at either the mid-ventral or mid-dorsal meridian (Suhama, 1983).

As in all other mirror-image ciliates, the local aspects of geometry are *not* mirror-imaged. The membranelles of the sOA are either 180-degree rotational permutations of normal membranelles, or are disorganized and fragmented (Suhama, 1982). The oral primordium is always formed to the *left* of a stomatogenic ciliary row, reflecting the invariant local asymmetry of the ciliary rows. As in *janus Tetrahymena*, although the geometry of ciliary rows remains normal, the arrangement and curvature of ciliary rows around the OA and the anterior pole are mirror imaged on the two sides of the cell (Suhama, 1982), making it unmistakably clear that a large-scale mirror-image organization is being superimposed on the unchanging and probably unchangeable local asymmetry of the ciliary rows. The plane of bilateral symmetry in this case coincides with the two meridians along which CVPs can be formed.

The mirror-image doublets of *Glaucoma* resemble the *janus* forms of *Tetrahymena* in all essential features except one. The one major difference is in the arrangement of CVPs relative to the OA. In *janus* tetrahymenas, CVPs are always located to the right of the pOM and to the left of the sOM. If one took the *janus* form of Figure 8.8a and brought the sOM and pOM closer together, one would obtain the basic geometry of the *Glaucoma* mirror-image doublets, *except* for the absence of the mid-dorsal CVP location. In Section 9.2.2 I show how both kinds of CVP arrangement can be generated by the operation of a single rule.

Do the mirror-image *Glaucoma* doublets, like *janus Tetrahymena,* arise from expression of a genic mutation, or are they a nongenic variant that propagates the consequence of some cortical accident? Because Suhama's studies were conducted on a single aberrant clone (Suhama, 1982) and as *Glaucoma* genetics is as yet undeveloped, this question has no certain answer. Several closely related facts, however, indicate that the mirror-image *Glaucoma* doublets did not arise from a genic mutation. The doublet condition was "maintained . . . by transferring only perfect doublets into fresh culture medium" (Suhama, 1982, p. 53); this suggests that the doublet condition itself rather than its genetic basis is self-propagating. In accord with this conclusion, normal singlet clones were selected from this stock

(Suhama, 1983), and clones of both normal and reverse singlets were generated by surgical transection (Suhama, 1985).

8.3.2 Organization of Self-Propagating Reverse Singlets

When reverse singlets of *Tetrahymena* were first seen on slides made from cultures of regulating parabiotic doublets, I predicted that such cells could not be self-propagating because ". . . they are presumably in the same morphogenetic dead-end as are stentors with reversed contrast zones" (Frankel, 1984). This prediction had been proven incorrect by the time that it appeared in print; as of this writing clones of reverse singlets of *Tetrahymena thermophila* and *Glaucoma scintillans* have been maintained for upward of 5 years in Iowa City and Hiroshima, respectively.

The reverse singlets in the two closely related genera are so similar that they can be described together. They were obtained in different ways: in *Glaucoma* by longitudinal transection of mirror-image doublets (Suhama, 1985) and in *Tetrahymena* by a process of spontaneous regulation from doublets (Nelsen and Frankel, 1989). The cytoplasmic basis of the difference between *Glaucoma* singlets and mirror-image doublets is self-evident from their mode of origin; in *Tetrahymena*, where the precise origin must be inferred, a nongenic basis of the reversal was demonstrated through genetic analysis (Nelsen et al., 1989a).

The organization of the reverse singlets is shown diagrammatically for *Glaucoma* (Fig. 8.9c) and photographically for *Tetrahymena* (Fig. 8.10b). The sole oral apparatus is now the secondary OA. As is characteristic of sOAs, the oral structures

Figure 8.10 Scanning electron micrographs of a normal (**a**) and reverse (**b**) singlet *T. thermophila* fixed after deciliation. Note that the membranelles of the OA in (**b**) are inverted relative to those of (**a**); corresponding locations in the two are indicated by *arrows*. The anterior suture, extending between the anterior pole of the cell *(asterisks)* and the anterior end of the OA, is much longer in the normal cell than in the reversed cell. Scale bars, 10 μm.

are sometimes inverted, sometimes superficially reversed (in *Tetrahymena*), and sometimes fragmented. CVPs are now located to the *left* of the oral meridian, at the same proportional distance around the cell circumference as in cells of normal large-scale asymmetry, despite the fact that the absolute number of ciliary rows differs from that observed in normal singlets (Suhama, 1985; Nelsen and Frankel, 1989). Although the local asymmetry of ciliary rows remains normal, all global features are reversed: the pattern of anterior basal-body pairs in *Tetrahymena* (Nelsen and Frankel, 1989), the choice of a *left* postoral meridian for oral development in both *Glaucoma* (Suhama, 1985) and *Tetrahymena* (Nelsen and Frankel, 1989), the location of the dividing micronucleus in *Glaucoma* (Suhama, personal communication), and the general arrangement and curvature of ciliary rows. The superimposition of global reversal on local normalcy applies even to the sculpturing of the cell surface. Although the general configuration of surface ridges is the same in normal and reverse singlets of *T. thermophila*, the highest and steepest ridges are found to the cell's right of the OA in normal cells (Fig. 8.10a) and to the left of the OA in reverse cells (Fig. 8.10b). Every cortical feature that is not uniformly repeating is globally reversed when compared with the corresponding feature of normal cells.

This leaves us to explain how the reverse-singlets of *Glaucoma* and *Tetrahymena* can escape the morphogenetic dead end of comparable *Stentor* and *Stylonychia* cells. The reason for the difference is that whereas in the latter two ciliates the location of the mouth parts (cytostome and cytopharynx) is determined extrinsically by a global basal-apical gradient (see Section 7.4.2), in *Glaucoma* and *Tetrahymena* the location of these structures is determined intrinsically "... in relation to the spatial orientation of the three membranelles regardless of the large-scale cellular polarity" (Suhama, 1985, p. 459). Many of the sOAs of reversed cells have well-formed buccal cavities (Fig. 8.10b), and can feed (Suhama, 1985; Nelsen and Frankel, 1989). In inverted OAs, the cytostome is clearly at the *intrinsic* posterior end of the 180-degree rotated oral structures, juxtaposed on the *anterior* end of the cell (Suhama, 1985). Thus, success of propagation of reversed *Glaucoma* and *Tetrahymena* cells is based at least in part on the intrinsic control of the location of the oral opening, which allows inverted OAs to develop their cytostomes at the correct internal location in defiance of general cellular polarity, a feat that is evidently impossible for *Stentor* or *Stylonychia*. In *Tetrahymena*, routine cultivation on a very rich bacteria-free medium may also contribute to success in propagating cells with defective OAs, by greatly reducing or perhaps even eliminating the requirement of phagocytosis for growth (Orias and Rasmussen, 1976).

When reversed *Tetrahymena thermophila* cells from an otherwise fertile stock were mated with normal cells, they conjugated in an abnormal heteropolar orientation, but nonetheless could both carry out micronuclear meiosis and form zygotic nuclei that can develop further, up to the formation of macronuclear anlagen (Nelsen and Frankel, 1989). We have not found viable progeny, however, even under selective conditions in which a tiny proportion of such progeny would be detected. The upside down configuration of conjugating pairs is probably a consequence of an inherent asymmetry of the union of conjugants such as has been demonstrated in *Blepharisma* (Honda and Miyake, 1976) and is strongly suspected in *Tetrahymena* (Nanney, 1977, pp. 378); the failure of formation of viable progeny might then be due to some form of damage to the zygotic nuclei during exchange through an abnormal conjugation bridge. [Reversed *Tetrahymena* cells of different mating types could produce viable progeny following mating (Nelsen et al., 1989a), although with a difficulty that probably is a consequence of the underdevelopment of the anterior suture region at which conjugating cells normally prepare to fuse (Wolfe and Grimes, 1979; Suganuma et al., 1984).]

8.3.3 Development of Secondary Oral Structures

Although the development of secondary oral structures is always basically similar in mirror-image forms and reverse singlets, it is better studied in the latter, for two reasons. The first and obvious reason is that all OAs in reverse singlets are secondary OAs, whereas the second and more interesting reason is that in both *Glaucoma* and *Tetrahymena* the secondary OAs are on the average larger and better developed in reverse singlets than in mirror-image doublets.

In both *Glaucoma* and *Tetrahymena,* there is a tendency for membranelles to appear as mirror-images of the normal ones at early stages of their development, and as rotational permutations at later stages. The basic reason for this changing orientation has already been explained in the previous chapter (see Section 7.2.6 and Figure 7.7); ciliary units often assemble into promembranelles as if in conformity with a vector jointly determined by the cell's polarity and asymmetry. As a consequence, when cellular asymmetry is reversed although polarity remains normal, the tilt of the promembranelles is a mirror-image of the normal, but the internal organization that becomes more evident at later stages is a 90-degree rotational permutation of the normal.

To make this partial description of membranelle assembly more complete, we must include three additional observations, rendered schematically in Figure 8.11

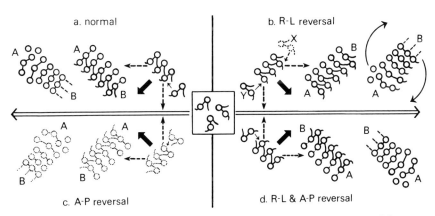

Figure 8.11   Diagrams indicating the influence of the four possible combinations of directions of asymmetry and polarity on the orientation of a developing membranelle of *Tetrahymena* cells. As in Figure 7.7, membranelle development is shown as proceeding from an initial condition of randomly oriented basal-body pairs in an anarchic field (box, center) to the left (**a**, **c**) or right (**b**, **d**) in the diagram. *Dashed arrows* represent the vector components of asymmetry (horizontal) and polarity (vertical), and *heavy solid arrows* show the imaginary resultant vectors. The ciliary elements in quadrant (**c**) are indicated by *dotted lines* because these configurations are not observed in cells. For further explanation, see the text.

(which is an amplified version of Fig. 7.7). The first observation concerns the *direction* in which promembranelle assembly proceeds. In normal oral primordia, assembly clearly proceeds from anterior to posterior, with basal-body pairs added at the cell's posterior-left (viewer's lower right) of each promembranelle, the future "B" end (Fig. 8.11a). In oral primordia formed under the influence of a reversed asymmetry vector-component (Fig. 8.11b), there is a choice. If the order of assembly were controlled by the intrinsic polarity of the promembranelle, then basal-body pairs should be added at the future "B" end of the promembranelle, which is now its *anterior*-left end (Fig. 8.11b,X). If, on the other hand, the order of assembly were controlled by the anteroposterior vector-component, then basal-body pairs would be added at the *posterior*-right end of each promembranelle, the future "A" end (Fig. 8.11b,Y). A systematic study (Nelsen et al., 1989b) indicates that the latter method is used predominantly whenever promembranelles are tilted as shown in Figure 8.11b. This means that the "A" end of the membranelle, where subsequent modification through displacement of basal bodies (see Section 7.2.5) is to take place, would be the latest-assembled part of the developing membranelle rather than being the first-assembled part as it is in normal OAs. Such a dissociation of assembly order from intrinsic organization might contribute to the frequent abnormalities in the final modeling of the membranelles of sOAs.

The second additional observation is specific to secondary oral primordia of *Tetrahymena*, and applies both to *janus* cells and to reverse singlets. The spatial orientation of membranelles becomes altered during late stages of development. In primary oral primordia, the membranelles seem to rotate about their axes to become almost transverse in orientation, and later rotate back to reassume their original diagonal pitch (Bakowska et al., 1982b). In secondary oral primordia, the initial rotation often continues in the same direction in which it started, so that membranelles that were originally assembled with a reverse tilt secondarily assume a normal orientation (Fig. 8.11b, curved arrows) (Frankel et al., 1984a; Nelsen et al., 1989b).

The third and most interesting observation, first made by Suhama (1982, 1985) on sOAs of mirror-image doublet and reverse singlet *Glaucoma* and then repeated on reverse singlet *Tetrahymena* (Nelsen et al., 1989b), was that sometimes *both* structural organization and assembly order can become inverted, i.e., permuted through 180 degrees rather than 90 degrees relative to the normal (Fig. 8.11d). Such an inversion requires a reversal of the polarity as well as the asymmetry vector-component. It should be noted that when this happens all axial conflicts within the oral primordium are resolved, as basal-body pairs are added at the anterior-right (B) end of the growing promembranelle in obedience both to the reversed polarity vector and to the inverted intrinsic organization of the preexisting promembranelle. Put differently, the polarity and asymmetry vectors are once again in their normal geometric relationship to one another (i.e., the polarity vector points to the left of a hypothetical observer who stands behind the asymmetry vector). Thus it is not surprising that in both *Glaucoma* and *Tetrahymena* the sOAs with the most nearly normal internal organization are those that are close to 180-degree rotational permutations of the normal structures.

The fact that the configuration shown in Figure 8.11c is not observed indicates that reversal of the anteroposterior axis within the tetrahymenid OA rarely, if ever, occurs by itself. This suggests that the polarity reversal shown in Figure 8.11d probably is a secondary consequence of the switch in asymmetry. The polarity reversal also appears to be more circumscribed than the reversal of asymmetry, as the former, unlike the latter, usually has no effect outside of the oral system itself. This fact can be appreciated if one recalls that the intrinsic posterior end of an inverted OA is juxtaposed on the anterior end of the cell (Fig. 8.10b). Owing to the cylindrical topology of ciliate growth (see Fig. 2.2), an anterior cell pole once was a fission zone. Hence, the fission zone is formed just anterior to the intrinsic *posterior* end of an inverted OP. If the reversal of the anteroposterior axis of the OP were associated with a reversal of the anteroposterior axis of the *entire* cell cortex, then the fission zone should have formed posterior to the OP, i.e., near the intrinsic *anterior* end of the inverted OP. In a few cases it does (Nelsen et al., 1989b), but these cases are exceptional. Thus the anteroposterior reversal of oral patterning appears to be another indication of the existence of a level of developmental coordination that is both relatively long range yet intrinsic to the oral system (see Section 7.2.7 and Table 7.1).

Whereas in normal oral primordia the order of formation of membranelles is clearly from the cell's left to right, with the outermost membranelle, M1 (see Fig. 8.9), appearing first and the innermost one, M3, last, in reversed and inverted oral primordia membranelles differentiate from the cell's right to left. This reversal is particularly clear in *Glaucoma,* in which the orientation of membranelles at early stages of development is nearly parallel to the anteroposterior cell axis, and results in M1 forming first, M2 next, and M3 last in inverted oral structures just as in normal ones (Suhama, 1982, 1985). A corresponding situation is observed in the more nearly diagonal membranelles of *Tetrahymena* (Nelsen et al., 1989b).

Before concluding, I would like to return to the puzzling observation, made in both *Tetrahymena* and *Glaucoma,* that secondary OAs tend to be larger and better developed in reverse singlets than in mirror-image forms. The simplest way to resolve this puzzle is to recall that in mirror-image forms the sOA is functionally gratuitous as the cell can (and does) take in its nourishment through the primary OA, whereas reverse singlets cells depend on their sOAs for their survival. Mirror-image doublets rarely go through oral replacement during the exponential phase of culture growth, whereas oral replacement is common in all cultures of reverse singlets (Suhama, 1985; Nelsen et al., 1989b). Poorly developed sOAs will, therefore, persist in mirror-image forms that also have pOAs, but will be replaced in reverse singlets that rely on their sOAs for survival. Therefore, there might be a positive relationship between completeness and persistence of sOAs in reverse singlets but not in mirror-image doublets. We doubt that this is an adequate explanation for the difference, however, as at least in *Tetrahymena* virtually all secondary oral primordia appear larger and better developed in reverse singlets than in *janus* forms. Development of secondary oral primordia in mirror-image forms might be influenced by the primary oral primordia, possibly by lateral interactions affecting timing of development (Tartar, 1966a) that force secondary oral primordia to proceed through development more rapidly than they would if left to themselves (Nelsen et al., 1989b).

8.4 CONCLUSIONS

Reversals in large-scale cortical patterns have been observed in a wide variety of ciliates, including all members of the cast introduced in Chapter 3 except for *Paramecium* and *Euplotes.* Despite this widespread occurrence, the phenomenon itself is remarkably uniform. The reversal is essentially a change in intracellular handedness that affects the arrangement of all cortical structures within its domain. The domain of the reversal may include either a part of a cell or an entire cell. In mirror-image cells, a subset of the usual cortical structures may be mirror-duplicated, whereas the remainder of the structures are totally missing. However, when an entire singlet cell is affected a complete set of structures is always formed in reversed spatial order. Wherever it is found, the reversal always affects only the *arrangement* of ciliary structures and never their internal organization. Probably as a consequence of this, ciliary structures within the domain of reversal often develop abnormally, with the abnormality generally being more severe the more complex the structure.

The reversals of large-scale asymmetry considered in this chapter offer clear evidence for the existence of a developmental hierarchy in the sense that goes beyond a mere nesting of smaller units within larger ones. There is a true "system of dual control" (Polyani, 1968), in which the "higher" (more global) level of the hierarchy is constrained by the rules of the "lower" (more local) level, but is not directly reducible to that lower level. At the local level, structures can undergo rotational permutation within the plane of the membrane, and hence may become inverted, but appear incapable of true reversal of asymmetry. Reversals of asymmetry can only occur at the more global level of arrangement of diverse cortical structures. Nevertheless, constraints imposed by the rules of the local level on the arrangements at the global level probably forbid a ciliate from generating a reversal of asymmetry sufficiently perfect to survive in nature and be encountered by ciliate taxonomists, who might then have given it a different species name (Nelsen et al., 1989a).

The complex maneuvers of oral development are severely affected by the clash between a reversed global asymmetry and an unchanged and (probably) unchangeable local asymmetry. It seems that the degree of reconciliation that is possible depends on the extent to which the oral system can become inverted. In oxytrichids, the elements of the oral apparatus—membranelles and undulating membrane—become rotationally permuted within the reversed territory as an automatic consequence of the spatial order of their assembly (see Fig. 8.3), and rotations on a larger scale may also occur. A total inversion of the entire oral apparatus, however, is impossible in these ciliates because of the extrinsic determination of the mouth parts (buccal cavity and cytostome) by the cellular basal-apical gradient. In tetrahymenids, however, the mouth parts are determined intrinsically, and virtually normal oral apparatuses can form within a reverse singlet cell because the whole structure can develop as a 180-degree rotational permutation of the normal arrangement. This occurs by a secondary reversal in polarity. The oral apparatus can thus regain internal normalcy at the expense of being upside down with respect to the remainder of the cell. This situation, however, poses no dilemma to the tetrahymenid cell because the reversal of oral polarity usually does not affect the cortex outside of the oral system itself and, therefore, does not interfere with propagation of the clonal cylinder. This fact reinforces evidence presented in Section 7.2.5 that there exists an intermediate level of spatial organization, beyond the range of macromolecular assembly processes but still far more circumscribed than the global domain within which reversals of asymmetry take place.

9

Origins of Reversals of Asymmetry

9.1 INTRODUCTION: ROUTES TO MIRROR-IMAGE ORGANIZATION

A change of handedness of a three-dimensional object could be achieved by reversing any one of its three axes while retaining the other two unchanged. A right hand would be changed to a left hand if one could reverse the anteroposterior axis while leaving the palm and the back of the hand in their original position. One would obtain the same final result by reversing the dorsoventral axis, so that the palm became the back of the hand and vice versa. Reversal of any of these axes would involve a respecification of every point on that axis, except one (Shinbrot, 1966).

If a pattern is expressed on a surface, as it is in ciliates, the number of relevant axes is reduced from three to two. In ciliates, these are most easily thought of as a circumferential (lateral) axis and an anteroposterior axis. Reversal of either of these axes could in principle change the handedness of the ciliate cell surface.

The known cases of reversal of handedness of the entire ciliate surface (reverse singlets) are derived secondarily from side-by-side mirror-image doublet forms, or from other more complex geometric situations such as Tartar's "quartered" stentors. In the best studied cases, the lateral axis or the anteroposterior axis becomes reversed in a portion of the cell surface, leaving the remainder in its original configuration. These modes of reversal are illustrated schematically in Figure 9.1.

The simplest type of transformation is the reversal of a lateral axis in one longitudinal half of the cell (Fig. 9.1b). This would create a side-by-side mirror-image form, several examples of which were described in Chapter 8.

Another equally simple transformation is a reversal of the anteroposterior axis in one latitudinal half of the cell. This would create a head-to-head (Fig. 9.1c) or tail-to-tail (not shown) mirror-image form. The tail-to-tail form would be inviable due to lack of feeding structures, whereas the head-to-head form might survive but would have difficulty reproducing its geometry unless it could create mirror-images of each cell half at the equator during each division. Mirror-image head-to-head forms have been generated in the heterotrich *Condylostoma* (Yagiu, 1951, 1952; Tartar, 1979), but these were not self-propagating (Yagiu, 1952).

Side-by-side mirror-image forms nonetheless can be derived by reversal of the anteroposterior axis, but by routes that necessarily are more complex than the one involving reversal of the lateral axis. One such route is through a folding fusion of a longitudinal fragment like the one in Figure 9.1d. After the completion of a 180-degree fold (Fig. 9.1e), one lateral half of the cell would be inverted with respect to the other half. The inverted half-cell would then have its lateral axis facing that of

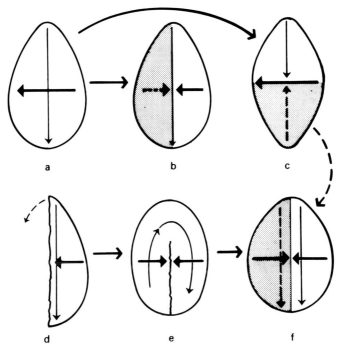

Figure 9.1 Possible routes by which reversal of an axis could lead to a mirror-image organization. The reverse domains are shaded. The lateral axis is indicated by the *heavy horizontal arrows*, and the anteroposterior axis is represented by the *lighter vertical arrow*. Reversed axes are *dashed*. A side-by-side mirror-image organization could arise directly from the normal organization (**a**) by a reversal of asymmetry in one-half of the cell (**b**), and a head-to-head mirror-image organization could arise from the normal organization by a reversal of the anteroposterior axis in the posterior half of the cell (**c**). A side-by-side mirror-image organization could also come about if removal of one lateral half of the cell (**d**) were followed by folding of the remaining portion (**e**) with subsequent reversal of the anteroposterior axis in the folded portion (**f**). The mirror-image organization shown in (**f**) could also arise from (**c**) by a clockwise rotation of the posterior half-cell relative to the anterior half.

the non-inverted half-cell, but the anteroposterior axis would be opposite to that of the complementary half. If, under these circumstances, the anteroposterior axis of the inverted portion underwent reversal in situ, one would then have converted a rotational permutation into a side-by-side mirror-image form (Fig. 9.1f).

The transformation sequence shown in Figure 9.1d to f is not the only way to get a mirror-image form through reversal of an anteroposterior axis. This could also be accomplished by reversing the anteroposterior axis first, as in Figure 9.1c, and then folding over the reversed half so that its anteroposterior axis comes to lie parallel to that of the normal (nonreversed) half.

The basic difference between the two types of routes to a side-by-side mirror-image form is in directness. If the transformation from a normal to a mirror-image pattern occurs without any major folding or rotation, then one can be reasonably confident that the lateral axis has undergone reversal. If this transformation is preceded, accompanied, or followed by a folding or shifting of parts relative to one another, then one may reasonably suspect that the anteroposterior axis has undergone reversal. In *both* cases, the end-result is that one half-cell is of reversed global asymmetry with respect to the other half-cell.

There is evidence that both types of routes can be followed in particular cases of the origin of heritable mirror-image doublets. Because the direct route through reversal of the lateral axis is at least nominally simpler than the more indirect route involving reversal of the anteroposterior axis, I describe cases of the former first and then proceed to the latter. The organization of the remainder of this chapter, therefore, is not taxonomic; although all mirror-image forms known in tetrahymenids seem to arise through reversals of the lateral axis, mirror-image configurations in heterotrichs and oxytrichids have been generated following both types of routes. Thus we encounter *Blepharisma* and *Stylonychia* in both major divisions of this chapter.

9.2 THE DIRECT ROUTE: REVERSAL OF THE LATERAL AXIS

9.2.1 How a *janus* Mutation Comes to Expression

The best way to find out how a reversal of asymmetry takes place is to observe a single living cell going through the entire process. In no case has such observation been made at the necessary level of resolution. The next best method is to cause the process to take place synchronously in a population of cells, and then study stages of the transformation in fixed and stained samples of that population. This can be accomplished either by carrying out an operation that reliably generates a reversal of asymmetry, or by bringing a *janus* allele to expression. The genetic technology available for *Tetrahymena* permits one to pursue the latter approach. The technique for doing so involves setting up a protocol in which old $janA^+/janA^+$ macronuclei are replaced synchronously by new $janA/janA$ macronuclei during conjugation (Frankel and Nelsen, 1986b). Exconjugants initially preserve their previous cortical organization intact, and then undergo a morphogenetic transformation from a normal to a mirror-image pattern during the division cycles subsequent to conjugation. Despite some variability both in the time of onset of expression of the *janA* allele and in the level of expression eventually obtained, a study of stained samples fixed at close time intervals has permitted us to infer a probable sequence of events for the onset of *janus* expression.

This sequence is summarized schematically in Figure 9.2. When exconjugant cells began to divide, there was a steady increase in the number of ciliary rows that were involved in formation of contractile vacuole pores (CVPs), with some indication of a transient shift of these CVP-rows to the right, away from the oral meridian (Fig. 9.2b). Afterward, CVP sets began to become divided into two subsets separated by one or two ciliary rows without CVPs. While this separation was beginning, about three to four generations after the genomic switch at conjugation, new secondary oral primordia began to appear (Fig. 9.2c, sOP), at a position that was close to (but not quite at) the definitive location of the sOM (Frankel and Nelsen, 1986b). About one generation after the first secondary oral primordia appeared, a few cells with secondary oral areas (sOAs) were seen. These cells undoubtedly were posterior division products of the first cells that earlier had

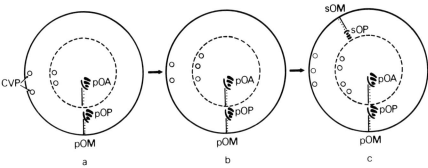

Figure 9.2   A polar projection showing how the *janA* mutation comes to expression in dividing cells. The pictorial conventions are the same as those introduced in Figures 8.6 and 8.7. (**a**) A normal singlet at the time of the *janA+* to *janA* genomic switch, with the primary oral area [pOA], primary oral primordium [pOP], primary oral meridian [pOM], and the contractile vacuole pore [CVP] set labeled. (**b**) A typical configuration one- or two-cell generations later, when the region of development of new CVPs has expanded and shifted somewhat to the cell's right. (**c**) The configuration at about four-cell generations after the genomic switch, when secondary oral primordia [sOP] defining the secondary oral meridians [sOM] have first appeared, and the newly developing CVP sets first became interrupted. Redrawn with slight modification from Figure 4 of Frankel and Nelsen (1986b).

formed secondary oral primordia. The proportion of cells with secondary oral structures rose subsequently, so that by 10-cell generations after the genic switch the full level of *janus* expression was attained (Frankel and Nelsen, 1986b).

As the mutant phenotype came to expression, the number and the asymmetry of ciliary rows remained the same. There was no evidence for any major deformation of cell structures as would be needed for generation of a mirror-image through reversal of the anteroposterior axis (Fig. 9.1d–f). Rather, it appeared as if the dorsal surface of a singlet cell was replaced by a ventral surface with reverse asymmetry, suggesting a direct in situ reversal of the lateral axis (Fig. 9.1b).

9.2.2 The Formation of Nongenic Mirror-Image Doublets

Tetrahymena *and* Glaucoma *mirror-image doublets*

No means has yet been devised to cause wild-type tetrahymenid cells to undergo a mass-conversion into mirror-image forms. Mirror-image doublets have not been produced by any form of mistreatment of normal singlet cells (such as exposure to drugs or heat shocks), *except* when that mistreatment first generates parabiotic doublets. Nongenic mirror-image doublets are generated by rare cases of aberrant regulation during the transition from the parabiotic doublet to the singlet condition.

A pictorial framework for the formation of mirror-image doublets from parabiotic doublets is presented in Figure 9.3. Imagine that a parabiotic doublet with two identically arranged cortical domains located directly opposite one another is bisected by a one-sided mirror (M) that is inserted lengthwise down its central axis (Fig. 9.3, top row), and that the domain located behind the mirror (bracketed) is then replaced by a mirror-image (shaded) of the domain in front of the mirror (Fig. 9.3, bottom row). The organization that results will then depend on the placement of the mirror relative to the major structures of the two component cells of the

Parabiotic
doublet

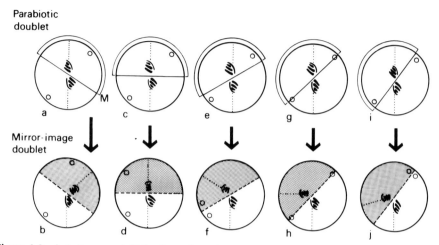

Mirror-image
doublet

Figure 9.3 A "mirror model" for formation of mirror-image doublets from parabiotic tetrahy-menid doublets. In the top row, parabiotic doublets are drawn as simplified polar projections, differing from each other only in the location of a mirror plane [M] relative to the oral and CVP meridians. In each case, the cortical domain on one side of the mirror plane (*bracketed in the top diagrams*) is replaced by a mirror-image of the region on the other side of the mirror plane (*shaded in the bottom diagrams*).

parabiotic doublet. If the mirror is positioned anywhere in the wide arc to the right of the CVP sets (Fig. 9.3a,c,e), the resulting mirror-image doublet would have a *janus*-like configuration, with a broad range of relative locations of OAs and CVP sets (Fig. 9.3b,d,f). If the mirror were placed at the plane that bisects *both* CVP sets (Fig. 9.3g), then the secondary oral structures of the ensuing mirror-image doublet would be positioned one-fourth of the doublet's circumference to the right of the primary oral meridian, and CVPs would be located halfway between the OAs in *both* hemispheres (Fig. 9.3h). If the mirror were then rotated slightly further so as to be located in the narrow arc to the *left* of the CVP sets (Fig. 9.3i), the secondary oral structures would shift even closer to the primary oral structures, and CVPs would be found *only* in the opposite hemisphere of the mirror-image doublet, the reverse of the *janus* configuration (Fig. 9.3j).

The patterns shown in Figure 9.3b,d,f correspond to the variety of *janus*-like *Tetrahymena* doublets that we have observed (Frankel and Nelsen, 1986a; Nelsen and Frankel, 1989), whereas the patterns shown in Figures 9.3f,h,j correspond to the stable mirror-image doublets of *Glaucoma* described by Suhama (1983, 1984). Suhama's observations fit this representation very closely: "When two OAs were sufficiently apart . . . two CVPs were located an equal distance from each OA between two OAs. When the distance became short . . . one CVP was situated near the middle point between two OAs. In cases of the shorter distance . . . one more CVP occasionally appeared near the middle point of the opposite (dorsal) side of the cell. In cases of the shortest distance, the CVP on the dorsal side was dominant" (Suhama, 1984). The "mirror-representation" thus is heuristically useful in showing that the major differences in arrangement of structures in mirror-image doublets of *Tetrahymena* and *Glaucoma* could result from a simple variation of a single basic process.

The inferred pathway of formation of nongenic mirror-image doublets in *Tetrahymena* is shown in Figure 9.4. This pathway is based on the transient appearance of *janus*-like mirror-image doublets in a culture of parabiotic doublets that were regulating slowly to the singlet state while maintaining oral meridians directly opposite to each other (which I call a "balanced" condition). Regulation from the doublet to the singlet state in *Tetrahymena* involves a progressive reduction in number of ciliary rows, with one of the oral systems disappearing when the number of ciliary rows is about midway between that of a typical doublet and a normal singlet (Nanney, 1966b). Our reinvestigation of this transition (Frankel and Nelsen, 1986a) uncovered cases in which loss of one OA was accompanied by a shift of the associated CVP set to the cell's left (Fig. 9.4b). Such combined loss and shift was probably what created the "two-CVP-set singlets" (Fig. 9.4c) that were observed in the same culture. These cells had only one OA but two distinct CVP sets in the hemisphere to the cell's right of that OA. The positions of these two CVP sets varied in a manner suggestive of gradual convergence to a normal singlet condition (Fig. 9.4e). In some regulating doublets, however, a secondary set of oral structures was interpolated to the right of the shifted CVP set, creating a mirror-image doublet (Fig. 9.4d).

Both the two-CVP-set singlets and the mirror-image doublets were observed in the same culture (with the former much more common), suggesting that there were two possible transformations affecting one of the two cortical domains of the original parabiotic doublet (the upper one in Fig. 9.4a). As in the *janus* case, the number of ciliary rows did not change during this transformation; the mirror-image doublets were of the same girth as parabiotic doublets, whereas two-CVP-singlets were skinnier. Thus, as in the *janus* example, a lateral axis was becoming transformed without interpolation of new material.

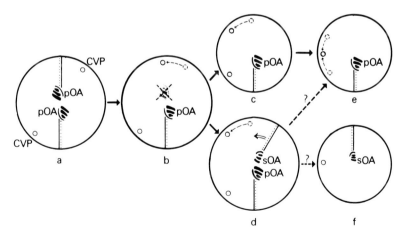

Figure 9.4 A simplified summary of inferred pathways of regulation of balanced doublets (**a**) of *Tetrahymena thermophila* in which (**b**) loss of one primary oral area [pOA] and shift of the accompanying CVP set leads to (**c**) two-CVP-set singlets or (**d**) mirror-image doublets with a secondary OA [sOA]. The two-CVP-set singlets almost certainly regulate to normal singlets (**e**), whereas the rare mirror-image doublets may give rise to reverse singlets (**f**) as well as to normal singlets (**e**). Redrawn with modifications from Figure 15 of Frankel and Nelsen (1986a).

Mirror-image doublets and two-CVP-set singlets were later observed repeatedly in cultures of reverse singlets that regulated spontaneously back to the normal singlet state, especially when such cultures were subjected to conditions that promoted fission arrest (Nelsen and Frankel, 1989). Regulation was probably through a pathway analogous to that shown in Figure 9.4a to d, except that the starting parabiotic doublet had to be a *reverse* doublet (i.e., a doublet in which *both* components were of reverse asymmetry). A mirror-image doublet would then be created by loss of one of the reverse domains and interpolation of a normal one, rather than by loss of a normal domain and interpolation of a reverse one as in Figure 9.4b to d. Reverse doublets have actually been seen in samples undergoing this type of regulation, but are very rare; Suhama (personal communication) observed a similar instability of reverse doublets in *Glaucoma*. Conversely, mirror-image doublets of the type shown in Figure 9.4d are common in cultures of reverse cells regulating to normal ones (Nelsen and Frankel, 1989).

The origin of the stable mirror-image doublets of *Glaucoma* is less clear. Suhama suggested that "... there is a possibility that these doublets arose through failure of dividing cells, because multiple monsters were simultaneously found together with doublets in cultures" (Suhama, 1982, p. 53). The culture that eventually generated the stable doublets also produced transient doublets some of which "had two OAs located 180 degrees apart"; these, however, "... soon returned to singlet forms" (Suhama, 1982, p. 53). Suhama's observations suggest a pathway of origin comparable to that of Figure 9.4, although with differences in detail due to a different location of the reversed domain.

These somewhat fragmentary observations do not show us clearly how to get beyond the heuristic "mirror-model" to a more detailed understanding of the reversal of asymmetry. The two-CVP-set singlets of *Tetrahymena* suggest that intermediate configurations are possible and indicate that the conversion of one type of cell domain to its mirror-image involves at least two separate steps. I will return to this in Section 10.3.4.

Mirror-image doublets from side-by-side juxtapositions in oxytrichids

To understand the origin of mirror-image doublets in oxytrichids, we must first review the nature of the nonmirror-image doublets in these ciliates. There are two basic kinds of oxytrichid doublets in which the component units are *not* mirror images. The type I doublets (see Fig. 4.10a) have "normal neighbors" at all points of the cell surface, and correspond to the parabiotic doublets of tetrahymenids. Tetrahymenid parabiotic doublets, however, regulate to singlets by an internal transformation, whereas their oxytrichid type I counterparts do so by progressive separation of components. The second type of oxytrichid doublet, the "side-by-side" doublet (type II; see Fig. 4.10b) is topologically very different from the type I doublets, because in type II doublets the right edge of one component abuts on the left edge of the other.

Evidence for formation of mirror-image doublets by reversal of the lateral axis in one component of type II doublets is available in *Paraurostyla weissei* (Jerka-Dziadosz, 1983) and in *Stylonychia mytilus* (Tuffrau and Totwen-Nowakowska, 1988). The starting points for the formation of mirror-image doublets in both cases were side-by-side (type II) doublets obtained either from arrested division (Fig.

9.5a,b) or from abortive conjugation (Fig. 9.5h,i). The precise sequence of events occurring after such arrested division is not yet fully clarified, and the two published accounts differ somewhat. The diagrams (Fig. 9.5a–e) are based on that by Tuffrau and Totwen-Nowakowska (1988) for *Stylonychia*. In this account, the anterior portion of the right component was spontaneously pinched off early in the sequence (Fig. 9.5c), whereas the posterior portion of the oral apparatus (OA) of the right component underwent a 180-degree rotation and joined to the distal end of the OA to its left to form the initial mirror-image oral structures of the right side of the future mirror-image doublet (Fig. 9.5d). After this, a sequence of cortical reorganizations (Fig. 9.5e,f) completed the formation of the mirror-image doublet.

In the account by Jerka-Dziadosz for *Paraurostyla*, the initial inversion of the posterior portion of the OA takes place in the same manner as was described later for *Stylonychia* by Tuffrau and Totwen-Nowakowska, but the anterior portion of the right component pinches off "during further reorganization." Critically, "It is not known whether the body ciliature of the right component becomes arranged in mirror-image symmetry during *one* cortical reorganization, or whether this process requires several rounds of ciliary replacement" (Jerka-Dziadosz, 1983, p. 186). Knowledge of just how these reorganizations proceed would be important for evaluation of the intercalation mechanism for reversal of global asymmetry that was proposed by Jerka-Dziadosz (1985) subsequent to her initial report (see Section 10.3.4).

Despite some uncertainty concerning details of the transformation, it is clear

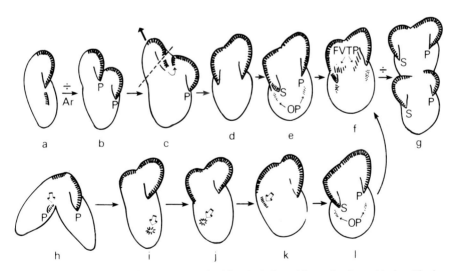

Figure 9.5 The formation of mirror-image doublets in *Stylonychia mytilus* from side-by-side doublets derived from arrested (Ar) cell division (÷)(**a–g**) and from abortive conjugation (**h–l, f–g**). The diagrams show oral structures, including oral primordia [OP], with structures of normal asymmetry labeled P (primary) and structures of reverse asymmetry labeled S (secondary). Two sets of frontal-ventral-transverse cirral primordia [FVTP] are shown in (**f**). The *open arrow* within each of the diagrams (**h**) to (**k**) points out the proximal remnant of the membranelle band of the right OA that migrates anteriorly to join the distal end of the right membranelle band. Redrawn with modifications from Figure 1 (sequences **d** and **e**) of Tuffrau and Totwen-Nowakowska (1988), with permission.

that generation of mirror-image forms requires that the two cell-units of the original blocked divider remain in a flat side-by-side configuration; when the two cell-units instead twisted around their common longitudinal axis to assume a back-to-back configuration, type IA homopolar doublets (see Fig. 4.9) were formed; these either maintained themselves as self-propagating type IA doublets, or split apart into singlets, but they *never* regulated into mirror-image doublets (Tuffrau and Totwen-Nowakowski, 1988).

During conjugation in *Stylonychia,* the proximal part of the membranelle band of the left conjugant abuts on the distal part of the membranelle band of the right conjugant (see Fig. 3.11). A small remnant of the posterior portion of the OA of the right component sometimes persists (Fig. 9.5h, arrow). When conjugation was aborted and the conjugating cells remained permanently fused, this persisting portion, including about 20 membranelles, sometimes rolled up into a circular palisade of membranelles within a subsurface vesicle (Fig. 9.5i), migrated anteriorly and to the cell's right (Fig. 9.5j), and then reemerged, straightened out (Fig. 9.5k), and joined with the former *anterior* end of the right membranelle band to produce a new *posterior* end of the OA of the reverse right component of a nascent mirror-image doublet (Fig. 9.5l) (Tuffrau and Totwen-Nowakowska, 1988).

It is hard to imagine that the unusual morphogenetic events described in the previous paragraphs could occur without some fundamental change in the underlying spatial organization in the cell cortex. What is this fundamental change and why does it come about? An anomalous form of pattern reorganization that was observed both in *Stylonychia* and in *Tetrahymena* suggests an answer to the second question and a hint toward resolution of the first.

9.2.3 Transient Triplet Forms

The "triplet" configuration was first discovered in *Tetrahymena patula* (then *Leucophrys patula*) by Fauré-Fremiet (1948). He was the first to notice that doublets that were regulating to singlets underwent a steady decline in number of ciliary rows. He observed that during the course of this decline, the two OAs often shifted to positions that no longer were directly opposite to one another, creating an illusion of bilateral symmetry (Fig. 9.6a). Such doublets, which I call *unbalanced,* were transitional between balanced doublets and singlets. Fauré-Fremiet, however, also observed something quite unexpected: the "induction" of a small additional OA in the narrower arc (semicell) between the two converging OAs (Fig. 9.6b). This additonal OA disappeared when the unbalanced doublets regulated to singlets.

This phenomenon was rediscovered by Kumazawa (1979) in *Blepharisma* cells that were grafted side-by-side to produce doublets. Balanced doublets, with normal "primary" OAs (pOAs) opposite one another, were propagated stably as such, but unbalanced doublets (Fig. 9.7a) often generated third OAs, again located in the narrower of the two semicells. These intercalated OAs sometimes were of normal asymmetry (Fig. 9.7b) and sometimes were of reverse asymmetry (Fig. 9.7c). A cell with one type of intercalated OA could readily produce the other during cell division or oral replacement, suggesting that these two types of OA were not as different as they seemed on inspection of living cells [which is why I have labeled both types of intercalated OAs as sOAs]. The "triplet" cells reverted to singlets by a conjoint

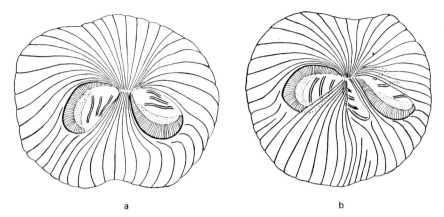

a   b

Figure 9.6   A polar view of unbalanced doublets of *Tetrahymena patula* (then *Leucophrys patula*) without (**a**) and with (**b**) an intercalated third OA. From Figures 7 and 8 of Fauré-Fremiet (1948), with permission.

loss of the two OAs on the cell's left, usually with an intermediate stage of fusion (fOA) of the sOA with the pOA to its left (Fig. 9.7d) (Kumazawa, 1979).

The third discovery of anomalous triplets was made by Nelsen among *Tetrahymena thermophila* doublets regulating to singlets. He observed, as Fauré-Fremiet had earlier in *T. patula,* that homopolar doublets generally became unbalanced as they regulated to singlets. No matter how the doublets were created originally, a third abnormal OA often appeared after progeny of the original (balanced) doublets

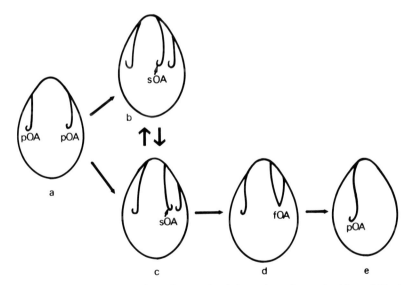

Figure 9.7   Schematic diagrams of regulation of unbalanced parabiotic doublets of *Blepharisma* produced by lateral grafting of two cells. The original graft complex is shown in (**a**), with two normal or primary oral areas (pOA). Secondary OAs (sOA) of either normal (**b**) or reverse (**c**) asymmetry were then intercalated. The triplet condition was resolved by fusion (fOA) of the sOA with the pOA to its left (**d**) and subsequent regression and disappearance of this fOA (**e**). Redrawn with slight modifications from Figure 3 of Kumazawa (1979), with permission.

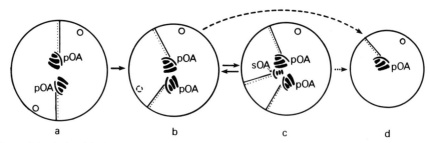

Figure 9.8 A simplified summary of the inferred pathway of regulation of unbalanced doublets of *Tetrahymena thermophila*. The starting point, a balanced parabiotic doublet (**a**), is the same as in Figure 9.4, but in this case the doublet first became unbalanced (**b**) and then underwent inter-calation of a secondary OA [sOA] between the two primary OAs [pOA]. The probable end point of this regulation is a normal singlet (**d**), which may be derived either from a triplet that has lost an sOA together with an adjacent pOA, or perhaps directly from an unbalanced doublet. Redrawn with slight modification from Figure 6 (A–D) of Frankel et al. (1987).

(Fig. 9.8a) became sufficiently unbalanced (Fig. 9.8b). Exactly as Fauré-Fremiet first described, this third OA *always* was formed in the narrower semicell, never in the wider one (Fig. 9.8c) (Nelsen and Frankel, 1986).

Nelsen and Frankel (1986) extended the initial discovery by Fauré-Fremiet with five further observations. First, unbalanced doublets had the same number of cili-ary rows irrespective of whether or not third OAs were intercalated, indicating that this intercalation involves reorganization rather than growth of the circumferential dimension. Second, the intercalated third OAs showed the same range of internal abnormalities as did secondary OAs of *janus* cells, whereas the flanking original OAs remained normal. The central OAs, therefore, were designated sOAs, the flanking ones pOAs (Fig. 9.8c). The sOAs commonly were reversed in overall geometry during development, and became very variable when mature; this vari-ability parallels Kumazawa's (1979) observation that the sOAs of unbalanced *Blepharisma* doublets were sometimes of reverse and sometimes of normal asym-metry. Third, the location of the sOA relative to the two pOAs was also quite vari-able, with fairly common slippage to the right or left during development of oral primordia. Just as in the *Blepharisma* triplets, however, the sOA was generally closer to the pOA to its left than to the pOA to its right, and primary and secondary oral structures sometimes fused (Nelsen and Frankel, 1986). Fourth, development of sOAs was sporadic, and sometimes even pOAs failed to develop at the expected times and places. Fifth, no CVP set was present in the narrower semicell of the most highly unbalanced doublets, irrespective of whether or not an sOA had been intercalated. Section 10.3.3 shows how these characteristics "make sense" if one considers the triplet form as an unstable transitional stage between a doublet and a singlet.

Most cells of clones initiated from balanced doublets (Fig. 9.8a) became unbal-anced (Fig. 9.8b), almost certainly due to a random slippage of the oral meridian superimposed on a steady loss of ciliary rows (see Appendix of Frankel and Nelsen 1986a). Secondary OAs (Fig. 9.8c, sOA) appeared in the narrower semicells of unbalanced doublets that were near the threshold of transition to the singlet state. Although doublets might become singlets by simple loss of one of the two primary oral meridians (Fig. 9.8b directly to d), an alternative route is through the simultaneous loss of a secondary oral meridian together with one of the two primary oral meridians (Fig. 9.8c to d). We suspect that in *Tetrahymena* as in *Blepharisma*, this dual loss may be preceded by a transient fusion of an sOA with a pOA to its left (see Section 10.3.3).

Finally, the triplet configuration has also been observed in *Stylonychia*. Sometimes, the anterior fission product did not pinch off after side-by-side fission arrest in *Stylonychia mytilus*, as was shown in Figure 9.5c, but rather slipped posteriorly and came to lie next to the posterior division product (Fig. 9.9c). In these cases, *three* distinct oral "V"s were subsequently formed: normal (primary) ones on the right and left sides (Fig. 9.3d, P) and a *reverse* (secondary) one in between (Fig. 9.9d, S) (Tuffrau and Totwen-Nowakowska, 1988). In subsequent reorganizations, the reverse central pattern became expanded to include the F-V-T cirri as well as oral structures (Fig. 9.9f). The newly developed central and right oral structures failed to link up with any parts of the original OA, and thus all of the oral systems except the one on the left remained fragmentary and presumably were resorbed shortly after differentiating (Fig. 9.9f). According to Tuffrau and Totwen-Nowakowska (1988), such triplets did not regulate either to mirror-image doublets or to singlets, but instead became increasingly monstrous and eventually died. Such *Stylonychia* triplets have been observed by others as well (Pang, personal communication; Yano, personal communication), and deserve further study.

9.2.4 A Provisional Summary

Reversal of asymmetry is accomplished somewhat differently in *janus* mutant tetrahymenas and in their wild-type counterparts. In *janus* cells, the starting point is an ordinary singlet cell, whereas in wild-type cells the path to reversal starts with a parabiotic doublet. In *janus* cells, the dorsal surface of the singlet is transformed into a ventral surface that is the mirror image of the original ventral surface. In wild-type cells, one cell-unit of a doublet is replaced by the mirror-image of the other cell-unit.

In both situations, the number of ciliary rows and the cell circumference remain the same during the reversal. Any growth that is taking place is longitudinal, at right angles to the axis that is undergoing reversal. Thus, in both wild-type and *janus* tetrahymenas the reversal is accomplished by a true internal reorganization of cir-

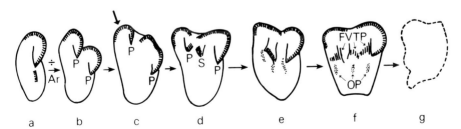

Figure 9.9 The formation (**a–f**) and demise (**g**) of triplets in *Stylonychia mytilus*. The conventions of illustration and labeling are the same as in Figure 9.5. Redrawn with modifications from Figure 1 (sequence c) of Tuffrau and Totwen-Nowakowska (1988), with permission.

cumferential (latitudinal) organization. The same is almost certainly true in the oxytrichid cases as well.

Regulating wild-type doublets may give rise either to mirror-image doublets (Figs. 9.3 to 9.5) or to triplets in which a reverse cortical domain appears *between* two persisting normal ones (Figs. 9.6 to 9.9). The triplets have been observed in a wide variety of ciliates: *Tetrahymena, Blepharisma, Stylonychia,* and, as we shall see in Chapter 10, in a vestigial manner in *Paramecium* as well. Although the mirror-image doublet configuration sometimes is stable, the triplet organization is always unstable and generally is resolved by transformation to the singlet condition. The challenging and paradoxical fact that such a triplet can be transitional between a parabiotic doublet and a normal singlet was the original motivation for the model to be presented in Chapter 10.

9.3 THE INDIRECT ROUTES: REVERSAL OF THE ANTEROPOSTERIOR AXIS

9.3.1 The Cut-and-Fold Experiment

The earliest surgically derived mirror-image doublets were obtained in *Blepharisma* by experiments involving abnormal polar juxtapositions (Suzuki, 1957). The simplest such experiment was a removal of the dorsal (aboral) half of the cell (Fig. 9.10a) followed by a folding of the anterior end of the remaining ventral (oral) half

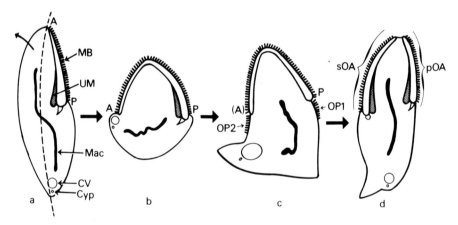

Figure 9.10 Transformation of a folded oral half of *Blepharisma japonicum* into a mirror-image doublet. The operation involved (**a**) removal of the aboral half of the cell followed by (**b**) immediate folding over of the oral half-cell to juxtapose the anterior [A] end of the oral apparatus against the posterior end of the cell, opposite to the original posterior [P] end of the oral apparatus. The structures labeled in (**a**) are the membranelle band [MB], undulating membrane [UM], contractile vacuole [CV], cytoproct [Cyp], and macronucleus [Mac]. (**c**) Two oral replacement primordia [OP1, OP2] were formed 12 to 15 hours after the operation. OP1 developed into part of a normal set of oral structures, whereas OP2 contributed to the formation of a reverse set of oral structures. After a second similar oral replacement, a mirror-image doublet (**d**) with a normal primary OA (pOA) and a reverse secondary OA (sOA) was generated 30 hours after the operation. Redrawn from Figure 43a, b, e, and h of Suzuki (1957), with modifications in orientation.

onto the posterior. As a result, the anterior half of the elongated oral apparatus (OA) was inverted, and its anterior tip was juxtaposed against the posterior region of the cell, including the CV and the cytoproct (Fig. 9.10b). The cell then produced *two* oral primordia: one at its usual location behind the original posterior end of the OA (Fig. 9.10c, OP1) and the other at a new site between the original anterior tip of the folded OA and the posterior structures of the cell (Fig. 9.10c, OP2). *Both* OP1 and OP2 produced mouth parts (cytopharynx and cytostome) at the ends nearest the cell's posterior end, a position consistent with numerous observations indicating that heterotrichs always form mouth parts in the portion of the developing OP that is nearest to the basal end of a global basal-apical gradient (see Sections 7.4.2 and 8.1). After two successive oral replacements, "...a bistomial monster with normal and reversed oral structures was obtained" (Suzuki, 1957, p. 163). This came about because the original membranelle band eventually was broken into two halves. The former posterior half became connected with the normally oriented OP1 and formed the normal or primary OA (Fig. 9.10d, pOA). The inverted former anterior half was joined to OP2, and acquired a polarity of large-scale organization that was the reverse of that of the old membranelle band. It is this reversal of the anteroposterior axis in the inverted portion of the cell that converted the folded complex into a mirror-image form, as had been shown schematically in Figure 9.1d to f.

The *Blepharisma* mirror-image doublets manifested a cortical organization analogous to that of the mirror-image doublets of *Stylonychia* described in Section 8.2. In the reverse OA of *Blepharisma* (Fig. 9.10d, sOA) "...The membranelles retained their intrinsic movement and beat in a reverse direction, and moreover the cytopharynx was incomplete" (Suzuki, 1957, p. 163). The only difference from the oxytrichid doublets was in durability; *Blepharisma,* unlike *Stylonychia* and *Paraurostyla,* maintained a mirror-image configuration for only 2- or 3-cell generations, after which the reverse structures "gradually dedifferentiated" (Suzuki, 1957).

9.3.2 The Central-Disc Experiment

The previous account implies that the logical way to obtain viable mirror-image doublets through reversal of the anteroposterior axis is by folding of a left, or oral, half, as is illustrated for *Blepharisma* in Figure 9.10a. Unfortunately, *Stylonychia* is a stiffer cell than *Blepharisma,* and longitudinal halves generally retain their original shape (Grimes and Adler, 1978). Out of 6000 longitudinal halves, both right and left, Grimes and co-workers obtained only 15 fully folded right halves and *no* fully folded left halves (Grimes and L'Hernault, 1979). The folded right halves underwent a regulation similar to that presented above [described in detail by Grimes and L'Hernault (1979)], but created inviable type IV mirror-image doublets (see Fig. 4.10c) with incomplete OAs in the center of the ventral surface.

The Chinese workers (Tchang et al., 1964) circumvented this difficulty in a surprising way. They removed the central region of a dividing cell at a stage just before the formation of the division furrow (Fig. 9.11a). This central region included the posterior end of the OA, the anterior end of the OP, and the entire macronucleus,

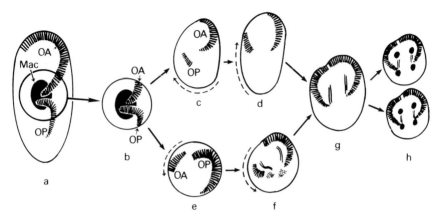

Figure 9.11 The origin of mirror-image doublets in *Stylonychia mytilus* from an excised central disc. The excision of a disc from the center of a predividing cell, with the condensed macronucleus [Mac] and portions of both the oral area [OA] and the oral primordium [OP], is shown in (**a**) and (**b**), the clockwise rotation of the OP that was postulated by Tchang and Pang (1982) is shown in (**c**) and (**d**), the counter-clockwise rotation of the OA that was inferred by Shi (personal communication) is shown in (**e**) and (**f**), and the resultant mirror-image doublets are illustrated in (**g**) and (**h**). (**c**) and (**d**) were redrawn from Figure 10 of Frankel (1984) (originally from Tchang and Pang (1982)), and the remainder were redrawn from originals kindly provided by Dr. Xinbai Shi.

which at that time was in the stage of maximal condensation (Fig. 9.11b). This experiment is geometrically analogous to cutting out a coin-shaped disc from the center of a pancake.

Only a small proportion of such central discs actually became mirror-image doublets (Tchang and Pang, 1965, 1982). As in the blocked-dividers considered earlier (see Section 9.2.2), whenever the anterior and posterior components succeeded in twisting around their common longitudinal axis so as to end up on opposite surfaces, they subsequently either pulled apart to form singlets or developed into stable type Ia doublets, but never formed mirror-image doublets (Tchang and Pang, 1982). When both OA and OP remained on the same side of a flattened disc, however, a rotation then took place *within the plane of the disc,* and mirror-image doublets sometimes formed subsequent to the rotation. Tchang and Pang (1982) interpreted this rotation as clockwise, with the original OP tending to be swept around to the right anterior part of the fragment by clockwise cytoplasmic flow (Fig. 9.11c,d). In a subsequent more detailed reinvestigation, Shi (personal communication) showed that this process instead involved a counter-clockwise rotation in which the original OA, rather than the OP, became displaced to the opposite side of the fragment as a consequence of the growth of the OP (Fig. 9.11e, f). The displacement of the original OA resulted in its becoming nearly parallel to the OP but in an inverted orientation, a situation remarkably reminiscent of Suzuki's cut-and-fold experiment. One thus may think of the central-disc experiment as a serendipitous discovery of conditions under which *Stylonychia* will occasionally undergo an internal fold-back, with consequences similar to those already presented for *Blepharisma.*

9.3.3 The Tandem-Graft Experiment

Results of the experiment

Several investigators have sought to devise an experimental protocol that would generate mirror-image doublets in a large proportion of the operated cells, so that the origin of these doublets could be analyzed in detail. Considering the observations summarized earlier, one might conclude that an axial reversal would require either a heteropolar state or the juxtaposition of anterior and posterior cell poles or perhaps both conditions simultaneously. Thus, the most promising experiment should be a heteropolar grafting of cells or cell parts. Although Suzuki had succeeded in obtaining some short-lived mirror-image doublets in *Blepharisma* by such means (Suzuki, 1957, Fig. 47), this strategy has thus far failed in *Stylonychia*. The components of heteropolar complexes either failed to interact, each developing more or less independently of the other, or else one of the components was resorbed (Grimes, personal communication; Shi et al., 1990a).

Shi and his coworkers eventually succeeded in reliably provoking mirror-image reversals by creating an anteroposterior juxtaposition *without* heteropolarity, as shown in Figure 9.12. This tandem tail-on-head juxtaposition (Fig. 9.12b) is remarkably similar to the tandem *Stentor* graft described in Section 5.3.1 (see Fig. 5.9). Shi's graft complex, like Tartar's, did not divide. Although the *Stentor* tandem

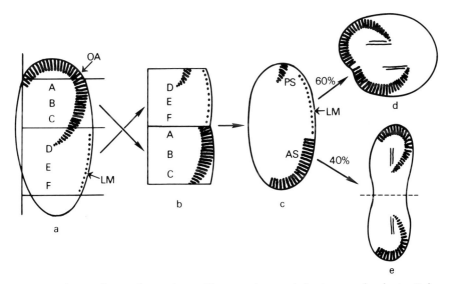

Figure 9.12 Shi's tandem-graft experiment. The operation, carried out on nondeveloping *Stylonychia mytilus,* is shown schematically in (**a**) and (**b**), with sequential anterior to posterior locations labeled A B C D E F. The healed complex is shown in (**c**) with the posterior segment of the original membranelle band [PS] well separated from the anterior segment [AS], and the left marginal cirri [LM] located anteriorly rather than posteriorly. This graft complex reorganized itself to form either a self-propagating mirror-image doublet (**d**) or a mirror-image heteropolar complex (**e**) that subsequently pulled apart to produce a reproducing normal cell and an inviable reverse cell. Redrawn with slight modifications from original drawings provided by Dr. Xinbai Shi, and used with permission.

graft-complex resolved the positional discontinuity by interpenetration of cortical material from both components (see Fig. 5.9), the stiffer *Stylonychia* graft-complex did not interpenetrate but instead often underwent a reversal of the anteroposterior axis of its posterior component.

In some of the *Stylonychia* graft complexes, the reversed posterior component separated from the anterior component (Fig. 9.12e). In others, the anterior ends of the normal and reversed components folded together to create a mirror-image doublet (Fig. 9.12d) (Shi et al., 1990a). Thus, mirror-image doublets were obtained by a combination of reversal of the anteroposterior axis and geometric reorientation, but in this case the polarity reversal took place very early and the reorientation occurred at the same time or later. These dramatic changes were carried out with remarkable speed, requiring only 4 to 5 hours after construction of the graft for completion (Shi et al., 1990a).

The transformation from tandem graft to mirror-image doublet is documented by an extensive series of photographs of protein-silver preparations (Shi et al., 1990a), a few of which are reproduced in this book (Fig. 9.13). After the fusion of the two components of the tandem graft (Fig. 9.12c), an oral primordium was formed (Fig. 9.13a1, OP) a short distance posterior to the posterior segment (PS) of the original OA. Membranelles differentiated at *both* ends of this OP (Fig. 9.13a2), rather than just at its anterior end as in normal development. Subsequently, the original composite oral primordium split into two separate oral primordia, the anterior OP remaining close to the PS of the old OA as in normal development, the posterior OP migrating toward the AS of the old OA (Fig. 9.13b), a process that has no counterpart in a normal cell. The linkup of membranelles of the anterior OP to the PS of the original OA was normal. In contrast, the posterior OP *rotated* 180 degrees before docking on the anterior end of the AS. Thus, in the stage shown in Figure 9.13c the membranelles of the anterior OP had already made a smooth juncture with those of the PS, whereas the membranelles of the posterior OP were still at a 90-degree angle relative to the those of the AS, having not yet completed their rotation.

Figure 9.14 illustrates why the membranelles of the posterior OP must rotate to fit onto the AS. Because we know that the internal organization of ciliary structures is the same everywhere in a mirror-image oxytrichid doublet (see Section 8.2.1), the two sets of membranelles in the bipolar OP of Figure 9.13a2 must also be internally similar, with the posterior set organized as a rotational permutation of the anterior set, as shown diagrammatically in Figure 9.14a. The anterior set of new membranelles conforms in both its topology and spatial orientation to the membranelles of the old PS, whereas the posterior set of new membranelles conforms

Figure 9.13 Photographs of stages in the reorganization of Shi's tandem-grafted *Stylonychia* fragments into mirror-image doublets. (**a1**) One hour after the operation. The two components have healed and an oral primordium [OP] has developed just posterior to the posterior segment [PS] of the original OA, to the right of the left marginal cirri [LM]. Two sets of frontal-ventral-transverse cirral primordia [FVTP] have developed, one to the right of the posterior oral segment, the other to the right of the posterior end of the anterior oral segment. (**a2**) An enlargement of the OP from (**a1**). Membranelles have begun to develop at both ends; the posterior membranelles never are observed in normal cells. (**b**) One and one-half hours after the operation. The single bipolar OP has separated into two portions, with the posterior portion approaching the AS of the original

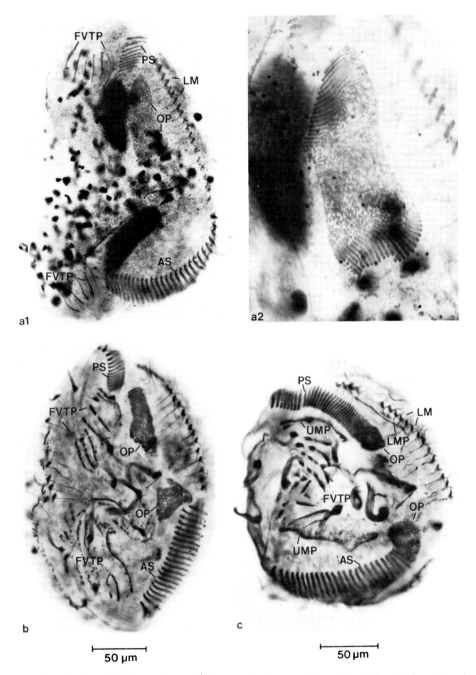

50 µm

50 µm

membranelle band. (**c**) Two and one-half hours after the operation. Both of the OPs have joined with the original oral segments, undulating membrane primordia [UMP] and left marginal cirral primordia [LMP] have developed, and cirri have formed within the FVTP. The graft has undergone internal rotation, so that the former anterior and posterior ends have begun to converge at the viewer's left side of the photograph. The anterior end of the future mirror-image doublet is thus at the viewer's left side of the photograph, the posterior end at the right side. Scale bars, 50 µm. Previously unpublished photographs kindly supplied by Dr. Xinbai Shi, and used with permission.

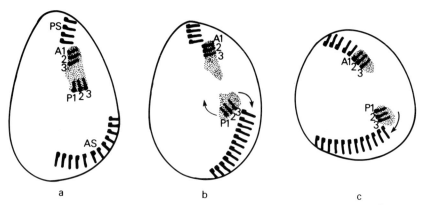

Figure 9.14 Schematic diagrams corresponding to photographs (**a1**), (**b**), and (**c**) of Figure 9.13, designed to illustrate membranellar organization. The anterior segment [AS] and posterior segment [PS] of the original membranelle band are labeled. The oral primordia are shaded. Each membranelle is drawn in a schematic manner that emphasizes its polarity and asymmetry. The membranelles within the oral primordia are numbered in their anatomical order, with the first membranelle of each respective set "tagged" as A1 (anterior) or P1 (posterior). The *arrows* near the posterior oral primordium in (**b**) and (**c**) indicate the direction of rotation of the oral primordium postulated by Shi et al. (1990a).

in topology but *not* in spatial orientation to the membranelles of the old AS (Fig. 9.14b). The new posterior membranelles, therefore, *must* rotate through nearly 180 degrees to fit onto the old AS (Fig. 9.14c). The physical mechanism by which a requirement for conformity of local organization brings about the rotation of a massive cortical structure is totally mysterious.

Two sets of frontal-ventral-transverse cirral primordia (FVTP) appeared concurrently with the onset of differentiation of oral membranelles, one set at its normal position next to the posterior segment (PS) of the original OA (Fig. 9.13a1, top), the other near the posterior end of the anterior segment (AS) of the original OA (Fig. 9.13a1, bottom). Although the polarity of the F-V-T streaks and associated ciliary primordia initially conformed to the original polarity of the transposed anterior segment, the development of these structures became highly disorganized at subsequent stages, with increasing indications of reversal of polarity. The actual process of reversal of the polarity of these ciliary primordia was sufficiently complex and variable to defy concise description (see Shi et al., 1990a for details); the reversal, however, was generally well established by the end of the first round of cortical reorganization and was completed during the second round (not shown here).

The reorientation that converted a bipolar form to a mirror-image doublet was already well underway in the cell shown in Figure. 9.13c. This reorientation was clearly a separate phenomenon, dissociable from the reversal of organizational polarity, as it often did not occur (Fig. 9.12e), and when it did occur its direction could be controlled by the details of the initial operation: if the right margin of the cell was cut off, as shown in Figure 9.12a, then the two mirror-image components bent to the right, yielding the type III mirror-image doublet with the membranelle bands on the outside as shown in Figure 9.12d and 9.13c; on the other hand, if a similar operation was carried out with the *left* margin excised, then the two components bent to the *left,* producing a remarkably complete type IV (see Fig. 4.10) mirror-image doublet with the membranelle bands near the midline (Shi et al., 1990b).

Interpretation of the experiment

The description thus far has skirted the central issue of how the posterior component of the graft complex could have undergone a reversal of its anteroposterior axis. If one thinks in terms of the intrinsic polarity of the membranelles, then the membranelle of the posterior OP that is labeled P1 in Figure 9.14 *ought* to have become the most anterior membranelle of the new OA. Instead, it became the most posterior membranelle. How?

One could imagine two rather different mechanisms of reversal of the antero-posterior axis. The first, an induction mechanism, pictures the reversal as spreading sequentially from a single restricted region, whereas the second, a respecification mechanism, postulates a more global reversal of cell-surface polarity.

The induction hypothesis has two main elements. The first element is that the formation of membranelles is induced to begin in those regions of the "common oral primordium" (of Fig. 9.13a) that are closest to preexisting membranelle bands, irrespective of the polarity of these preexisting bands. The second element is that the OP retains its original polarity (Fig. 9.15, heavy arrow within the OP) through-out the transformation process. This means that membranelles that had become initiated ectopically at the posterior end of the OP (star in Fig. 9.15a) would have had to differentiate in the "wrong" direction (posterior-to-anterior rather than ante-rior-to-posterior) relative to the pervasive polarity of the OP. Such an inverted direction of differentiation implies an oppositely directed membranelle asymmetry. To visualize this, imagine yourself first at the north end of a field facing south. If you stretch out your right arm, it points west. Now walk to the south end of that field and turn around to face north. Your outstretched right arm will now point to

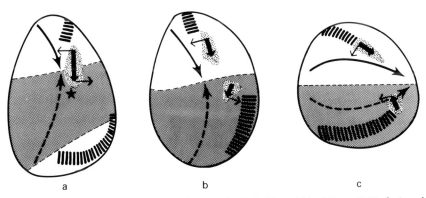

a   b   c

Figure 9.15 Diagrams corresponding to photographs (**a1**), (**b**), and (**c**) of Figure 9.13, designed to illustrate models for reversal of polarity in the tandem-graft experiment. The *arrows* that cover the entire cell surface indicate the pervasive anteroposterior polarity as postulated by the respecification hypothesis, with the *dashed* portions indicating regions of reverse polarity *(shaded).* The *heavy arrows* within the OPs indicate the intrinsic polarity of these OPs postulated by the induction hypothesis. The unique posterior site of initiation of membranelle formation is indicated by a *star* in (**a**). *Light arrows* extending from the OPs indicate the intrinsic asymmetry of the membra-nelles that were formed at the two ends of the bipolar OP. For further explanation, see the text.

the east. This geometric situation is shown in Figure 9.15a by thin arrows for the membranelles that are formed at the opposite ends of the bipolar OP. If the posterior OP retains its original polarity vector throughout its separation, migration, and rotation, the combination of the inverted polarity vector of the rotated OP and the asymmetry vector of each individual membranelle within that OP will create a mirror-image configuration (Fig. 9.15c). One could then imagine these vectors influencing the polarity and asymmetry of the remaining cortical structures through a sequence of secondary inductions.

The respecification hypothesis postulates a simultaneous and general reversal of the anteroposterior axis of the posterior portion of the cell surface (Fig. 9.15, dashed long arrow), brought about by the juxtaposition of extreme posterior on extreme anterior cortical positions (F on A in Fig. 9.12b). One could think of the ensuing reversal as being due either to a reorganization of a basal-apical gradient to extend in *both* directions from a posterior organizing center that has been relocated to the center of the complex (F in Fig. 9.12b), or to a process of intercalation, in which a DEFABC sequence would be reorganized into a DEFEDCBABC sequence (intercalated values underlined). The large-scale polarity of the posterior half of the bipolar OP as well as of all other newly developed cell-surface structures would thus be reversed from an early stage, with all of these ciliary primordia responding to a common global influence, much as in Suzuki's folded blepharismas.

Each hypothesis is better suited to explaining some of the observations. The induction hypothesis accounts better for the internal organization of the oral primordium, whereas the respecification hypothesis accommodates the indications that other ciliary primordia undergo reversal before the OP completes its rotation. It might be that both hypotheses are partially correct, despite certain incompatible features; on the induction hypothesis, the posterior half of the original OP retains its initial polarity although it is situated within the territory that is postulated by the respecification hypothesis to be globally reversed (Fig. 9.15).

The mirror-image doublets obtained from the tandem-graft operation were transmitted for hundreds of generations and passed through conjugation as well as encystment and excystment (Shi et al., 1990a). Nonetheless, such an operation could hardly have brought about a genic change. To prove the absence of a genic change, Shi and his co-workers obtained reverse singlets by cutting mirror-image doublets in half longitudinally (as in Fig. 8.5), and then carried out the same tandem-graft operation on the *reverse* singlets that they had earlier carried out on normal singlets. The result was that these tandem-grafted reverse singlets underwent a polarity reversal of the posterior portion of the cell surface, to produce a normal cortical domain (Shi et al., 1990a). Therefore, a large-scale reversal of cellular asymmetry that was produced at will could also be reverted at will. The reversed pattern is heritable but is not a result of a genic change.

9.4 THE ORIGIN OF REVERSE SINGLETS

Reverse singlets have been generated microsurgically by various means. By far the simplest is longitudinal transection of a mirror-image doublet, as mentioned earlier and in Section 8.2.3 for *Stylonychia.* The self-perpetuating reverse singlets of *Glaucoma* (Suhama, 1985) were obtained in the same way. The operations used by Tartar and by Uhlig (1960) (see Section 8.1) to obtain reverse stentors were more complex, and those pathways to the reversed condition were not elucidated in detail.

The heritable reverse singlets of *Tetrahymena* were obtained without microsurgery, by selection from cultures of regulating mirror-image doublets (Nelsen and Frankel, 1989). Genetic tests established that the origin of these reverse singlets could not have been based on mutations in nuclear genes (Nelsen et al., 1989a), indicating that in *Tetrahymena,* as in *Stylonychia,* the heritable reversal of cellular handedness was not due to a genic change. Because the origin of the *Tetrahymena* reverse singlets was rare, the precise pathway by which they arose is uncertain. Indeed, it is not definitely known whether reverse singlets originated from the transient triplet forms, shown in Figure 9.8c, or from the mirror-image doublets of Figure 9.4d, or from both. The motivation for placing reverse singlets at the end of the doublet pathway shown in Figure 9.4 rather than the triplet pathway of Figure 9.8. is largely theoretical, and will be covered in Chapter 10.

9.5 CONCLUSIONS

The genesis of mirror-image ciliates involves the sudden appearance of new cell-surface configurations, at the "wrong" places, with reverse large-scale asymmetry or polarity. Such unusual appearances imply a preceding reversal of the lateral or anteroposterior axes. Thus, the first and perhaps most basic conclusion to be drawn from these observations is that such in situ reversals really do occur. Although a change of handedness of a formed hand by reversal of one of its three orthogonal axes can be accomplished in thought but not in real life, the same feat can be carried out by a fully developed ciliate, sometimes in a matter of hours and with no intervening growth. A (re)specification event occurs in a portion of the existing cell cortex, and this event then brings about the reversal of spatial relationships of new structures that form in that region of the cortex. Thus the intracellular axes of the ciliate surface are labile, a lability that is shared by some early embryos (see Chapter 11).

The axial reversals themselves are invisible to our current means of detection. Thus, we have a long way to go before we really understand what is happening. Nonetheless, what we can see may help guide us toward what we cannot now see. In my view, the observations that have been made thus far give us two important clues.

The first clue is that reversal of asymmetry sometimes does and sometimes does not involve a genetic change. When it does, alleles at several gene loci (collectively called *janus*) each can bring about essentially the same transformation. When it does not, the transformation depends on new spatial arrangements and not on new physiological conditions; no known physiological trauma can by itself change a normal cortical domain into its mirror image. These observations suggest that although altered gene expression may sometimes create the conditions for the axial transformation, it is not fundamental to the transformation itself. A deeper analysis must begin rather than end at the cell surface.

The second clue is that although the geometric situations that result in mirror-image reversals are very diverse, with the starting points ranging down from whole parabiotic doublets to relatively small parts of a singlet cell, most of them involve abnormal proximity of different regions of the cell surface that usually are located far apart. This is particularly pronounced in *Stylonychia* and *Paraurostyla,* in which *all* of the starting points for formation of mirror-image doublets involve juxtapositions of right-on-left or anterior-on-posterior, but it also applies to the unbalanced parabiotic doublets of *Blepharisma* and *Tetrahymena.*

One can think of the axial reversals that bring about mirror-image transformations as ways in which a state of topological continuity that has been severely disturbed may be restored. For this idea to be understood clearly and evaluated critically, however, it first must be formulated precisely. Such a precise formulation must then be assessed in the various circumstances in which reversals of asymmetry take place, to find out whether the formulation can make sense out of situations that appear inexplicable or even paradoxical in the absence of that formulation. Finally, the formulation must make testable predictions, indicating what should and what should not be observed if the formulation is correct. Chapter 10 attempts to carry out this program.

10

A Cylindrical-Coordinate Model
for Pattern Formation in Ciliates

10.1 GLOBAL REVERSALS IMPLY INTRACELLULAR NONEQUIVALENCE

The previous two chapters showed that ciliates can propagate reversals of global asymmetry. Every such case involves a reversal in the spatial arrangement of structures whose internal asymmetry remains normal (Fig. 10.1). This dissociation of global from local asymmetry indicates that a system that controls the "handedness" of the cell surface is independent of the asymmetry of the microscopically visible cortical architecture. Furthermore, the demonstration that the contractile vacuole pores (CVPs) of *Tetrahymena* and *Glaucoma* are located at the same proportion of the cell circumference irrespective of handedness (see Section 8.3.2) indicates that whatever is subject to reversal of handedness is a global system of positional values that can be used for measuring and/or counting. Therefore, different regions of the same cell must be nonequivalent in a sense that is comparable logically to the positional nonequivalence of different cells of a multicellular organism.

The demonstration that the reversed condition can be transmitted nongenically to progeny cells (Nelsen et al., 1989a) makes sense if one assumes that the contours of the graded system are propagated longitudinally as cells grow. To visualize this, we first label the contours of the circumferential system in the same way that the gradation of time is labeled on a traditional clockface: by numbers around a circle (Fig. 10.1). If the winding of these numbers in cells of normal handedness (arbitrarily called right-handed, or RH) is clockwise, then that of cells of reverse handedness (left-handed, or LH) must be counter-clockwise. Next, we represent this ring of positional values as a set of longitudes that is propagated along the ciliate clonal cylinder (Fig. 10.2). Because no continuous deformation can possibly change the clockwise winding of an RH cell into the counter-clockwise winding of an LH cell, the geometry of the system insures propagation of its handedness.

One can characterize a continuum of values placed around a circle by an integral "winding number." The winding number is defined as "the net number of times the output value runs through a full cycle around a ring, as the input value is varied once forward along its ring" (Winfree, 1980, p. 14). In this definition, the "input value" is a member of the actual array of positional values around the physical circle (Fig. 10.3, left column), and the "output value" is the mapping of that value onto a "standard clockface" that represents the full set of possible values (Fig. 10.3, right column). A single complete circular set of values runs one full cycle, and thus it has a winding number of 1. If an RH cell (Fig. 10.3a) is assigned a winding

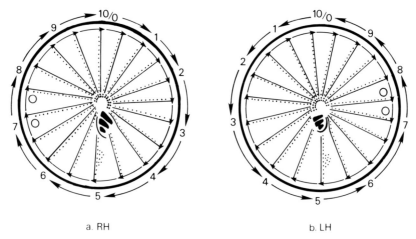

Figure 10.1 A polar representation of (**a**) normal right-handed (RH) singlet cells and (**b**) globally reverse left-handed (LH) singlet cells of *Tetrahymena thermophila*, emphasizing the dissociation of the local asymmetry of the ciliature (*arrowheads* inside the circumference of the polar map) from the global asymmetry of arrangement of nonrepeated structures (the sequence of numbers and large arrows outside the polar map). For further explanation, see the text.

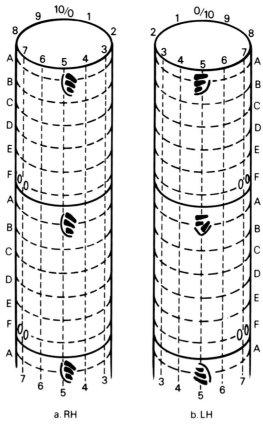

Figure 10.2 Orthogonal coordinates superimposed on the clonal cylinder for (**a**) right-handed [RH] and (**b**) left-handed [LH] cells. For further explanation, see the text.

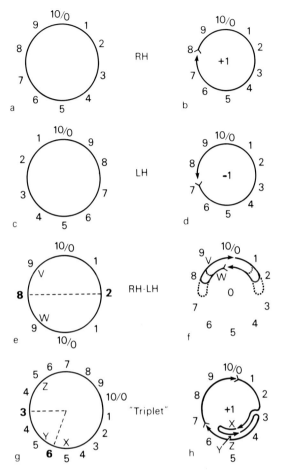

Figure 10.3   Topological mapping of different arrangements of positional values around the ciliate circumference. Each diagram on the left represents an actual cell circumference, with numbers indicating the positional values expressed on that circumference. Numbers representing longitudes of symmetry in positional values are indicated in boldface. Each curve on the right is a mapping of the set of positional values shown in the left onto a standard clockface that always has the same complete set of positional values displayed in the same stereotyped order. Construction of the map on the right requires reading the positional values on the left in a uniform direction (clockwise). The winding number of each map is given in its center. (**a, b**) A normal right-handed [RH] singlet maps as one complete clockwise turn within a standard clockface. (**c, d**) A reverse left-handed [LH] singlet maps as one complete counter-clockwise turn within the standard clockface. (**e, f**) A mirror-image (RH–LH) form (see Fig. 10.4g) maps as a closed loop within a portion of the standard clockface. The letters V and W show how two points with the same positional value [9] on the actual cell circumference map to the same value on the standard clockface. The loop can be extended or retracted at the edges without affecting continuity *(dashed lines)*, indicating the possibility for gain or loss of positional values at lines of symmetry. (**g, h**) A triplet (see Fig. 10.5b) maps within the standard clockface as one complete clockwise turn with extra length within a folded region. The letters X, Y, and Z show how three points with the same positional value [5] map to the same value on the standard clockface.

number of +1 (Fig. 10.3b), the winding number of an LH cell (Fig. 10.3c) is −1 (Fig. 10.3d).

Although circular sets with the same winding number can be transformed continuously into one another, circular sets with different winding numbers cannot be converted into one another without creating a topological discontinuity [this statement is derived from the index theorem of topologists, which, although usually stated for vector fields on planar regions of surfaces, also applies to one-dimensional phase-maps (Glass, 1977) such as are used here]. To effect such a conversion, one needs to cut into the circle and remove and/or add values. Such a drastic operation must be carried out once for each integral change in winding number; thus to convert a circle with a winding number of +1 to one with a winding number of −1, or the converse, one must commit not one but *two* topological transgressions. It is thus not surprising that LH cells fail to switch back to the RH condition even when the penalty for this failure is death by starvation (*Stentor:* Section 8.1; *Stylonychia:* Section 8.2.3). When LH cells can feed and grow, however, switches do occasionally occur. Much of the remainder of this chapter deals with the problem of *how* they occur. This question is addressed in the context of a "cylindrical-coordinate model," developed in Section 10.2 and applied in Sections 10.3 and 10.4.

10.2 A CYLINDRICAL-COORDINATE MODEL

In the cylindrical-coordinate model (CCM), *positional information* is provided by an orthogonal coordinate grid on the surface of the ciliate clonal cylinder (Fig. 10.2). In this model, the ciliate has two sets of positional values, longitudinal and circumferential. These two sets of values are treated as independent of each other. The coordinates are "interpreted" to promote formation of specific cortical structures. For example, we can let the pair of positional coordinates 5A specify the location of an oral apparatus, whereas 7F will specify the location of contractile vacuole pores. The specific numbers used are arbitrary but their spatial relations and the requirement for consistency are not.

The *circumferential system* defines the positional longitudes of the cell. The system is represented by labeling points around the cell circumference with numbers indicating different positional values, much as positions around the circumference of a traditional clockface are labeled with numbers indicating different times. An apparent discontinuity (10/0 in Figs. 10.1 and 10.2) must exist somewhere around the circle, but this discontinuity is an artifact resulting from the mapping of a number-line onto a circle; the distance between 10/0 and 1 is exactly the same as the distance between 1 and 2, just as the interval between 12 and 1 on a clock is exactly the same as the interval between 1 and 2 o'clock. A genuine discontinuity would be created if one were to remove a number from the circular set of positional values, just as one would have a real discontinuity—and a useless clock—if one removed one number from the face of a clock.

As is evident from Figure 10.1, the circumferential system of positional values is quite distinct from the ciliary rows and associated cytoskeletal structures. The numbers in the positional system mark off contours of an underlying gradation in some substance or condition, whereas the ciliary rows mark off a structural repeat.

The number of ciliary rows can and sometimes does change during growth of a clone, but the number of positional values arrayed around the cell circumference cannot change without a major topological transition (discussed below). In the cell shown in Figure 10.1, there are 10 values and 19 rows, so the average value-interval per row-interval is 10/19 or 0.53. If the number of ciliary rows became larger, the denominator of the fraction would become greater while the numerator necessarily remained the same, and then the value-interval per row-interval would decrease. If the number of ciliary rows became smaller, the value-interval per row-interval would increase. From now on, I will replace the cumbersome phrase "value-interval per row-interval" with the simple word "slope," understanding that the unit for assessment of this slope probably is a structural repeat rather than a physical distance.

Apart from the underlying idea of intracellular nonequivalence and a cylindrical coordinate system, the CCM as applied to the circumferential dimension has three basic postulates: the first is that the ensemble of circumferential positional values is stable when its slope is normal, whereas it is unstable when its slope is sufficiently steeper than normal. This "rule of sufficiently spaced neighbors" could be thought of as an extension of Mittenthal's rule of normal neighbors (Mittenthal, 1981), as a direct confrontation of nonnormal neighbors (a positional cliff) is a limiting case that occurs when the positional slope becomes increasingly steep in a local area.

The second postulate of the CCM is that the circumferential positional system tends to readjust itself to achieve a continuum of normally spaced positional values. When alternative routes of adjustment are available, as when a microsurgical operation results in juxtaposition of nonadjacent positional values, restoration will generally proceed by the shortest route. This aspect of the CCM is a restatement of the familiar shortest distance intercalation rule of the polar-coordinate model (see Section 1.5 and Fig. 1.8).

The third postulate is not strictly a part of the CCM but is essential to make it work: this is that the clonal cylinder has some specification of an optimal number of structural repeats, which is *independent* of the positional values arrayed around it. If cells are forced to deviate from this optimal number, they tend to return to it. This postulate is a restatement of Nanney's concept of a "stability center" (see Section 4.2.1).

The *anteroposterior system* is represented by letters in Figure 10.2. The values of this system mark cell latitudes. The anteroposterior system is more labile than the circumferential system, because as the ciliate clonal cylinder grows, original longitudes are preserved but latitudes must change. When the cell switches from the state of inhibition to activation for division (see Section 5.2.3) an ABCDEF system becomes subdivided into a tandem ABCDEFABCDEF system. Such subdivision implies a repetitive creation of discontinuities (between F and A). The CCM postulates that when cells are activated for division the presumptive anterior and posterior ends of daughter cells behave as "normal neighbors" and can be intercalated after removal (see Section 5.3.1), whereas at other times they do not behave as normal neighbors.

Two topological assumptions pervade the CCM. The first assumption is that the topology appropriate to cortical development in ciliates is indeed cylindrical. This ignores the obvious fact that real ciliates look much more like spheres than cylinders. If one changed each cylinder-segment of Figure 10.2 into a sphere, one would have to pull together all of the circumferential values of the cylinder at the two ends, creating two points (analogous to the north and south poles of the earth) that express all longitudes simultaneously, and are thus examples of what topologists call critical points or singularities (Firby and Gardiner, 1982, Chapter 7). The possible developmental effects of these two singularities are ignored in this treatment, a neglect that I justify by assuming that positional longitudes, like ciliary rows, do not really meet at the ends, thereby preserving a cylindrical topology. The second assumption is that the control of patterning on the ciliate surface is a resultant of two separate one-dimensional processes, one of them working on the longitudes and the other on the latitudes of the "clonal cylinder." This treatment requires that the two components of cortical location (the latitudes and the longitudes) are orthogonal and independent in the sense that they influence the morphogenetic end result but not each other. This postulate is consistent with virtually all of what is known about ciliate development, with the important qualification that events occurring along the anteroposterior dimension may be required to "turn on" the expression of the circumferential system, i.e., regulation of positional longitudes can take place only at latitudes where development has been initiated.

The cylindrical coordinate model (CCM) differs topologically from the classic polar coordinate model (PCM) (see Section 1.5) because the topology of the CCM is that of the surface of a cylinder, whereas the topology of the PCM is that of the surface of a cone. The "coordinate-free" simplification of the PCM, which reduces a cone to its topological equivalent, a planar surface (Lewis, 1981, 1982; Mittenthal, 1981; Winfree, 1980), therefore, cannot apply directly to the CCM because a cylinder is not topologically equivalent to a planar surface. But the ideas of the PCM, as well as the topological insights of the authors who generalized it, nonetheless contributed heavily to the formulation of the CCM.

10.3 THE CIRCUMFERENTIAL SYSTEM

10.3.1 Overview

My goal here is to use the CCM to make sense out of the varied situations in which lateral (circumferential) axes undergo reversal. To do this, I will first (Section 10.3.2) apply the CCM to some simple microsurgical experiments in *Stentor*, in part to show how an emphasis on continuity changes one's perspective on ciliate pattern formation. I then apply the CCM to endogenous reversals of asymmetry in various ciliates, first analyzing transitions from the doublet to the singlet condition in which reversals of asymmetry may occur transiently (Section 10.3.3) and then considering how the more stable reversals of cellular handedness may be generated (Section 10.3.4). These sections deal only with wild-type cells, in which one may presume that the positional system and the rules for its intracellular interpretation remain unaltered. Finally (Section 10.3.5), I consider how and to what degree the CCM can be applied to those genic mutants in which the distribution of cell structures around the circumference is known to be changed.

10.3.2 Regenerating the Contrast Zone in *Stentor*

I start by applying the CCM to some simple experiments in *Stentor*. Consider the consequence of removing wedges of varying width, all of them including the contrast zone (Fig. 10.4). If narrow wedges are removed (Fig. 10.4a,N), the contrast zone regenerates and an oral primordium is formed (Fig. 10.4b,c). As the wedge gets wider, approaching a bisection of the cell, regeneration of the contrast zone generally occurs (Fig. 10.4d,e), but it may fail, especially if one-half or more of the cell circumference is removed (Fig. 10.4f,g). In the latter case, a healthy *Stentor*

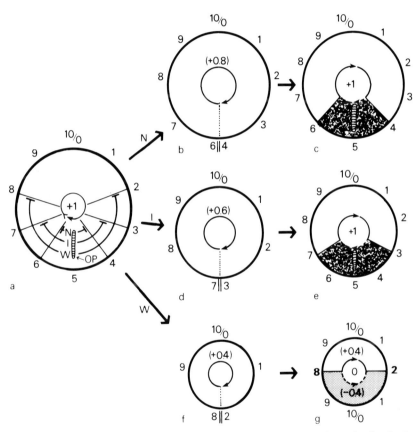

Figure 10.4 Regulation after removal of wedges of different sizes from the oral side of a *Stentor* cell, interpreted according to the CCM. The circles represent schematic polar maps labeled with positional values; the contrast zone with an oral primordium (OP) is placed at the value of 5. (**a**) A diagram showing the size of the three excised wedges, narrow [N], intermediate [I], and wide [W], respectively. (**b, d, f**) Cells immediately after the closing of the wounds resulting from removal of narrow (**b**), intermediate (**d**), and wide (**f**) wedges, with the proportion of the cell circumference remaining after the excision indicated in parentheses above the central circle [the place in the center is reserved for winding numbers, which must be integers]. (**c, e, g**) The result of intercalation by the shortest route, with (**c, e**) restoration of the missing positional values or (**g**) mirror-image duplication of the extant values. Regenerated regions are *pebbled,* duplicated regions *shaded.*

fragment fails to develop a contrast zone, cannot form oral primordia, and eventually dies (see Section 6.3.1).

In terms of the CCM (Fig. 10.4), the excisions remove positional values as well as cytoplasm. After these excisions, nonnormal neighbors are juxtaposed. This juxtaposition triggers restoration of normal neighborhoods by intercalation of positional values following the shortest path. This results in restoration of the original set of circumferential values, including the contrast zone, after narrow and intermediate excisions. The crowding of positional values in the resultant abnormally small stentor is probably relieved by the growth back to normal size that ensues after the fragment begins to feed.

In the widest excisions, however, more than half of the positional values have been deleted, and intercalation by the shorter route will lead to the formation of a mirror-image duplicate of the extant subset of values (8-9-10-1-2). The winding number of the cell circumference is, thereby, reduced from $+1$ to zero: if you move in a clockwise direction around the circumference of the abnormally regulated fragment (Figs. 10.4g and 10.3e), you first move "up" by four steps and then return to the starting point by moving back "down" by the same four steps; this tour is shown in Figure 10.3f in the form of a map onto a standard clockface. The new zero-winding-number situation allows for positional continuity at all points without requiring that the cell possess a complete set of positional values; therefore, the system can now obey all of the rules of the CCM without ever being required to generate the positional value (5) that specifies an oral-primordium site. Considered from this perspective, the cell's morphogenetic failure is a consequence of its falling into a topological trap.

This particular treatment of regeneration in *Stentor* departs from the traditional view of pattern restitution in *Stentor*. The unorthodoxy lies in postulating that the major visible discontinuity at the *Stentor* "locus of stripe contrast" is an expression of continuity at a more fundamental level. This new view is different from previous interpretations [including my own (Frankel, 1982)] but is consistent with most of the experimental results on *Stentor*. This new interpretation also makes it unnecessary to postulate invisible contrast zones in the many ciliates that lack visible ones, and can explain situations of "conditional totipotency" in other ciliates, such as the failure of narrow lateral fragments of *Urostyla grandis* to regenerate except when certain unusual spatial relations are achieved (see Fig. 6.3).

10.3.3 The Transition from Doublet to Singlet
Transient triplets

RH–RH doublets formed by parabiotic union (see Fig. 4.7) generally can perpetuate their doublet condition (see Section 4.3) but tend to revert to the RH singlet state. The simplest way for doublets to regulate to singlets is by separation of the two components; this actually happens in oxytrichids (see Section 4.3.3). The next simplest is by a loss or regression of one set of cortical structures, as has been reported for *Paramecium* (Sonneborn, 1963) and *Tetrahymena* (Nanney, 1966b). We also know, however, that some unanticipated geometric configurations may appear along the route from doublets to singlets. Perhaps the most provocative of these is the transient triplet organization, in which a third, reverse set of cortical structures appears in the narrower of the two semicells of an unbalanced parabiotic doublet (see Section 9.2.3 and Fig. 10.5). The formation of this third set of oral structures is paradoxical because, in *Tetrahymena* and *Blepharisma* at least, it is a prelude to simplification to the singlet state. How can three ever lie between two and one?

The CCM provides a solution. A parabiotic (RH–RH) doublet has two complete sets of positional values, arranged in tandem; its winding number is +2 (the mapping is not shown but is the same as that of the RH singlet shown in Figure 10.3b, except that it goes twice rather than once around the standard clockface). Given that the number of ciliary rows is reduced progressively to the value typical for a normal singlet (the stability center), yet continuity of positional values is maintained, it follows that the average slope of the positional gradation must become steeper. If the doublet becomes unbalanced, either as a deliberate consequence of microsurgical operation (*Blepharisma:* Kumazawa, 1979) or as a result of random slippage of the oral meridians (*Tetrahymena:* appendix of Frankel and Nelsen, 1986a), the slope of the positional gradation will become especially steep in the narrower semicell. When the system becomes too unstable to maintain itself, it can reduce the slope in only one way: by reverse intercalation. A subset of values that is wound in one direction is replaced by a smaller complementary subset that is wound in the opposite direction. In the example of Figure 10.5, +0.7 of a complete set of positional values (7 units worth, from 6 through 10/0 to 3) is removed and −0.3 of a complete set (from 3 back to 6), is substituted (Fig. 10.5a,b). Reverse-substitutions are, of course, possible at other places within the narrower semicell. Any substitution of a subset of positional values by a complementary subset in reverse order will reduce the winding number of the doublet from 2 to 1. To convince yourself of this, make a tour of the new circle immediately after the substitution (as in Fig. 10.5b): if, for example, you start at the 3 at the left side of Figure 10.5b and move in a clockwise direction, you will go up through a complete set of values to the opposite 3 (+1), then go up another 3 units to 6 (+0.3), and then

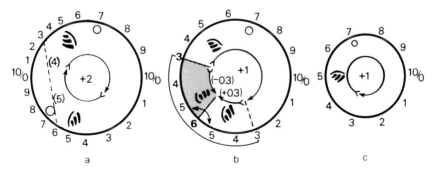

a   b   c

Figure 10.5 Formation and resolution of transient triplets, interpreted according to the CCM. In these and subsequent maps of tetrahymenids, the oral structures are placed at the periphery to emphasize their position around the cell circumference. (**a**) An unbalanced RH–RH doublet, with the route of subsequent reverse intercalation shown by a *dashed line*. (**b**) The immediate consequence of reverse intercalation. The reversed region is *shaded*. The positional values at the two resulting meridians of symmetry are indicated in *boldface*. Incremental loss of positional values is indicated by the *double-headed arrow*. The region that will eventually be deleted is *bracketed*. (**c**) A normal RH singlet after the completion of incremental loss of positional values.

back down from 6 to 3 (-0.3). The algebraic sum is $+1$. This tour is mapped onto a standard clockface in Figure 10.3h. The topology will be the same no matter where one places the reverse-substitution.

Although the winding number has been reduced from 2 to 1, the number of oral structures has gone up from 2 to 3. Why? Because a reverse intercalation that lies entirely within the narrower of the two semicells *must* include the positional value of 5, which is interpreted by the cell cortex as an oral apparatus (OA). This simultaneously accounts for the appearance of the third set of oral structures, its abnormality (due to the reverse gradation of positional values) and its restriction to the narrower of the two semicells.

This triplet configuration, however, is unstable. It can resolve itself to a singlet without any further topological transgressions. At both edges of the region of reverse asymmetry (shaded in Fig. 10.5b) there is now a longitude of reversal in the slope of the positional gradation. Positional values are arranged symmetrically on both sides of this longitude. As the circumference of the cell undergoes further reduction, positional values can be lost incrementally at either (or both) of these meridians of symmetry while maintaining continuity at every step. Imagine that the loss occurs at the lower **6**. One starts with 3-4-5-**6**-5-4-3, proceeds to 3-4-**5**-4-3, then to 3-**4**-3, and finally one is left with just the **3**, at which point one has a normal (RH) singlet. To visualize this in another way, consider the mapping in Figure 10.3h and imagine this as an elastic string that is progressively shrinking: the region of triple-overlap will progressively shorten until it disappears and the string is stretched taut around the circumference of the standard clockface.

During this process of incremental loss, a topologist would say that nothing of consequence was happening: the winding number was $+1$ when the process began, remained at $+1$ at every instant as it was occurring, and was $+1$ at the end of the process. The observer of the cell, however, would see dramatic changes. Through successive cycles of oral development one normal and one abnormal OA would approach one another, the two would fuse, and then the two would both disappear as if they had annihilated each other. There is direct evidence for this mode of loss in transient triplets of *Blepharisma* (Kumazawa, 1979), and inferential evidence in *Tetrahymena* (Nelsen and Frankel, 1986).

Blepharisma and *Tetrahymena* are not the only members of our ciliate cast in which a triplet configuration is resolved smoothly into a singlet state. In *Stentor,* a graft of a wide-stripe region into the center of the fine-stripe zone commonly resulted in the formation of three adjacent oral primordia, the central one reversed (see Fig. 6.4). During subsequent rounds of cortical development the number of developing oral structures was reduced to one, always of normal asymmetry (Tartar, 1956c, Uhlig, 1960). This reduction had earlier been interpreted as a result of direct mutual interference of activated oral primordium (OP) sites (Frankel, 1974); it now can be reinterpreted as a consequence of incremental loss of positional values across longitudes of reversal in slope of positional values.

In *Paramecium*, a trace of the transient triplet phenomenon has been observed despite the incapacity of this ciliate to form an OP anywhere except at the right margin of the old OA (see Sections 3.2.2 and 4.3.2). When *Paramecium* doublets regulate to singlets by a convergence of the two oral meridians (each possessing a cytoproct) and eventual disappearance of one of them, a third cytoproct appears transiently near the center of the narrow semicell of the doublet (Sonneborn, 1963; Kaczanowska and Dubielecka, 1983). According to Sonneborn (1963, p. 196) "This suggests that cytopyge (=cytoproct) formation is determined by interaction between . . . two kinety fields or the cortical regions in which they are located." I would reinterpret this phenomenon as this cell's maximal response to a positional instruction to form a third oral meridian.

Two-CVP-set singlets

When *Tetrahymena* doublets lose ciliary rows while remaining balanced, the slope of the positional gradation might increase uniformly around the circumference of the cell, so that when the critical slope is attained reverse intercalation could, in principle, occur at any location. One possible route would lead to the deletion of one of the two oral meridians as well as a shift of the nearby CVP set to the cell's left (Fig. 10.6a,b). Subsequent incremental loss of positional values at longitudes of symmetry would bring about a return to the normal RH singlet condition (Fig. 10.6b,c) as in the case of transient triplets described above. This would account for the two-CVP-set singlets that we have observed in our study of the regulation of balanced doublets to singlets (see Section 9.2.2 and Fig. 9.4). A rather similar alternative pathway of regulation from doublet to singlet, with early loss of one OA but preservation of CVP sets, was observed by Kaczanowska and Dubielecka (1983) in *Paramecium.*

Can one perhaps manipulate the possibilities of reverse intercalation to obtain virtually any arrangement of structures? One cannot. The topological restrictions on a single intercalation event that obeys the postulates of the CCM permit only a few structural configurations in addition to those depicted in Figures 10.5 and 10.6. Most of these are relatively minor variations of the two configurations described earlier, some of which have been observed and others not. One configuration, however, is quite different, and special because of its simplicity. If the cell were to intercalate between the values 2 and 8 in a balanced RH–RH doublet, replacing (2)-3-4-5-6-7-(8) with (2)-1-10/0-9-(8), it would have removed both an OA (5) and a CVP set (7) and substituted "dorsal" values, which have no special morphological expression. This would be seen cytologically as a direct transformation of an RH–

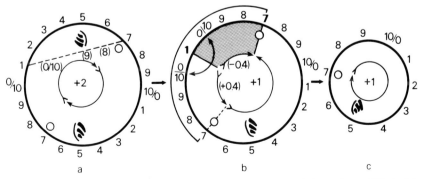

Figure 10.6 Formation and resolution of two-CVP-singlets from balanced RH–RH doublets, interpreted according to the CCM. The illustrative conventions are the same as in Figure 10.5, and comparable steps are shown.

RH doublet to an RH singlet. Such transformations undoubtedly occur, and may even predominate in balanced doublets (Nanney, 1966b). They are consistent with the CCM but by themselves do not count in its favor, as an apparent direct transition from a doublet to a singlet condition could be explained by other more traditional ideas, such as loss of an oral meridian through lateral inhibition.

Speaking more generally, the CCM works better when accounting for gain of structures (or major shifts in location of structures) than for loss of structures. An *absence* of a structure at a location predicted by the CCM may be rationalized as resulting from an incapacity to respond to an appropriate positional signal, but the *presence* of a structure will always count, in favor of the CCM if it appears at a location predicted by the model, against the CCM if it develops at a location excluded by the model. Thus far, patterns unequivocally excluded by the CCM, such as CVPs in cell sectors to the right of a reversed OA, or a third abnormal OA in the wider of the two semicells of an unbalanced doublet, have not been observed, but some configurations not predicted by the CCM did appear, especially in the *hypoangular* putative-mutant (see Section 10.3.5).

10.3.4 Reversals of Global Asymmetry in Wild-Type Cells

Routes to reversal of asymmetry

We return now to the problem of how to get from a singlet cell of normal (RH) to one of reverse (LH) asymmetry. Because the former has a winding number of $+1$, the latter a winding number of -1, the cell must go through two topological discontinuities to make this transition. It is thus not surprising that the two operations (both in *Stentor*) that have succeeded in converting RH into LH singlets created at least two major positional discontinuities. In all other cases, LH singlets probably arose from mirror-image (RH–LH) doublets (see Section 9.4). Because a mirror-image doublet has two components of equal size and opposite handedness, its winding number is zero (its map would be similar to Fig. 10.3f, but with the loop extending around virtually the entire circumference of the standard clockface before turning back on itself). Thus, the transition from a parabiotic (RH–RH) to a mirror-image (RH–LH) doublet still requires that the winding number must go down by two units, in this case from $+2$ to 0. Doublets, however, possess a built-in instability that may provide a "driving force" to generate these transitions. Because both the specific instability and the path to the final outcome appear to be different in oxytrichids and tetrahymenids, I describe the two cases separately. The discussion of oxytrichids at this point deals only with reversals of the lateral (circumferential) axis; the radically different mode by reversal of the anteroposterior axis is presented later in this chapter (Section 10.4).

Reversal in oxytrichids

The specific version of the CCM presented here (Fig. 10.7) is a slight modification of the model originally proposed by Jerka-Dziadosz (1985, Fig. 24) for *Paraurostyla*. The initial instability is a major one, as the sequence begins with side-by-side (type II) RH–RH doublets, joined at points that are 180 degrees out of phase in the positional circle (3 and 8 in Fig. 10.7a,b); it is as if the little finger of one right hand became joined to the thumb of another right hand. The simplest way to restore

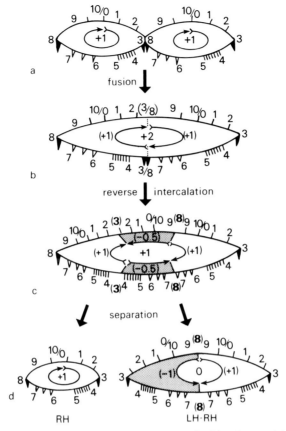

Figure 10.7 A hypothesis for formation of mirror-image doublets in oxytrichids, based on the CCM. These diagrams represent cross sections seen from the anterior end (see Figs. 4.9 to 4.12), and hence are comparable to the other figures in this chapter. (**a**) Side-by-side fusion of two RH singlets results in (**b**) a type II RH–RH doublet. Then reverse intercalation at the lines of discontinuity creates two (**c**) LH regions *(shaded)* between the two RH domains. A complete normal cell then buds off on the right to give rise to (**d**) an RH singlet plus a mirror-image (LH–RH) doublet.

continuity is to intercalate the equivalent of a left hand at *both* the dorsal and the ventral surfaces (Fig. 10.7c). This reverse intercalation reduces the winding number by 1, from +2 to +1 (Fig. 10.7c). (Intercalation by the alternative route, without reversal, would have increased the winding number from +2 to +3 and also would have interpolated ventral structures on the dorsal surface and dorsal structures on the ventral surface.)

At this stage we have a geometric situation not unlike that of the *Blepharisma* and *Tetrahymena* triplets, in which incremental loss of positional values across lines of symmetry (3 and 8 in Fig. 10.7c) would relieve crowding of positional values and eventually return the cell to the normal (RH) singlet state. Although this probably does occur (discussed later), the compound system sometimes appears to do something else, namely fragment into a RH singlet (on the right side) and a mirror-image doublet (on the left side). This physical removal of a RH singlet cell could convert the RH–LH–RH complex to a LH–RH mirror-image doublet. (Note that in terms of relative position the RH component is the *left* component and the LH component is the *right* component of the mirror image doublet, as described in Chapters 8 and 9; the RH–LH terminology used here refers to internal asymmetry, not relative placement.)

The evidence for each individual step shown in Figure 10.7 is considerably better than for the process as a whole. All three authors who have investigated this process (Jerka-Dziadosz, 1983, 1985; Tuffrau and Totwen-Nowakowska, 1988; Yano, personal communication) have reported an origin from type II (side-by-side) doublets, all have reported the budding off of singlet cells on the right side, and all have observed an interposition of a third, reverse set of ventral cortical structures, although to varying degrees (the dorsal surface has not been analyzed in these studies, therefore, its configuration remains conjectural). The timing of the budding-off of the RH singlet component, however, remains in doubt (see Section 9.2.2). Furthermore, Tuffrau and Totwen-Nowakowska (1988) view the triplet configuration as part of an alternative pathway (Fig. 9.9) rather than a step in the formation of mirror-image doublets from side-by-side doublets. Also, the transformation that they observed after side-by-side conjugation does not appear to involve any budding process (Fig. 9.5h–l). Finally, the regulation of the ciliature of the dorsal and ventral surfaces is closely coordinated in some circumstances (Grimes and Adler, 1978; Jerka-Dziadosz, 1989), but not in others (Hammersmith and Grimes, 1981). Additional work is needed before the interpretation shown in Figure 10.7 can be considered either as firmly established or unequivocally refuted.

Mirror-image doublets of oxytrichids, however generated, usually lack marginal structures at the meridian of symmetry (see Section 8.2.1). This can be understood in terms of the CCM by recalling that loss (or gain) of positional values can occur at this meridian with no loss of continuity. A . . . 6-7-**8**-7-6 . . . configuration can change to a . . . 6-7-6 . . . and back again. Such losses and gains (shown by dotted lines in the mapping in Fig. 10.3f) do not change the winding number, which remains zero. The topological situation, however, changes drastically when the two components are separated physically by transection across the mirror plane, creating complementary RH and LH singlets. If there has been a loss of positional values at the line of symmetry of the LH–RH doublet (Fig. 10.8a,b), then transection at that line will result in a positional gap at the newly formed margins (Fig. 10.8c). The subsequent intercalation (Fig. 10.8d) should specify restoration of the marginal cirral row and formation of a normal F-V-T cirral set. In agreement with this prediction, such restoration occurs not only in the viable RH component but also in the LH component that dies soon afterward because it is unable to feed (Grimes and Goldsmith-Spoegler, 1990b; Shi and Frankel, 1989b). The fact that a moribund cell draws on its dwindling energy reserves to carry out a functionally useless cortical restoration suggests that it is following a rule that genuinely constrains its development.

As noted earlier (see Section 8.2.3), the two components of a mirror-image doublet may separate spontaneously into a viable RH singlet and an inviable LH singlet; this is consistent with the CCM, especially as the one published photograph of such separating cells (Fig. 18 of Pang et al., 1984) indicates that marginal structures had been restored at the midline before separation. The observation, however, reported in the same paper that RH–LH mirror-image doublets may revert

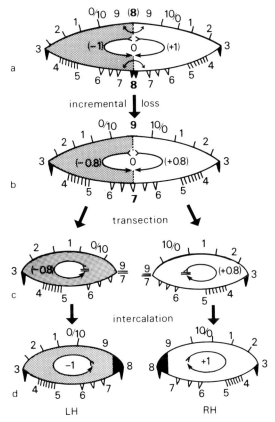

Figure 10.8 A hypothesis for loss of structures at the line of symmetry in mirror-image doublets of oxytrichids, based on the CCM. A hypothetical complete mirror-image doublet (**a**) may lose positional values incrementally at the meridians of symmetry (*double-headed arrows*) and, therefore, (**b**) fail to form structures specified by the "marginal" values. When such an incomplete mirror-image doublet is transected at the midline, both halves (**c**) will have a positional discontinuity after healing, which will trigger intercalation to form (**d**) new marginal cirri plus other structures expected near the margin of the cell (*blackened region*).

to RH singlets by resorption of the LH component is less easily reconciled with the CCM.

Reversal in tetrahymenids

The path outlined previously for oxytrichids probably is not taken by tetrahymenids, because they do not appear to bud off singlet cells from triplet complexes. The appearance of a few mirror-image doublets in the same culture that contained many more two-CVP-singlets (see Section 9.2.2 and Fig. 9.4) suggests that the route leading to formation of mirror-image doublets is related to that leading to two-CVP-singlets. The single reverse intercalation event postulated for the origin of two-CVP-singlets (Fig. 10.6) is not sufficient, however, to generate mirror-image doublets. *Two* separate reverse intercalation events must take place to bring the winding number down from +2 to 0. The location of these intercalation events determines the initial mirror plane.

The way in which two reverse-intercalation events could generate a mirror-image doublet of *Glaucoma* is shown in Figure 10.9a,b (for *Tetrahymena,* see Fig. 16 of Frankel and Nelsen, 1986a). Each reverse-intercalation event subtracts one complete circular set of positional values; subtraction of two adjoining sets converts one of the two RH semicells of an RH–RH parabiotic doublet into an LH sector of an LH–RH mirror-image doublet.

This scheme has an improbable feature: why intercalate twice if, as shown in Figures 10.5 and 10.6, a single intercalation event is sufficient to relieve all of the "crowding problems" of a shrinking parabiotic doublet? The concept of the zero-winding number condition as a topological trap offers a way out of this dilemma. Suppose that reverse intercalation events were triggered at random around the surface of a shrinking balanced RH–RH doublet. Simultaneous double occurrences might occur but would be rare. Mirror-image doublets would then originate much less commonly than two-CVP-set singlets or triplets. We have seen, however, that a triplet or two-CVP-set singlet is topologically equivalent to an ordinary RH singlet and is readily converted to such a singlet by incremental loss of positional values. A mirror-image doublet, on the other hand, is topologically different from either an RH or LH singlet and, therefore, should remain a doublet even while

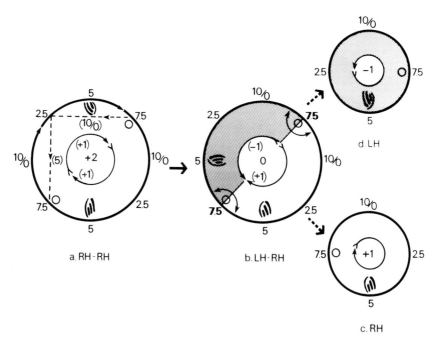

Figure 10.9 A hypothesis for the formation of mirror-image [RH–LH] doublets and left-handed [LH] singlets from a balanced RH–RH doublet of *Glaucoma.* The illustrative conventions are the same as in Figures 10.5 and 10.6, but to avoid clutter fewer positional values are shown. (a) The presumed original parabiotic doublet may have undergone reverse-intercalation at two sites *(dashed lines)* to convert the winding number of one semicell from +1 to −1, generating (b) a stable mirror-image [LH–RH] doublet, which then may undergo reversible loss of positional values (and, therefore, of a CVP set) at either of the two meridians of symmetry *(double-headed arrows).* Transection at a plane of symmetry could convert such an LH–RH doublet into one RH singlet (c) and one LH singlet (d).

losing positional values across one or both of the lines of symmetry. This explanation is particularly appropriate for *Glaucoma,* in which a particular clone (HR-2) has maintained the mirror-image doublet configuration with great stability despite substantial variation in the number of ciliary rows (Suhama, 1982, 1983, 1984), and temporary loss of either of the two sets of CVPs. The mirror-image doublet configuration might preserve and amplify the outcome of an exceedingly rare event.

This leaves us with the question of how LH singlets might originate. For *Glaucoma,* the answer is simple: Suhama (1985) obtained LH singlets by transecting mirror-image doublets. For *Tetrahymena,* the answer is less obvious, as LH singlets were obtained not by surgery, but by selection from cultures of regulating parabiotic doublets. This suggests that a mirror-image doublet of *Tetrahymena* either could transect itself, which would conform with the CCM, or might occasionally suffer regression of the RH component, which would not. It is not known how this conversion actually occurs.

If cells can, as shown in Figure 10.9, proceed from the RH to the LH singlet condition through the intermediate states of a parabiotic doublet (RH–RH) and then a mirror-image (RH–LH) doublet, we would predict that they should be able to go in the other direction, from an LH singlet to a RH singlet by way of a LH–LH parabiotic doublet and then a mirror-image (RH–LH) doublet. Consistent with this prediction, the frequency of reversion of LH singlets back to RH cells is greatly increased by conditions that promote fission blockage, which should create LH–LH doublets (Nelsen and Frankel, 1989; Suhama, personal communication). The predicted LH–LH doublets have been observed but they are surprisingly rare in both *Tetrahymena* and *Glaucoma;* the LH–LH doublet condition clearly is extremely unstable. Probably because of this instability, an abundance of intermediate forms is observed in cultures regulating from the LH to the RH condition. Among these, mirror-image (RH–LH) doublets and triplet forms with a *normal* (primary) OA located between two *abnormal* (secondary) OAs have been observed in reverting cultures of LH *Glaucoma* (Suhama, personal communication) and *Tetrahymena* (Nelsen and Frankel, 1989), although the latter form is rare in *Tetrahymena.* Additional forms that have been observed in both ciliates are consistent with the CCM if one assumes that OAs sometimes fail to develop in LH regions. Thus, pathways comparable to those worked out for the "forward reaction" from RH to LH singlets appear also to apply to the "back-reaction" from LH to RH singlets (Nelsen and Frankel, 1989).

The studies on regulating LH cells have, however, led to two further conclusions, one entirely new and the other anticipated by earlier results. The new conclusion arose from the unanticipated instability of LH–LH doublets, which indicates that there is no perfect parity between LH and RH forms. This, in turn, suggests that the interaction between the as-yet-invisible positional system and the visible microtubular cytoskeleton involves more than just an assessment of the slope of the positional system; there may be a preference for intercalary reorganization in the direction in which the asymmetry of the positional system is in harmony with that of the cytoskeleton. We already have encountered another possible case of bias in favor of a "normal" direction of intercalation in Uhlig's segment-exchange experiments in *Stentor* (see Section 6.3.1).

The second additional conclusion also concerns a preference, although of a somewhat different nature. Whenever a cell possesses both a normal and a globally reversed positional system, there must necessarily be *two* meridians of symmetry, at the two borders where reverse and normal systems abut on one another (indicated in boldface in Fig. 10.3 through 10.11). As explained earlier (see Section 10.3.3), the CCM allows loss of positional values to occur incrementally at *either* of these meridians. There appears, nonetheless, to be a clear preference for loss at the meridian at which the vectors of decrease in positional values face *away from* the line of symmetry. Thus, in triplets, the LH OA preferentially fuses with the RH OA to its *left,* both in *Blepharisma* (see Fig. 9.7) and *Tetrahymena* (Fig. 10.5; see also Fig. 10.11). For example, in the case shown in Figure 10.5, incremental loss of positional values occurs mainly at the 5-**6**-5 meridian, not at the 4-**3**-4 meridian. The same kind of preference seemed to operate in the rare RH–LH doublets that appeared during regulation of balanced RH–RH doublets, in which the secondary oral meridian appeared to shift in a counter-clockwise direction, bringing the CVP sets closer together (see Fig. 9.4d). This directional shift was even more obvious in the numerous RH–LH doublets found in our recent study of regulating LH cells, leading Nelsen to recognize that the system has a general capacity to distinguish positional "ridges" (e.g., 7-**8**-7 from "valleys" (e.g., 7-**6**-7) (Nelsen and Frankel, 1989). How it might do so is totally unknown.

10.3.5 Mutations Affecting Circumferential Positioning
janus *mutations and pattern maintenance*

The effects of *janus* mutations of *Tetrahymena thermophila* can be explained by the CCM if one assumes that these mutations prevent the expression of subsets of positional values (Frankel and Nelsen, 1986b). Suppose that the wild-type allele of the *janA* locus (*janA*$^+$) makes a product that is required to maintain the positional arc (8)-9-10/0-1-(2) (Fig. 10.10a). If, during conjugation, cells that were originally

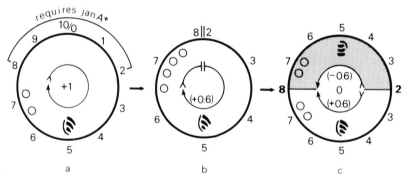

a   b   c

Figure 10.10 An interpretation of the conversion of a normal (RH) *T. thermophila* singlet into a *janus* mirror-image cell after a change of genotype from *janA*$^+$/*janA*$^+$ to *janA*/*janA*. Diagrammatic conventions are the same as in Figures 10.5, 10.6, and 10.9. After the genomic switch, an RH singlet (**a**) loses the positional values that require *janA*$^+$. This loss generates a major positional discontinuity (**b**) that provokes intercalation of the remaining permitted values to form (**c**) a cell with a mirror-image duplication of the ventral organization. The reverse-duplicated region is shaded.

of genotype *janA⁺/janA⁺* become *janA/janA,* dilution of the presumed *janA⁺* gene product during subsequent divisions would result in a progressive loss of the dorsal positional values, so that eventually the values 8 and 2 would confront each other, whereas the remaining values would spread out over the cell circumference (Fig. 10.10b). The confrontation of the noncontiguous positional values would trigger an intercalary reorganization by the only permitted route: (8)-7-6-5-4-3-(2) (Fig. 10.10c). This reverse-intercalation would result in a double-half cell with two ventral surfaces in mirror-image, and hence a winding number of zero.

A zero-winding number topology permits but does not force incremental loss of positional values at either or both of the two longitudes of symmetry. In *janus* cells, this loss presumably is limited by the fact that the cell circumference is nearly normal to begin with. Incremental elimination of positional values, however, should become activated if one could bring about an increase in the number of longitudes of symmetry. This was done by fusing *janus* cells side-by-side (Fig. 10.11a). Such *janus* "doublets" still have a winding number of zero (0 + 0 = 0), but now have *four* meridians of reversal in slope of positional values, i.e., two positional "ridges" and two "valleys." Under these conditions, when incremental loss of positional values occurs, the *janus* "doublet" should be converted to a *janus* "singlet" through collapse across at least one of the four longitudes of symmetry. One route of collapse, across a positional "ridge," is shown in Figure 10.11b,c. This situation allows a test of several predictions based on the CCM, most of which were confirmed (Frankel and Nelsen, 1987). These confirmed predictions include an increased variation in the relative width of complementary sectors, the loss of CVP sets in the narrowest sectors (Fig. 10.11b lower-left), and the convergence and probable fusion of secondary oral meridians with primary oral meridians to their left (however, there was no evidence for fusion in the opposite direction). These observations are all consistent with the CCM, although the preference for fusion and loss

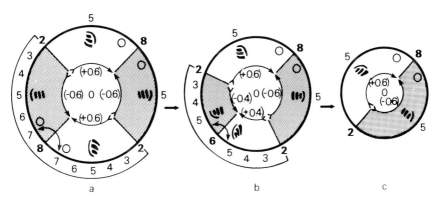

Figure 10.11 Interpretation according to the CCM of the conversion of *janus* "doublets" to *janus* "singlets." (**a**) A polar map of two *janus* cells that had been fused side-by-side to form a *janus* doublet. The *open* regions are of RH asymmetry, the *shaded* regions of LH asymmetry. When incremental loss of positional values occurs at the meridian of symmetry that is shown by the 8 at the lower left of the diagram, a reduction in the distance between one primary set of oral structures and the secondary set to its right will ensue. This will lead first to loss of the intervening CVP set (**b**), then to fusion and subsequent loss of both oral structures (not shown), and finally to reversion to the *janus* "singlet" condition (**c**).

of LH and RH oral meridians due to incremental intercalation along positional "ridges" rather than "valleys" (see Section 10.3.4) was not predicted by the original CCM and, therefore, is an extension of the model.

The problem of absence of predicted structures, already mentioned at the end of Section 10.3.3, seriously affects the application of the CCM to *janus* cells. Secondary oral structures always are expressed incompletely in *janus* clones. The simplest way to tackle this difficulty is to assume that "the postulated morphogenetic field underlying the secondary oral axis must be continuously propagated in a manner that is independent of its expression" (Jerka-Dziadosz and Frankel, 1979, p. 196). This rationale, however, implies that the presence of an appropriate positional value does not provide a *sufficient* condition for the actual formation of an oral structure. This limits both the explanatory power and the testability of the CCM. If the rationale is correct, then one should be able to find a molecular marker that labels all secondary oral meridians, whether expressed by secondary OAs or not.

The hypoangular mutant: a novel member of the janus family

If the CCM is correct, then viable mutations yielding a mirror-image *(janus)* global phenotype should be recovered only if the "prohibited" subset of positional values did *not* include positional values that specify "vital organs" such as the OA or the CVPs. This idea is compatible with the organization of *janA* and *janC* mutants (Fig. 10.12a,b) and with the somewhat different organization of *janB* mutants at 39°C (Fig. 10.12c,d), and also with that of *bcd,jan* double homozygotes (Cole et al., 1988). Imagine, however, that the positional values of most of the right half of the cell were removed (Fig. 10.12e). The result of reverse-intercalation of the remaining (permitted) values should be a secondary OA a short distance to the right of the primary OA, with *no* CVP set located anywhere in the cell (Fig. 10.12f). Such a cell would fail to osmoregulate and would certainly die, at least under normal culture conditions. Hence, no *janus*-type mutant with secondary oral structures located a short distance to the right of the primary oral structures should be recovered. We have, however, recently isolated such a putative mutant after our conventional mutagenesis procedures (Frankel and Jenkins, unpublished observations). Unfortunately, it is sterile and, therefore, its true genetic status remains unknown. In this putative mutant, typical *janus*-type secondary oral structures appear at about 20 percent of the cell circumference to the right of the primary oral meridian, yet a CVP set *is* present. This set is located about 10 percent to the right of the primary oral meridian instead of the normal 20 to 25 percent (Fig. 10.12h); on the average, the CVP set is midway between the two oral meridians in cells where both are present, thus preserving the *janus*-type topology. We have called this putative mutant *hypoangular* because it has a reduced angle of CVP determination relative to the primary oral meridian.

The hypoangular phenotype could be explained by assuming a "local nonautonomy" similar to that first observed by Stern (1954) in mosaic studies on the expression of the *achaete* mutation of *Drosophila*. If the region that normally is specified to produce a structure either is missing or rendered incompetent to perform its normal developmental task, a nearby region might substitute. We know,

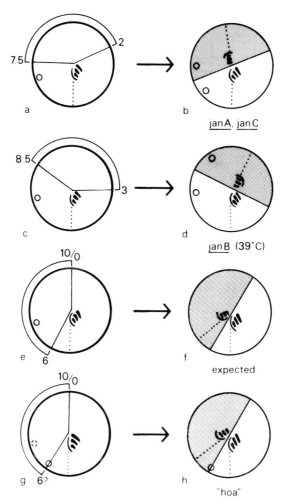

Figure 10.12 Polar maps showing consequences of elimination of a subset of positional values (*bracketed* on left) and subsequent reverse-intercalation of the remaining values (*shaded* on right) in *Tetrahymena thermophila*. (**a, b**) Elimination of a dorsal subset of values from approximately 7.5 to 2 brings about the average phenotype of cells homozygous for *janA* and *janC*. (**c, d**) Elimination of a dorsal subset of values from approximately 8.5 to 3 generates the average phenotype of *janB* cells (at 39°C) and of *bcd,janA* double homozygotes. (**e, f**) The elimination of a right-lateral subset of values from 6 to 10/0 should bring about loss of the CVP set and formation of a secondary oral meridian a short distance to the right of the primary oral meridian. (**g, h**) The hypoangular "mutant" brings about *janus*-like cells with a secondary oral meridian at the location predicted in **f**, but CVP sets, nonetheless, are formed in the narrow sector between the two oral meridians.

however, that CVPs are *not* formed within similarly narrow cortical sectors in wild-type "transient triplets" and in regulating parabiotic doublets of *janA* (note the absence of a CVP at the "6" positions in the narrow sectors between primary and secondary OAs in Fig. 10.5b and Fig. 10.11b). Thus, it appears as if the putative *hypoangular* mutation has altered either the locations of certain positional values or the way in which they are read.

bcd and mlm: *destabilizing mutations?*

A pair of allelic mutations in *T. thermophila* and a set of genic segregants in *Paraurostyla weissei* bring about comparable modifications of the distribution of cortical structures around the cell surface. The two *broadened cortical domain (bcd)* mutant alleles of *T. thermophila* broaden the circumferential arcs within which specialized cortical structures develop; oral primordia often develop near two or even three adjacent ciliary rows instead of the normal one, whereas CVPs are produced near posterior ends of four, five, or more ciliary rows instead of the usual two (Cole et al., 1987). The broadening of these domains of expression is not associated with any major shift in relative locations of the midpoints of the domains (Fig. 10.13). Surprisingly, when multiple oral primordia form, sometimes one of them manifests a reverse tilt of membranelles, much like the secondary oral primordia of *janus* mutants (Cole et al., 1987).

Cells made doubly homozygous for *bcd* and *janA* appear for the most part as if the *janA* mirror-reversal were superimposed on the *bcd* widening, although there is an additional shift of the secondary oral meridian to the right. Although the CVPs are still located within the sector to the right of the primary oral meridian and to the left of the secondary oral meridian, their location within that sector is less precisely regulated than it is in *janus* cells without *bcd* (Cole et al., 1988).

A cortical aberration that may be similar to *bcd* was described earlier in certain

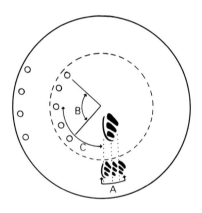

	wt	bcd	bcd/wt
A. OP domain	5.3%	7.2%	1.4
B. CVP domain	10.8%	24.1%	2.2
C. CVP location	22.9%	25.9%	1.1

Figure 10.13 A summary of the phenotype of cells homozygous for the *bcd* mutation of *T. thermophila*. The diagram is a polar map of a dividing *bcd* cell, showing oral primordia along several adjacent ciliary rows, making up the oral domain [A], and a broadened region of expression of contractile vacuole pores, the CVP domain [B]. The distance in ciliary row-intervals between the right postoral ciliary row and the midpoint of the CVP domain is the CVP location [C]. The table compares these values in wild-type [wt] and *bcd* cells, measured in ciliary rows and given as a percentage of the cell circumference. Data from Cole et al., 1987.

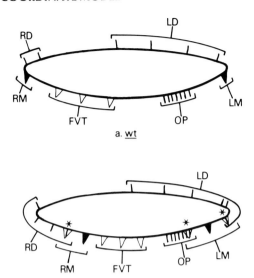

Figure 10.14 A schematic summary of the distribution of ciliary structures around the cell cir-cumference in (**a**) wild-type (wt) and (**b**) enhanced multi-left-marginal (*mlm*)(clone 95) *Paraurostyla weissei*. The labeled structures are the oral primordium (OP), frontal-ventral-transverse cirri (FVT), right marginal (RM) and left marginal (LM) cirri, and right dorsal (RD) and left dorsal (LD) ciliary rows. The longitudes of overlap of ciliary structures of different kinds are indicated by *stars*.

abnormal clones of *Paraurostyla weissei* (Jerka-Dziadosz and Banaczyk, 1983). Genetic analysis showed that this aberration was caused by a recessive allele at a single locus, called *multi-left marginal (mlm)* (Jerka-Dziadosz and Dubielecka, 1985). The *mlm* phenotype is highly variable, with expression dependent on genetic background. *mlm* by itself alters the distribution of cortical structures on the left half of the cell surface (Dubielecka and Jerka-Dziadosz, 1989); when combined with a suitable enhancer, *mlm* can bring about abnormalities in the distribution of all classes of cortical structures (Jerka-Dziadosz, 1989). The enhanced phenotype, summarized schematically in Figure 10.14, includes a varying combination of three anomalies: (1) changes in width of cortical domains, with widening most pro-nounced for the left marginal cirri, whereas the right marginal domains sometimes are contracted; (2) an overlap in the longitudes of expression of different structures, commonly observed for marginal cirri and adjacent dorsal bristles, and also for left marginal cirri and the oral primordium; and (3) occasional reversals of orientation of structures, especially within the most expanded zone, the left marginal cirri. These anomalies were explained by Jerka-Dziadosz (1989) as consequences of instability in the system of positional values combined with a lag in the interpre-tation of these values by developing ciliary primordia. A comparable instability might also account for the *bcd* phenotype in *Tetrahymena*, although the paucity of structures that are differentially expressed at different cell longitudes makes this elaboration of the CCM operationally meaningless until more cortical markers are found.

10.4 THE ANTEROPOSTERIOR SYSTEM

Unlike the circumferential system, the postulated anteroposterior system of positional values cannot propagate itself but must instead undergo periodic subdivision. It, therefore, must be far more labile than the circumferential system. In view of this major difference, one could reasonably ask whether the CCM makes any sense when applied to the anteroposterior organization. There are three kinds of observations that suggest that it does; these are summarized in a highly abstracted form in Figure 10.15.

The first of these observations is Tartar's (1967a) demonstration that the fission zone can rapidly be regenerated by predividing stentors after the presumptive fis-

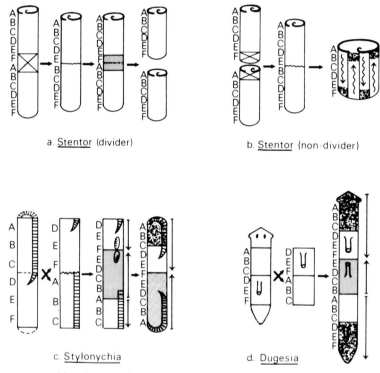

Figure 10.15   Highly schematic diagrams representing interpretations according to the CCM of experiments on the anteroposterior organization of (**a, b**) *Stentor* and (**c**) *Stylonychia*, compared with (**d**) Chandebois' interpretation of an experiment by Okada and Sugino in the flatworm *Dugesia. Wavy arrows* inside the diagrams represent direction of migration, *solid-tipped arrows* outside diagrams indicate direction of polarity. Regions of terminal reorganization or regeneration are *pebbled*, regions of intercalation *shaded*. (**a**) Removal of the fission zone of *Stentor*, with subsequent intercalation of a new fission zone and ensuing cell division. (**b**) Tandem fusion of two stentors after removal of cell extremities, with subsequent telescoping to form a stentor of near-normal length and double width. (**c**) Exchange of anterior and posterior regions of *Stylonychia*, with subsequent reverse intercalation near the line of fusion accompanied by completion of the anterior end and probable posterior respecification of global polarity. (**d**) Exchange of anterior and posterior regions of *Dugesia*, with subsequent reverse intercalation near the line of fusion, and anterior and posterior regeneration of head and tail, respectively.

sion zone is removed (see Section 5.3.1 and Fig. 5.8). This can be interpreted as an intercalation process (Fig. 10.15a) if one assumes that when a cell is activated to divide the extreme values of the anteroposterior system ("A" and "F") become topologically continuous.

A second observation, also by Tartar (1964), that juxtaposition of posterior on anterior regions in *non*dividing stentors leads to interpenetration of cortical wedges (see Section 5.3.1. and Figs. 5.9), shows that extremities of cells not activated for division cannot intercalate a fission zone. In terms of the CCM, this indicates that in cells not activated for division "F" and "A" are not normal neighbors. If one accepts this special postulate, the outcome of this experiment supports the idea of anteroposterior intracellular nonequivalence very strongly, because it shows that parts of the cell cortex will literally go to extraordinary lengths to meet their positional matches. This is presumably made possible by the flexibility of the cortex of *Stentor.*

The third experimental result is the outcome of Shi's anteroposterior juxtaposition in *Stylonychia,* in which there was no interpenetration of cortical regions but rather a reversal in polarity of cortical structures that were newly formed in the zone between the two grafted components (see Section 9.3.3). The general reversal of polarity was best explained by the "respecification hypothesis" presented in Section 9.3.3. When translated into the language of the CCM, the respecification hypothesis postulates that the juxtaposition of discrepant anteroposterior values in a nondividing cell with a cortex that presumably is more rigid than that of *Stentor* triggers intercalation of a cell-domain with reverse polarity (Fig. 10.15c).

A comparison of Shi's experiment with Okada and Sugino's (1937) topographically similar transplantation experiment in the flatworm *Dugesia* (see Section 1.5) reveals the generality of intercalation in the anteroposterior dimension but also highlights a possible difficulty in application of the CCM to the *Stylonychia* transplant. The regeneration of reversed structures at the site of positional disharmony in the worm (Fig. 10.15d) corresponds to the formation of reversed structures at the corresponding location in *Stylonychia.* The regeneration of a head at the anterior end of the "DEF" piece of *Dugesia* is paralleled by the less conspicuous remodeling of the anterior portion of the truncated "DEF" piece of *Stylonychia* to produce the most anterior structures (membranellar collar, etc.). However, the regeneration of a tail by the posterior end of the "ABC" piece of *Dugesia* has no parallel in *Stylonychia,* where there is instead a respecification of polarity to correspond to the reversed polarity of the middle region. The difference might be related to the relative morphogenetic passivity of latitudes that lie outside of the "proliferative zones" (Grimes, 1976), in contrast to the ability of all latitudes of flatworms to undergo both intercalation and terminal regeneration.

10.5 SUMMARY AND EVALUATION

Summary of the CCM

The cylindrical coordinate model (CCM) is basically an application of borrowed ideas of nonequivalence and continuity to the topology of the ciliate clonal cylinder. The CCM can be summarized in six statements: first, the appropriate topological framework is an indefinitely elongating cylinder. Second, two orthogonal and independent sets of positional values, one circumferential and the other anteroposterior, are located in the surface layer of this cylinder. Third, the set of circumferential values is propagated as the cylinder elongates, and the set of anteroposterior values is periodically subdivided. Fourth, positional values are "interpreted" to call forth production of specific intracellular structures. Fifth, the stability of a set of positional values depends not only on continuity but also on sufficient spacing of positional values. Sixth, reorganization by intercalation is triggered when continuity is violated or spacing is insufficient.

Application of the CCM

The CCM accounts for a substantial number of patterning phenomena in a variety of ciliates, including several that are unexpected or paradoxical. In *Stentor*, the CCM helps to explain some of the "insuperable dilemmas" (Tartar, 1966d), in which cells provided with both nucleus and cortex failed to regenerate. In *Stylonychia* and *Paraurostyla*, the CCM accounts for the contrast between the instability of side-by-side doublets and the stability of mirror-image doublets and also for the fact that mirror-image doublets lose structures at the line of symmetry, whereas singlets derived from these doublets regain such structures. Most strikingly, the CCM accounts for globally reversed arrangements of structures that appear spontaneously between juxtaposed normal ones, especially in the transient triplets encountered in several ciliates. Finally, the CCM accounts for most of the pathways of formation of mirror-image doublets and reverse (left-handed, or LH) singlets.

Testing of the CCM

The CCM has been deliberately tested in *Tetrahymena*. These tests were carried out in three situations, all of which involved major changes in cortical organization: first, the regulation of *janus* "doublets" to *janus* "singlets," second, the regulation of wild-type reverse (LH) singlets to normal (RH) singlets, and third, the phenotypes of new mutants affecting circumferential patterns. In the first test, the important predictions of the CCM were confirmed. In the second test, the cortical configurations predicted by the CCM have been observed and configurations "forbidden" by the CCM have not turned up. In the third test, although all mutations that generate additional oral meridians have displayed the predicted *janus*-type topology of cortical arrangements, the expected relative distances between different structural domains were altered in one such "mutant" (hypoangular). Another mutant *(bcd)* manifested an unanticipated loosening of positional constraints; Jerka-Dziadosz (1989) has accounted for the similar but more extensive changes observed in the *mlm* genic segregant of *Paraurostyla* by postulating a destabilization of positional values.

From the perspective of the CCM, "errors of omission" (a structure not forming where one is expected) have been observed very frequently, but "errors of commission" (a structure forming where it is not expected) have been rare, except in certain mutants. Both of these difficulties can be explained away. The errors of omission have almost all involved the failure of globally-reversed structures, usually oral apparatuses, to form when and where they were expected. These failures may be attributed to impediments in development of complex ciliary structures in LH regions due to the conflict between locally normal and globally reversed asymmetry. As for the errors of commission, as the "rules of the game" are presumably specified by genes, one might expect that genic mutants might occasionally modify these rules. Nonetheless, the CCM would be on stronger footing if we had a means of assessing positional organization independently of the structures specified by that organization, and if we could find out exactly how the mutant genes affect the underlying positional organization.

How good is the CCM?

The CCM, although fairly comprehensive and to some degree predictive, is at best a starting point for understanding the complex global dynamics of the ciliate surface. It clearly does not meet Lewontin's (1974, Chapter 1) tests for a good theory: dynamic and empirical sufficiency. Empirically, there are, as indicated earlier, major gaps in testability of the CCM. We do not yet know whether these gaps reflect flaws in the CCM or insufficient methods for detecting reversals when they exist. Even when results predicted by the CCM were obtained, these results revealed other features consistent with the CCM but not predicted by it. Dynamically, the situation is even worse. The CCM is in no sense a dynamic model, because it does not specify the nature of either the entities or the forces that underlie its formal rules.

In view of these major defects, I make three claims for the CCM: (1) it provides partial explanations for otherwise incomprehensible phenomena; (2) it suggests some additional types of observations for pushing our understanding further; and (3), most important, it calls attention to the fact that some dynamic system(s) controlling large-scale topological order *must* exist within the ciliate cortex.

How can better models be built?

I am aware of two possible directions toward construction of models that might potentially meet Lewontin's criteria. One is a line of theoretical investigation initiated by Goodwin (1980) that involves an application of Laplace's equation to the ciliate surface. The idea of continuity is embedded within this equation. Furthermore, Goodwin's model, especially in a more recent elaboration (Goodwin, 1989), deals seriously and explicitly with the topology of the ciliate surface, although in a manner entirely different from that of the CCM. The ciliate is dealt with as a compact solid, a topological sphere. Orthogonal fields, characterized by dual ("conjugate") solutions of Laplace's equation, are expressed on this surface. At specific times and places (determined in part by a third "solid field") these dual fields develop paired singularities: a region of converging field lines (sink) forms adjacent to a region where no field lines meet (saddle point). These two types of local singularity have an opposite topological index ($+1$ and -1, respectively), allowing for a major local perturbation while preserving smoothness at distant regions. The location of the paired singularities determines the position of the new oral apparatus (OA), the distance between the sink and its associated saddle determines the size of this OA, and the field vectors between the sink and the saddle specify the orientation of the oral ciliature. The asymmetry of the oral structures is controlled by "some weak chiral influence" (Goodwin, 1989), which in turn is dependent on the direction of curvature of the more longitudinal of the dual laplacian surface fields.

There are at least three potential strengths of this model. First, it deals with the physical surface of the ciliate cell, which is topologically spherical and not cylindrical. Second, it dispenses with the mysteries of interpretation of positional values and instead postulates that new structures are generated directly through the operations of the fields. Third, it can account not only for the location and asymmetry of oral structures, but also for the detailed organization of membranelle sets, a feature that is not touched on by the CCM. Nonetheless, this theory must meet the major empirical challenge of accounting for the phenomena summarized in the previous two chapters, including especially the coordinated maintenance and propagation of arrangements of many separate cortical structures. There is also the quite different question of dynamic sufficiency: it is not clear to me to what degree the Laplacian equations as used in Goodwin's theory can provide a mathematical foundation for a dynamic physicochemical system.

A class of mathematical models that certainly does supply a "dynamic" is the reaction–diffusion class, first introduced by Turing (1952). These models show that standing waves in concentrations of chemical substances (which Turing called "morphogens") can be produced by autocatalytic and crosscatalytic processes involving two or more components with different diffusion constants. Such standing waves can be generated by different "families" of reaction–diffusion models (Harrison, 1987). In the best-known family, the more slowly diffusing component is an autocatalytic activator, the more rapidly diffusing component, an inhibitor. The formation or release of the inhibitor is stimulated by the activator, and the inhibitor feeds back on the activator to inhibit its formation or release (Gierer and Meinhardt, 1972; Meinhardt, 1982). In another family, an autocatalytic activator depletes a substrate necessary for its own production (Lacalli, 1981; Meinhardt, 1982, p. 35). It is important that although the dynamics of these reaction–diffusion models are specified by the Turing equations, the actual components are not (Harrison, 1982).

At first glance, a reaction–diffusion model could readily be applied to the circumferential component of ciliate organization. Turing's original mathematical application was to a ring, and there have been several promising recent applications to the formation of periodic structures around the circumference of unicellular algae (Lacalli, 1981; Harrison and Hillier, 1985) and of cellular slime molds (Byrne and Cox, 1987). There are, however, two major problems with application of this class of models to circumferential patterning in ciliates. One is that although reaction–diffusion models predict uniform spacing around a ring, the transient-triplets described in Section 9.2.3 seem deliberately to be choosing nonuniformity. The second major problem is that the periodic patterns that are excellently explained by reaction–diffusion models are radially symmetrical, whereas ciliate patterns are highly asymmetrical. If any model of the reaction–diffusion type is applied to ciliates, it must account not only for nonuniform spacing but also for the differences in handedness of multiple patterns.

A dynamic theory that can account for asymmetrical periodic patterns must have components with phase relations other than 0 or 180 degrees. Unfortunately, of the well-known two-component reaction–diffusion models, the activator–inhibitor family has the two sets of standing waves (of activator and inhibitor, respectively) in phase, whereas the activator-depleted substrate family has the two sets of waves symmetrically (180 degrees) out of phase. The asymmetrical out of phase systems that have been proposed, first by Goodwin (1976, Chapter 6) and later in more detail by Russell (1985), are not in the form of a dynamically coupled reaction–diffusion equation; indeed Russell (1985, p. 279) comments that the two waves in his model belong to two "entirely independent systems." To my knowledge, however, a two component reaction–diffusion system with the required asymmetry is not theoretically impossible; alternatively, a model with four components (two for each of the asymmetrically out of phase waves) might do the trick. Either type of dynamic model is likely to be rather complex. Another alternative might be a mechanochemical model in which one of the components might be a structural feature of the membrane skeleton (Goodwin, 1989). Such models are conceptually related to reaction–diffusion models (Harrison, 1987), but their properties have not been as fully explored.

11

Can Ciliates Help Us to Find "Nontrivial Universals"?

11.1 INTRODUCTION: UNIQUENESS AND GENERALITY

Ciliates are peculiar organisms. Phylogenetically, they make up a group parallel to and of roughly the same status as animals, plants, and fungi (Sogin et al., 1986b). Organizationally, major aspects of both their nuclear and their cortical makeup are unique: the chromosomal complement undergoes a wholesale reconstruction whenever new macronuclei are formed, and a complex structural organization is perpetuated longitudinally as the ciliate clone grows.

Unique features are sometimes very useful for uncovering general biological principles. Thus, synchronous reconstruction of chromosomes during formation of the macronucleus provided a special opportunity for elucidating the organization of chromosome ends (telomeres). This organization, originally worked out in the ciliate *Tetrahymena thermophila* (Blackburn and Gall, 1978), has more recently been demonstrated in chromosomes of a wide variety of other organisms as well, including humans (Roberts, 1988). In this application and others, the contributions of research on ciliates to general molecular biology are now widely known. What is not as generally known is the degree to which such important discoveries depended on unique aspects of ciliate nuclear organization.

This chapter is devoted to demonstrating how equally unique aspects of ciliate cortical organization may contribute to a general understanding of pattern formation in multicellular, as well as unicellular organisms. I first summarize the way in which the principal ideas of ciliate pattern formation are related to the special features of ciliate cortical organization, and then highlight three ideas that are particularly important in a wider context. This leads to a more detailed exploration of one of these three ideas, that of intracellular nonequivalence, where I argue that a ciliatological perspective might help us in the search for "nontrivial universals" of positional organization (Lawrence, 1985). This argument is focused on one nonciliate system, the insect (particularly *Drosophila*) embryo, where I show how a viewpoint related to ciliate studies and models can help in the search for a unified explanation of phenomena that ordinarily are viewed as separate and unrelated. Finally, I consider the possible molecular mechanisms that might underlie intracellular nonequivalence.

11.2 GENERAL IDEAS DERIVED FROM CILIATE CORTICAL ORGANIZATION

11.2.1 Continuity and Hierarchy

For a ciliate, geometry is destiny. The critical geometric feature of the ciliate cortex, first explicitly recognized by Tartar (1962), is its potentially endless longitudinal growth. The "clonal cylinder" (see Section 2.1.2 and Fig. 2.2) grows continuously along its longitudinal dimension and becomes subdivided periodically at right angles to the longitudinal axis of growth. This geometry lends itself to description in terms of longitudes and latitudes (see Fig. 10.2). The dynamic properties of longitudes and latitudes differ greatly, however, as the longitudes can extend themselves endlessly, whereas the latitudes must undergo reorganization in every cycle of growth and subdivision. This contrast immediately suggests two ideas: that structural inheritance prevails along longitudes, and that regulative gradients govern the reorganization of latitudes.

Although these ideas are suggested by descriptive geometry, they must be established by experiment. As shown in detail in Chapter 4, the important principle that supramolecular organization can be inherited directly has been demonstrated rigorously in a number of cases. Although the experimental paradigm for this principle is the propagation of the preexisting symmetry and polarity of inverted ciliary rows (see Section 4.2.2), such self-propagation has been demonstrated not only for ciliary rows but also for structural systems at levels of organization ranging from microtubule bands to global fields.

In Chapter 5, we saw that cell division is preceded by a reorganization of an original single system of latitudes into a new tandem dual system; the cell-equator of the parental cell is transformed into the posterior and anterior poles of the daughter cells before cytokinesis begins. This situation lends itself to visualization of anteroposterior organization as a highly labile gradient. The idea of gradients was introduced into ciliate biology by Uhlig (1959); as we saw in Chapter 6 (see also Section 7.4.2), the evidence for the actual existence of such gradients is best for the organization expressed along the anteroposterior axis.

The remaining central ideas of ciliate spatial organization depend not just on the cylindrical geometry of ciliate growth, but also on the nested complexity of structural organization. Much of this organization depends on the integration of ciliary units into compound structures. Such multiple ciliary units (membranelles and undulating membranes) themselves are further compounded into yet more complex structures such as the oral apparatus (see Section 2.2, also Fig. 7.1).

Once again, an impression of an organizational hierarchy that is gained from descriptive work is reinforced by analytic studies. We have seen, in Chapter 7, that the formation of the oral apparatus involves much more than assembly mechanisms working over short distances; distinct longer-range integrative processes coordinate the organization of separate oral membranelles in a manner reminiscent of the operation of morphogenetic fields of embryos.

Spatial integration is also at work at still more global levels of organization. The nature of this integration, presented in detail in Chapters 8 and 9, is best summa-

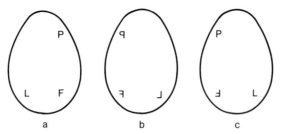

Figure 11.1 Diagrams illustrating hypothetical configurations of organelle sets in a ciliate. For explanation, see the text.

rized by some simple diagrams. Imagine a projection of the ciliate surface with three structures, represented by the letters P, F, and L (Fig. 11.1a). Each letter indicates an organelle system, such as a membranelle-set or a group of cirri, that has its own, often complex, internal organization. The positions of these letters represent the locations of these organelle systems in the ciliate cortex. Now imagine that we could *completely* reverse the handedness of the cell. This would convert it to the cell shown in Figure 11.1b, in which *both* the arrangement of structures and the internal organization of each structure are mirror-images of those in Figure 11.1a. Such total reversal has *never* been observed in any ciliate. What has been observed repeatedly is reversal of the *arrangement* of structures, whereas the details of internal organization of *each* of these structures have not been reversed. The individual structures retain their normal polarity and asymmetry, with a spatial organization that either is completely normal (the P and L in Fig. 11.1c) or is rotationally permuted (the inverted F in Fig. 11.1c). When the cell forms a new set of structures and divides, the arrangement of structures is perpetuated, but the specific orientation of each structure often is not.

The emergence of a new phenomenon (reversal of asymmetry) at a global level of organization in itself suggests that mechanisms are at work at that level that differ from those at the more local levels. The most appropriate, although not perfect, expression to describe this situation is that of hierarchical organization. This use of the term "hierarchy" is meant to convey the idea that the larger organization is more than a mere extrapolation of the smaller units nested within it. A political analogy would be to a national government that is more than a mere confederation of local governments. The term as used here, however, implies only the distinctness of organizational systems working over different scales, not the dominance of one system over another; again using the federal analogy, I am not implying that the "national government" of a ciliate is more powerful than the "local government," only that it is different.

Having summarized the basic ideas presented in this book, next I single out three specific areas in which a ciliate perspective can provide insights into analogous (possibly homologous) events in other organisms.

11.2.2 Three Ideas at Three Hierarchical Levels
Structural inheritance

The idea of structural inheritance is probably the best known contribution of ciliate biology to general concepts of pattern formation, as it has been brilliantly demonstrated and powerfully advocated by Sonneborn and his students (e.g., see Sonneborn, 1964, 1970; Nanney, 1968b; Grimes, 1989). As we have seen in Chapter 4 and summarized above, the topology of ciliate growth facilitates the rigorous demonstration of the inheritance of differences in supramolecular organization in the presence of a constant genotype. In essence, ciliate geometry allows a structural scaffold to perpetuate its own organization in a simple and obvious way. Although the geometry of ciliates is special, the principle of heritable intracellular organization, which this geometry helps us to recognize, may be much more general (see Section 4.5 for examples). This principle also has an interesting application to the propagation of polarity at the tissue level (see Section 11.3.2). Its relevance to embryonic development is more doubtful, as embryos appear to wipe their parental structural slate clean and start building their organization anew.

Batesonian fields

The examination of the details of the modeling of ciliary structures within the most complex of ciliate organelle systems, the oral apparatus, has led us to a rediscovery of William Bateson's principle of integration of serially repeated parts, 90 years after Bateson originally proposed it (see Section 7.2.5). The idea is that the specific configuration of each member of a series of repeated parts, such as digits or teeth, depends on its position within the whole series (Bateson, 1892, 1894). In Bateson's view, there exist no separate specifications for the anatomy of the individual elements within the series; instead, the anatomy of each element depends on its location relative to a continuously varying property that spans the entire field; a structure is told what it is according to where it is. This idea is illustrated in Figure 11.2, showing how spatially coordinated variation could lead to systematic differences in a series of imaginary structures depending on whether there were three such structures (Fig. 11.2a) or four (Fig. 11.2b).

The detection of underlying continuities spanning a set of serially repeated structures requires, first, that each element of the normal series has its own distinguishing characteristics, and second, that the number of elements within the series varies naturally or can be made to vary experimentally. The best documented vertebrate example is the pattern of cusps in teeth (Bateson, 1894, Chapters IX and

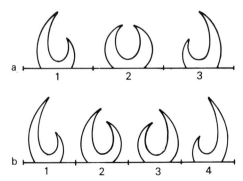

Figure 11.2 The principle of a batesonian field illustrated with a hypothetical linear array of structures, in (a) an organism with three such structures and (b) another organism of the same species with four such structures. For further explanation, see the text.

X). In the ciliate *Tetrahymena,* the unique arrangement of basal bodies within each membranelle, created by the final modeling of these structures, provides an even clearer example, as one can map the location and condition (ciliated or unciliated) of each individual basal body in suitably prepared specimens. In ciliates, this principle operates at the intermediate level of organization of an organelle complex, with dimensions too large for macromolecular assembly processes but still substantially smaller than the entire cell surface (see Section 7.2.5).

From my reading of the embryological literature, it appears that Bateson's principle of integrated organization has not so much been rejected as forgotten. Thus it is difficult to decide, when vertebrate digits or insect segments are labeled by number in illustrations within published experimental studies, whether these numbers reflect a truly digital (in the information sense) underlying organization, or whether the real organization might follow the analog mode postulated by Bateson. If the latter were true, investigators might unconsciously be partitioning a true continuum into artificially discrete states by seeking best matches between their observations and members of a set of platonic mental images. Bateson's idea may, of course, really be false in many systems, particularly those in which the discreteness of organization of the elements is necessary for normal function. In the ciliate case, the set of membranelles operates as an integrated entity, and their joint conformation is probably more important than their individual architecture. This might well also be true for the dentition or even for a foot or a wing; Bateson's principle at least deserves explicit attention.

Intracellular nonequivalence

The analysis of mirror-image reversals that occur uniquely at the most global level of ciliate organization has led us to the conclusion that the Lewis-Wolpert principle of nonequivalence (see Section 1.6) applies not only to separate cells of multicellular organisms but also to regions within cells (see Section 10.1). Although the case for intracellular nonequivalence was presented in Chapter 10 entirely from evidence in ciliates, this idea was originally based on a resemblance of certain paradoxical findings in ciliates to similar counter-intuitive observations made in insects. One such insect observation was that the juxtaposition of anterior and posterior portions of a body segment of a bug *(Oncopeltus)* led not to regeneration of the expected midregion of the segment but rather to the formation of a reversed segment border (Fig. 11.3) (Wright and Lawrence, 1981). The parallel ciliate observation was of an equally unexpected formation of a globally reversed third set of cortical structures between two normal structures that were placed abnormally close together (see Section 9.2.3). In both cases, the observations made sense if one assumed the existence of a set of underlying positional values with a circular topological organization that can be restored to a required state of continuity through intercalation by the shortest available route around that circle. These ideas were derived from the polar coordinate model (PCM) of French et al. (1976).

The striking parallel between reverse-intercalation in ciliates and in insects led to the formulation of a cylindrical-coordinate model (CCM) for the organization of positional values in ciliates. The formal structure of the CCM is largely parallel to the PCM except for some topological details (Chapter 10). There are two major

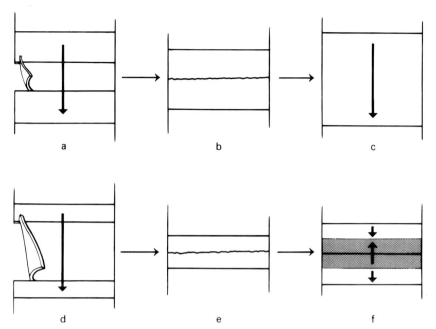

Figure 11.3   Diagrams illustrating results of excisions of strips of cuticle and underlying epidermis in third-stage larvae of *Oncopeltus fasciatus*. *Heavy vertical arrows* show cuticular polarity. (**a**, **d**) The original segments, showing the excised regions. (**b**, **e**) The bringing together of the wounded edges immediately afterward. (**c**, **f**) The result observed at the fifth stage. (**a**, **b**, **c**) An experiment in which less than half of the width of the segment was removed. The excised central region regenerated, with retention of the original polarity of cuticular sculpturing. (**d**, **e**, **f**) An experiment in which more than half of the width of the segment was excised. An extra segment border regenerated *(heavy horizontal line)*, and polarity of cuticular sculpturing in its neighborhood *(shaded)* was reversed. Drawn from the description of Experiment 7 of Wright and Lawrence (1981).

differences, however, in the way in which these models are applied to biological systems: first, in ciliates positional values must label parts of cells, not whole cells; second, pattern regulation in ciliates is by reorganization (morphallaxis), not by localized growth (epimorphosis). In the PCM, the restoration of continuity of positional values after juxtaposition of cells bearing nonadjacent values requires cell proliferation at the line of discontinuity. Within ciliates, there is no juxtaposition of separate cells; reorganization of large-scale pattern commonly occurs in ciliates that are growing and preparing to divide but does not appear to take place by means of local insertion and/or expansion of new surface. These differences are so large that they suggest that the similarities between ciliates and animals might exist only at a formal level and that the underlying molecular mechanisms are altogether different.

This last conclusion, however, rests largely on the assumption that intercalation in animals is always local and epimorphic. Although the bulk of the evidence favors the epimorphic view for intercalary regeneration in the postembryonic insect systems to which the PCM was originally applied (Anderson and French, 1985; Kiehle and Schubiger, 1985; Bryant and Fraser, 1988), morphallactic respecification may sometimes occur (Truby, 1986). In certain other organisms, notably the flatworm *Dugesia,* intercalary regeneration appears to include substantial respecification of existing tissue (see Section 1.5).

The "intercellular-local-epimorphic" view of positional reorganization was formulated from studies on postembryonic development and regeneration in insects and vertebrates, whereas the "intracellular-global-morphallactic" view was conceived largely (although as we shall see not entirely) from work on ciliates. The question is, which view better describes positional regulation during embryonic development? The following section shows how far the intracellular-global-morphallactic perspective can take us in interpreting mirror-image reversals in insect embryos.

11.3 A CILIATOLOGIST'S VIEW OF EARLY *DROSOPHILA* DEVELOPMENT

11.3.1 Why Concentrate on Insect Embryos?

I should first pause to explain why a major portion of the final chapter of a book on pattern formation in ciliates should be devoted to *Drosophila* and its close relatives. The reason can be put in the form of an answer to a hypothetical challenge: what can research on global patterning in ciliates tell us about pattern formation in a different system, particularly one in which genetic and molecular studies are far more advanced than they are in ciliates? The remainder of this section is an answer, centered around an analysis of mirror-image duplications in *Drosophila* and related dipterans. My basic arguments are: first, that although molecular–genetic findings contribute enormously to an understanding of pattern formation in embryos they do not, and perhaps cannot, provide sufficient explanations for the rules of global patterning, and second, that studies on pattern formation in the ciliate cortex can point us in the correct direction for seeking a full solution. To make the first argument, I review a particular set of phenotypes observed during early development in dipteran (particularly *Drosophila*) embryos (Sections 11.3.2 to 11.3.5); then for the second argument I provide a point-by-point comparison of pattern regulation in insects and ciliates (Section 11.3.6), drawing primarily, but not exclusively, on material covered in the immediately preceding review.

11.3.2 A Patterning Hierarchy May Exist in the Insect Cuticle

In the original formulations of gradients in the insect segment (Stumpf, 1966; Lawrence, 1966), the pattern of differentiation and the orientation of structures such as bristles and hairs were believed to be separate manifestations of the same underlying gradient. This is most easily visualized in Lawrence's (1966) "sand model" of morphogenetic gradients, in which altitude on an imaginary sand slope determines the nature of the structures, whereas the direction of the slope determines the orientation of these same structures. However, patterns of differentiation and of polarity can be affected separately in mutants of *Drosophila* (Gubb and Garcia-Bellido, 1982; Held et al., 1986) and can be dissociated in rotated cuticular grafts in the bug

Dysdercus (Nübler-Jung and Grau, 1987). The latter authors have concluded that these two aspects of pattern are under separate control (Nübler-Jung and Grau, 1987; Nübler-Jung, 1987). Although the differentiation pattern is dependent on some property that is graded across the entire segment (Locke, 1960; Nübler-Jung, 1977), the polarity pattern is strictly dependent on local cell-to-cell interactions, possibly related to concordance in orientation of microtubule organizing centers (Whitten, 1973; Nübler-Jung et al., 1987). This view is consistent with the local (and directional) nonautonomy of some mutants at the *frizzled* locus that affect the polarity of bristles and hairs (Vinson and Adler, 1987). Nübler-Jung et al. (1987) have compared the hierarchy of control in *Dysdercus* to the situation encountered in *janus Tetrahymena,* in which dissociation of local and global asymmetry was first reported (Jerka-Dziadosz and Frankel, 1979).

11.3.3 Mirror-Image Duplications Can Be Generated at Three Stages of Insect Development

In what follows, we deal exclusively with embryos of Diptera (flies), primarily *Drosophila*. These are holometabolous insects with four discrete life-cycle stages (egg–larva–pupa–adult). Early mitoses are syncytial; in *Drosophila,* nuclei destined for somatic regions divide near the center of the egg during the first seven mitotic cycles, migrate to the periphery during the eighth and ninth cycles, divide there for four more cycles (10th to 13th) without creating cell boundaries (syncytial blastoderm stage), and produce cell boundaries during a prolonged 14th cycle. Thus, blastodermal cells become sealed off from each other at the end of the 14th cycle, coincident with the onset of gastrulation and of germ band elongation (Foe and Alberts, 1983), although they retain cytoplasmic connections with the underlying yolky region until the middle of germ band elongation (Rickoll and Counce, 1980). The process of visible segmentation begins later, when the germ band has completed its elongation (Turner and Mahowald, 1977). In this embryo, which is of the "long germ" type (Sander, 1976), segmentation is simultaneous over the entire germ band; there is no localized growth zone (Turner and Mahowald, 1977). Indeed, between the cellularization of the blastoderm and the beginning of the first larval stage, there are only two further rounds of cell division in the ectodermal region that secretes the larval cuticle (Szabad et al., 1979; Campos-Ortega and Hartenstein, 1985).

At the blastoderm stage, certain regions are set aside that later invaginate to form imaginal discs (Gehring, 1978). During pupation, the larval integument is resorbed, and is replaced by an adult cuticle secreted by the evaginated epithelium of the imaginal discs.

Mirror-image reversals in dipterans are of three types: whole-embryo, intrasegmental, and imaginal disc. Whole-embryo reversals involve replacement of blocks of one or more segments, and have been induced by experimental manipulations of noncellularized embryos or by effects of mutated genes that act in the nurse cells, the oocyte, or in the embryo before cellularization. Intrasegmental reversals involve absence of a portion of the pattern of each larval segment with replacement by a mirror-image of the remaining portion, and have been induced by mutated genes that act after cellularization. Imaginal disc reversals involve partial pattern losses and complementary duplications (or occasionally duplications without losses) of parts or all of the derivatives of an imaginal disc, and have been induced by mutations acting during the embryonic or larval stages and by surgical operations carried out on imaginal discs of larvae. Because in the vast majority of these cases a part of a normal structure is missing and the complementary part duplicated in reverse, these mirror-image patterns are comparable topologically to those of *janus Tetrahymena* (see Section 10.3.5). We focus here on the whole-embryo and intrasegmental reversals observed in embryos, which are more uniform in their characteristics than are the reversals involving imaginal discs.

11.3.4 Whole-Embryo Reversals

The paradigm for whole-embryo reversals is the "double abdomen" phenotype, most intensively investigated by Kalthoff and collaborators in chironomid larvae, especially those of the genus *Smittia*. In brief, various types of damage applied to the anterior end of the embryo during preblastodermal cleavage stages resulted in a replacement of the head, thorax, and the first one to three abdominal segments by the remaining abdominal segments arranged as a mirror-image of the normal abdomen (Fig. 11.4). A wide variety of experiments indicated that formation of double abdomens was triggered by destruction of an anterior determinant (A in Fig. 11.4) that had the properties of an anteriorly localized RNA (reviewed by Kalthoff, 1979, 1983). This destruction was presumed to permit an ectopically located posterior determinant (P′ in Fig. 11.4) to reorganize the anterior half-embryo in the same way that the normal posterior center (P in Fig. 11.4) organizes the normal posterior half. Although this analysis was almost entirely conducted by manipulations of wild-type embryos, a maternally predetermined genic condition was

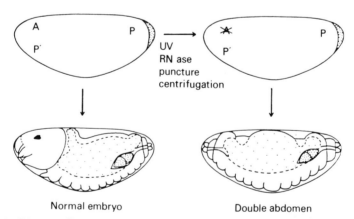

Normal embryo   Double abdomen

Figure 11.4 Diagrams illustrating normal development of first-stage larvae of the midge *Smittia* (*left*) and the induction of double abdomens (*right*) by various treatments, all of which are assumed to inactivate or displace anterior determinants [A]. The formation of reversed abdomens at the anterior end is presumed to occur under the influence of posterior or posterior-like factors [P′] localized at the anterior pole. The somatic primordia are fully reversed, but germ cells (*dark stippling*) form only in the original posterior abdomen. From Figure 2 of Kalthoff (1979), with permission.

recently found in one chironomid *(Chironomus samoaensis)* that greatly enhanced the propensity to form double abdomens (Kuhn et al., 1987).

Drosophila forms double abdomens less easily than do the chironomids. Induction of double abdomens in wild-type *Drosophila melanogaster* required extreme measures: cytoplasm had to be removed from the head region of early cleavage-stage embryos *and* cytoplasm derived from the tail region had to be injected in its place; either procedure by itself resulted only in head defects, without mirror-image reversals (Fröhnhofer et al., 1986). In *Drosophila,* however, a family of nonallelic maternal-effect mutations, known collectively as *bicaudal,* produced a similar double-abdomen phenotype (Nüsslein-Volhard, 1977; Mohler and Wieschaus, 1985, 1986). In all of these mutations, the phenotype of the embryo depended exclusively on the genotype of the mother, and temperature-sensitive periods during oogenesis (Nüsslein-Volhard, 1977; Mohler and Wieschaus, 1985) indicate that these genes exert their function before fertilization. Expression of all *bicaudal* mutations was variable; each female of an appropriate genotype produced embryos with a range of defects. This range, analyzed for the original *bicaudal (bic)* mutation by Nüsslein-Volhard (1977) and subsequently for the *BicD* mutation by Mohler and Wieschaus (1986), is presented in Figure 11.5.

The segmental layout of normal *Drosophila* larvae, projected back onto a rough approximation of the blastodermal fate map [see Fig. 8.1 of Campos-Ortega and Hartenstein (1985) for an accurate fate map], is shown in Figure 11.5a. The larva possesses 12 externally obvious segments plus two involuted posterior-head (gnathal) segments; these 14 segments are collectively labelled H4 to A8. The anterior head region (HR) is presumed to include the equivalent of three additional segments (H1 to H3) (Jürgens et al., 1986), whereas the tail region (TR) is believed to be made up of fused derivatives of four segments (the posterior half of A8 through A-11) (Jürgens, 1987).

The less severe effects of *bicaudal* mutations involved loss of some or all of the head structures (Fig. 11.5b). If all of the thoracic structures (and usually some anterior abdominal structures) also were lost, then the embryo was no longer simply truncated; instead the head and the thorax were replaced by a reversed tail region (Fig. 11.5c). As more of the anterior segments were lost, more and more posterior segments were duplicated in reverse (Fig. 11.5d), with the extreme and most common state being a perfectly symmetrical mirror-image duplication of the posterior abdominal segments that was complete except for the absence of the germinal pole cells at the ectopic posterior end (Fig. 11.5e) (Nüsslein-Volhard, 1977; Mohler and Wieschaus, 1986). In this phenotypic series, three rules are always obeyed: (1) when a given segment is missing, all of the more anterior segments are missing as well; (2) the embryo is either of normal polarity throughout (Fig. 11.5a,b) or has a duplicated tail-region at the anterior pole (Fig. 11.5c–e); and (3) in double-abdomen embryos, both sets of abdominal segments are continuous and all members of each set have an identical internal polarity.

The same rules are obeyed by embryos of *Chironomus samoaensis* possessing a maternal genotype predisposing toward double abdomens (Percy et al., 1986). One of the few differences from *Drosophila* is that in *Chironomus* fewer anterior structures were missing in association with each class of mirror-image duplications: a tail-region could appear at the anterior end even if the entire thorax was still

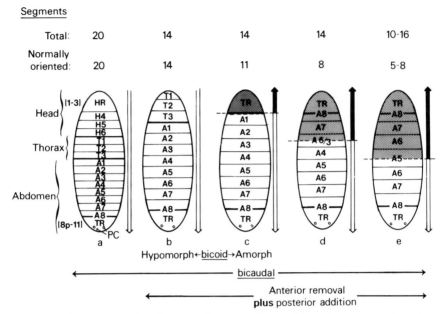

Figure 11.5 Sketches of the larval segmental pattern of *Drosophila melanogaster,* projected onto the blastodermal fate map. (**a**) is an approximate representation of the undisturbed condition, (**b**) to (**e**) of the disturbances after anterior losses resulting from mutations or operations. Each segment is indicated by a letter indicating the major body division [H, head; T, thorax; A, abdomen] and by a number indicating its position within that region. The segmental equivalents of the nonsegmented head-region [HR] and tail-region [TR] are given only in (**a**). Pole cells [PC] are indicated by *small circles* at the posterior end. The *dashed horizontal lines* in diagrams (**c**), (**d**), and (**e**) indicate planes of reversal of polarity, and the *vertical arrows* indicate the polarity on either side of these planes, with the *open downward-pointing arrows* representing the normal polarity, the *filled-in upward-pointing arrows* reversed polarity. Regions with reversed polarity are shaded. A "hypomorph" is an allele with partial loss of function, an "amorph" is an allele with total loss of function. For further information, including sources, see the text.

present, whereas in symmetrical double-abdomen embryos the average mirror-plane was in the third abdominal segment rather than the fifth abdominal segment as in *bicaudal Drosophila* embryos. This more anterior mirror-plane was also characteristic of double abdomens in *Smittia,* which were almost always symmetrical (Kalthoff, 1983).

Whatever the severity of the larval phenotype, *bicaudal* embryos of *Drosophila* were of normal size at the blastoderm stage. The positional losses and substitutions resulted from coordinated changes of specification of individual regions. For *BicD,* these changes of specification were determined during oogenesis and first became evident at the time of gastrulation (Mohler and Wieschaus, 1985).

The respecification of what otherwise would have been an anterior end to form a reversed posterior end generally has been interpreted as a direct consequence of a posterior influence coming to predominate at the anterior pole. In chironomids, a latent posterior influence presumed always to reside at the anterior pole is sufficient to bring about a complete double-abdomen phenotype (discussed above); in *Drosophila,* however, removal of an anterior determinant, using either a micropipette (Fröhnhofer et al., 1986) or a lack-of-function (amorphic) mutation of a gene *(bicoid)* that codes for an anterior determinant (Fröhnhofer and Nüsslein-Volhard, 1986; Driever and Nüsslein-Volhard, 1988a,b), can by itself do no more than to cause a posterior terminal region to appear at the anterior end (Fig. 11.5c). A symmetrical mirror-image embryo can form only if posterior determinants are added, either surgically by addition of posterior cytoplasm to the anterior pole of a wild-type (Fröhnhofer et al., 1986) or *bicoid* (Nüsslein-Volhard et al., 1987) embryo, or genetically by mutated *bicaudal* genes. [*BicD* is known to be an "antimorphic" mutation (Mohler and Wieschaus, 1986) that causes inappropriate expression of a posterior determinant in an anterior region (Lehmann and Nüsslein-Volhard, 1986; Nüsslein-Volhard et al., 1987).] This interpretation is a contemporary extension of the classical idea that the character of each structure formed along the anteroposterior axis of an embryo is determined by the relative levels of two graded substances that act antagonistically, one with a maximum near the anterior end, the other with a maximum near the posterior end (Hörstadius, 1973), an idea that was first applied to insect development by Sander (1959).

Mohler and Wieschaus (1986) have also noted a more general relationship: whenever one-half or more of the normal complement of segments was missing, parts or all of the remaining segmental array underwent mirror-image duplication. The conclusion that this relationship indeed is general is supported by the fact that it applies not only to double abdomens, but also to different mutant phenotypes in which other, nonterminal groups of segments were mutationally eliminated (Schüpbach and Wieschaus, 1986). In both of the cases cited by Schüpbach and Wieschaus (*Krüppel* homozygotes and *staufen-exuperantia* double homozygotes) mirror-image duplications were found only when eight or nine, or more, of the normal segments were missing, and was *not* found in milder cases in which fewer segments were missing. The same rule applies to the *hunchback* mutation, described later by Lehmann and Nüsslein-Volhard (1987). These cases, together with *bicaudal,* suggest that only the number and not the identity of the missing segments matters for the triggering of mirror-image reversals of entire segments or groups of segments.

Mohler and Wieschaus (1986), and also Schüpbach and Wieschaus (1986), have noted that the conditions for formation of these mirror-image reversals are reminiscent of the observations that inspired the polar coordinate model. One major difference, however, in the double-abdomen case is that initiation of reorganization does not involve an actual juxtaposition of disparate positional values (Mohler and Wieschaus, 1986). It is as if a loss of a sufficient portion of the system (or perhaps an expansion of the portion that is *not* lost?) itself can trigger a reorganization that mirror-duplicates some or all of the remainder of that system.

This general interpretation of whole-embryo mirror images in *Drosophila* also suggests a conceptual link between the greater ease of generating whole-embryo mirror images in chironomids (*Smittia* and *Chironomus*) compared with *Drosophila* and the fact that the plane of symmetry generally is more anterior in the chironomids than it is in *Drosophila.* The link may be a different structural threshold for reorganization (e.g., a difference in "elasticity" of structural elements or systems), or, as suggested by Percy et al. (1986), a different location of segmental anlagen on the blastodermal fate map, in which case deletion of fewer segments might be sufficient to cause the same physical expansion of the remainder. In either case, it seems unlikely that molecular signals that determine the identity of particular segments can provide a sufficient explanation for the occurrence (or nonoccurrence) of mirror-image reversals in the early embryo. The true decision probably is made by a separate structural system on which these signals impinge.

11.3.5 Intrasegmental Reversals

In *Drosophila,* a substantial number of recessive mutations in zygotically acting genes produce intrasegmental reversals of both differentiation and polarity patterns. These segment-polarity phenotypes are generated by mutations in ten genes that are primarily or exclusively zygotic in their action [*runt, armadillo, cubitus interruptus, fused, gooseberry, hedgehog, patched, wingless* (Nüsslein-Volhard et al., 1982), *odd-skipped* (Nüsslein-Volhard et al., 1985), and probably also *naked* (Martinez-Arias et al., 1988)]. In addition, at least two late-acting embryonic lethal mutations display a segment-polarity phenotype as a maternal effect [*disheveled* (Perrimon and Mahowald, 1987) and *costal-2* (Grau and Simpson, 1987)]. What all of these mutations have in common is loss of structures characterizing a portion of a segment and replacement of the missing structures by a mirror-image of the remaining structures. Analysis of temperature-sensitive periods (Gergen et al., 1986; Baker, 1987), of somatic mosaics (Wieschaus and Riggleman, 1987), and of changes in molecular expression (DiNardo et al., 1988; Martinez-Arias et al., 1988) all indicate that these effects are both specified and executed between the time of cellularization of the blastoderm and the formation of the cuticle of the first larval stage.

Two examples, *armadillo* and *patched,* can be used to illustrate the phenotypes of segment-polarity mutations. In abdominal segments of a wild-type first-stage larva (Fig. 11.6a), the segment border (horizontal dashed line) lies just behind the most anterior of six or seven transverse denticle bands, and the remaining denticle bands occupy the anterior one-third of the next segment. Each denticle band has its characteristic shape and orientation. Different bands have different orientations, but for simplicity the orientations are displayed in Figure 11.6 as if they were homogeneous (for a full description, see Gergen and Wieschaus 1986b). The remainder of each segment consists of naked cuticle.

In embryos expressing the most extreme *armadillo* phenotype, the segments were substantially shortened and the naked cuticle of each segment was replaced by a set of denticle bands, each with an orientation that is the reverse of that of the corresponding normal denticle band (Fig. 11.6c, shaded area) (Wieschaus and Riggleman, 1987). Effectively, the posterior two-thirds of each segment was replaced by the anterior one-third duplicated in reverse. This replacement was reflected at the molecular level by the premature loss of the product of the *engrailed* locus, which normally is expressed in approximately the posterior quarter of each normal segment (Fig. 11.6b, heavy dots) (DiNardo et al., 1988).

In the most extreme expression of *patched,* the larva was of nearly wild-type size, and the posterior half of each set of denticle bands was missing, whereas the anterior half was duplicated in reverse (Martinez-Arias et al., 1988). Thus, it could be inferred that the midregion of each segment was replaced by the reversed seg-

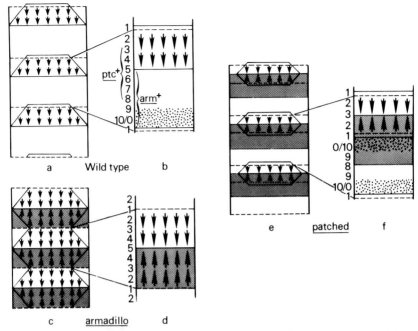

Figure 11.6 Diagrams of organization of the ventral cuticle of abdominal segments of first-stage larvae of *Drosophila melanogaster*. (**a**, **b**) wild-type, (**c**, **d**) *armadillo* homozygotes, and (**e**, **f**) *patched* homozygotes. The left diagrams (**a**, **c**, **e**) each show three successive segments, with the *dashed horizontal lines* indicating segment borders. The *trapezoids* indicate the regions occupied by denticle bands; the *small vertical arrows* represent the polarity of these bands in a simplified manner (see text). The regions in which reverse intercalary reorganization has taken place are *shaded* in (**c**) through (**f**). The right diagrams (**b**, **d**, **f**) show the midline region of one segment, with numbers indicating the presumed positional values. In (**b**), the *vertical brackets* indicate the positional values that are presumed to be dependent on the wild-type products of the *armadillo* [arm+] and *patched* [ptc+] genes. The *heavy dots* represent the localization of the *engrailed* [en+] product. Information for (**c**) and (**d**) from Wieschaus and Riggleman (1987) and DiNardo et al. (1988), for (**e**) and (**f**) from Martinez-Arias et al. (1988) and DiNardo et al. (1988).

ment-border region (Fig. 11.6e) (Nüsslein-Volhard and Wieschaus, 1980). This interpretation was confirmed by molecular studies that showed that the product of the *engrailed* gene appeared as an *extra* set of bands in each segment of *patched* embryos at the position expected for reverse-duplication (Fig. 11.6f) (DiNardo et al., 1988; Martinez-Arias et al., 1988).

Sander (1981) first noted the remarkable correspondence between the phenotype of *patched* mutant embryos and the reverse-intercalation resulting from microsurgical removal of the midregion of a segment of *Oncopeltus* (see Section 11.2.2 and Fig. 11.3). Therefore, he suggested that the *patched* phenotype might be generated by reverse intercalation of a subset of positional values subsequent to loss of the complementary subset. This is illustrated in Figure 11.6b and f. The anteroposterior dimension of a single wild-type segment is arbitrarily marked off into ten positional values (Fig. 11.6b). In homozygous *patched* embryos, the range of values normally found in the midregion of the segment (3.0-4-5-6-7-8.0) is lost and replaced by the complementary range of values arrayed in reverse order (3.0-

2-1-10/0-9-8.0) (Fig. 11.6f). If the same principle is applied to *armadillo,* the pos-terior range of values 5.0-6-7-8-9-10/0-1.0 is lost and is replaced by 5.0-4-3-2-1.0 (Fig. 11.6d).

This formal principle works for mutations in all of the ten loci with segment-polarity phenotypes. The actual portions of the segment potentially subject to loss are specific for each gene, and the extent of the actual loss varies among alleles of the same locus. Reverse duplication of the remaining portion, however, occurs *only* if one-half or more of the segment is missing (Nüsslein-Volhard et al., 1982). For example, Gergen and Wieschaus (1986b) observed that a series of increasingly hypomorphic *runt* alleles specifies a corresponding increase in extent of loss of seg-mental pattern elements, with an abrupt transition to a coordinated mirror-image duplication of remaining elements. The fact that this difference is observed between alleles indicates that it is the size of the missing region *per se,* and not the particular affected gene product, that decides whether or not reverse-intercalation takes place.

Nonetheless, as Sander (1981) also noted, the analogy of the segment-polarity phenotype to the microsurgically elicited reverse intercalation in *Oncopeltus* raises the following problem: if the analogy were exact, one would expect that regions corresponding to the "forbidden" positional values should be physically absent, so that the segmental anlagen, instead of being the usual four cells wide at the time of cellularization, should instead be one or two cells wide. This deficiency should then be followed by intercalary growth at the zones of juxtaposition of nonadjacent posi-tional values. As Sander noted, however, in *patched* (and other segment-polarity mutants) the blastoderm is of normal length. Subsequent studies did detect seg-mentally repeated regions of cell death in some segment-polarity mutants (Marti-nez-Arias, 1985; Perrimon and Mahowald, 1987, DiNardo et al., 1988; Martinez-Arias et al., 1988), but no such cell death was found in *patched* (Martinez-Arias et al., 1988). Even in those mutants in which it did occur, the localized cell death was observed shortly before visible segmentation (Perrimon and Mahowald, 1987) and after the onset of the molecular symptoms of respecification (DiNardo et al., 1988). Thus, localized cell death is probably a result rather than a cause of the abnormal pattern specification in these mutants. Further, no *extra* cell divisions in addition to the two that normally occur between the onset of gastrulation and the formation of the larval cuticle have yet been reported for any segment-polarity mutant. The evidence currently available thus is not consistent with the view that the segment-polarity phenotype involves compensatory growth initiated by the physical loss of certain regions.

Analyses of mosaics support the alternative view that intercalation involves in situ respecification. Zygotically acting mutations can be assessed for cellular auton-omy of action through production of somatic mosaics (Stern, 1968). Mutations at the three X-chromosomal loci with segment-polarity phenotypes (*runt, fused,* and *armadillo*) were assessed in this manner and all found to be cell-autonomous (Ger-gen and Wieschaus, 1985, 1986a; Wieschaus and Riggleman, 1987). The analysis of *armadillo* is particularly instructive (Fig. 11.7a). When mitotic recombination was induced at the blastoderm stage in embryos heterozygous for *armadillo* and for a linked recessive cell-marker mutation *(shavenbaby)* that affects denticle mor-phology, small patches bearing denticles appeared within the naked cuticle region. These patches were of the size expected for clones that had undergone the normal

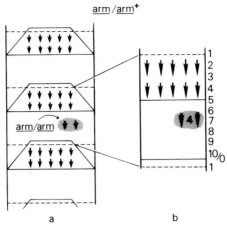

Figure 11.7 Schematic illustrations of *Drosophila melanogaster* larvae mosaic for *armadillo* as a result of mitotic crossing over induced at the blastoderm stage. Diagrammatic conventions parallel those of Figure 11.6. (**a**) A homozygous *armadillo* [*arm/arm*] patch *(shaded)* in a background of naked cuticle of heterozygous [*arm/arm*+] genotype. Note that the polarity of denticles in this patch is *not* reversed. Based on information from Wieschaus and Riggleman (1987). (**b**) An interpretation in terms of positional values. The "4" inside the patch indicates either that positional information within the patch is read as a 4, or is converted to a 4.

two divisions between their origin at the blastoderm stage and the time of cuticle formation (Wieschaus and Riggleman, 1987). The denticles within these clones expressed the linked cell-marker mutation, indicating that the clones were almost certainly also homozygous for *armadillo*. Hence *armadillo* can be expressed cell-autonomously, in the form of ectopic denticles within clonal patches. Interestingly, however, the orientation of these ectopic denticles was variable and generally *not* reversed (Wieschaus and Riggleman, 1987). Thus, although the differentiation pattern is cell-autonomous, the polarity pattern is not; this is reminiscent of the directional nonautonomy of *frizzled,* described in Section 11.3.2.

One way of thinking about the cell autonomy of the differentiation pattern of *armadillo* is to assume that "the wild-type *arm*+ product might serve as a cellular interpreter of external morphogen levels, transducing or internalizing the positional information within each cell" (Wieschaus and Riggleman, 1987, p. 182). The mutation might then alter the "perceived level of morphogen," for example causing a cell to read a 7 as if it were a 4. An alternative idea would be that the *arm/arm* cells, unable to create the 7 state that is emerging in the surrounding *arm/arm*+ cells, instead make the closest genically permitted state, in this case a 4 (Fig. 11.7b). The fact that the polarity pattern is not coordinately changed suggests that it may depend on separate information that requires physical contact between adjacent cells, as suggested by Nübler-Jung (1987). The ciliate analog would be the loss of local organization (extra marginal cirral rows) in cortically dedifferentiated *Oxytricha* cysts contrasted to the preservation of global organization (singlet versus doublet) in these same cysts (see Section 4.3.3).

Cell autonomy of the differentiation pattern in some segment-polarity mutations leaves open the possibility that other segment-polarity mutations might be nonautonomous. If specific positional values are dependent on products of genes of the segment-polarity class, then some of these products might not be able to move from one cell to another, whereas others might move freely. Loss-of-function mutations in genes coding for nonmobile products would yield cell autonomy, whereas loss-of-function mutations in genes coding for mobile products would yield nonautonomy because the necessary products could be supplied by neighboring cells. Wieschaus and Riggleman (1987) were unable to detect denticle-bearing clones anywhere in the naked-cuticle region of *wingless, gooseberry,* or *hedgehog,* all mutations with phenotypes resembling that of *armadillo,* suggesting possible nonautonomy (clones homozygous for nonautonomous mutations produced in the naked-cuticle area would be invisible because they would not produce any denticles). For *wingless,* the genetic and molecular evidence for nonautonomy is quite strong (Baker, 1987).

This section can be summarized quite simply by the statement that the bulk of the evidence available at the time of this writing (mid-1988) supports the conclusion that the reverse intercalation observed in the segment polarity mutants involves respecification (morphallaxis) and not local growth (epimorphosis).

11.3.6 Positional Specification in Insects and Ciliates

Insect embryos and ciliates are altogether different types of organisms, whose most recent common ancestor probably lived more than a billion years ago. Thus, I argue not for a point-for-point similarity, but rather for similarities in the "generative rules" (Green, 1987; Goodwin, 1988) governing formation of global patterns, correspondences that open the possibility of common underlying mechanisms. I outline the similarities under seven headings, drawing heavily on material reviewed in the preceding pages for insects and on the remainder of the book for ciliates.

Similar size

The sizes of the eggs of the three dipterans considered in this chapter are all within the ciliate range: 200 to 250 micrometers in length for *Smittia* (Kalthoff and Sander, 1968), 300 micrometers for *Chironomus* (Percy et al., 1986), and 450 micrometers for *Drosophila* (Campos-Ortega and Hartenstein, 1985). This makes the egg of *Smittia* roughly the size of *Paraurostyla weissei,* that of *Chironomus* close to *Blepharisma japonicus,* and the *Drosophila* egg smaller than *Stentor coeruleus* (see Chapter 3). In general, the germ-band of insect eggs varies in length from 0.2 to 1.5 mm (Fig. 4 of Sander, 1976), roughly corresponding to the lengths of the larger ciliates. Although similarity of size does not in itself indicate similarity of mechanisms, it shows that dimensions offer no impediment to the hypothesis that the mechanisms might be similar.

Hierarchical organization

The existence of a hierarchy of global superimposed on more local controls of ciliary patterns has been a major theme of this book (briefly summarized in Section 11.2.1). A comparable hierarchy exists in the insect cuticle (Section 11.3.2), with strictly local controls of polarity-pattern separate from more global controls of differentiation-pattern. A reader might be inclined to believe that this can be no more than a superficial analogy, as the polarity-pattern in ciliates is strictly intracellular, whereas the polarity-pattern in the insect cuticle must involve interactions between cells. As first suggested by Whitten (1973) and later shown in detail by Tucker (1981), however, cytoskeletal coordination is likely to exist between as well as within cells. Indeed, if we recall that in certain ciliates the membrane skeleton is found within alveolar sacs of the cell membrane, with connections spanning adjacent sacs (see Section 2.2.1), the nature of cytoskeletal continuity may sometimes be topologically the same in ciliates and in multicellular systems. Hence, there is no conceptual barrier to believing that the nature of the organizational hierarchy is not merely analogous, but also homologous, in ciliates and in insects.

Supracellular patterning

A century ago, the embryologist C. O. Whitman wrote an article entitled "The inadequacy of the cell theory of development," in which he noted the similarity of processes of development in embryos with very different cleavage patterns and concluded that "The plastic forces heed no cell boundaries, but mould the germ mass regardless of the way it is cut up into cells" (Whitman, 1893, p. 644). Subsequent experimental studies reinforced Whitman's conclusions by showing that considerable morphogenesis could occur in embryos of the annelid worm *Chaetopterus* (Lillie, 1906) and of the frog *Rana* (Holtfreter, 1943; Smith and Ecker, 1970), even when cleavage was suppressed; in *Drosophila,* embryos homozygous for a temperature-sensitive allele of *shibire* that prevents formation of cell membranes at the restrictive temperature could still undergo gastrulation movements even after cellular subdivision was prevented by heat shocks applied at the blastoderm stage (Swanson and Poodry, 1981).

With regard to pattern formation, it is clear that the original specification of segments in *Drosophila* must occur before cellular subdivision, as all whole-embryo pattern reversals are determined before cellularization (Section 11.3.4). Specification of the intrasegmental pattern, however, clearly takes place *after* cell boundaries have largely been completed (Section 11.3.5); it, therefore, could be quite distinct from the earlier events (DiNardo et al., 1988) and involve sequential cell-to-cell interactions (Martinez-Arias et al., 1988).

I believe that there are two lines of evidence, one direct and the other more indirect, which suggest that these later events are supracellular in Whitman's sense. The direct evidence that "the plastic forces heed no cell boundaries" is the independence of the width of gene expression bands from cell number in *Drosophila* embryos. This point has been demonstrated for the width of bands of the protein product of the *fushi tarazu (ftz)* gene in embryos mutated in other genes that affect nuclear size (ploidy) and hence cell density (Sullivan, 1987). These bands normally are 3 to 4 cells wide at the cellular blastoderm stage, but can be fewer (1 to 2) cells wide in embryos with reduced nuclear densities and more (5 to 6) cells wide in mutants with increased nuclear densities. The actual breadth of a band measured as a proportion of the length of the embryo is the same irrespective of the number of nuclei in the band; indeed, a given band retains its breadth even when crossing a borderline of markedly different nuclear densities. Sullivan (1987, p. 166) concluded " . . . that the pattern is not established through a cell counting mechanism . . . the molecules involved in the expression of *ftz* protein possess propagation properties which are independent of cell size and density."

This conclusion can be criticized because Sullivan counted *nuclei* yet drew conclusions concerning *cells*. Although the emergence of the *ftz* banding pattern coincides with the progress of cell-boundary formation during the 14th nuclear cycle (Carroll and Scott, 1985), this banding pattern can still be established even if cellularization is inhibited by injection of colchicine plus cytochalasin B at the beginning of the 14th nuclear cycle (Edgar et al., 1987). The *ftz* bands, therefore, could be an expression of an "early programme" that "operates largely in a syncytium," whereas the architecture within each segment could result from a distinct later program that is initiated after blastoderm cells have lost their direct cytoplasmic continuity with one another (DiNardo et al., 1988). In that case, Sullivan's conclusions might not apply to specification within a developing segment.

If a sequential cell-to-cell mode of interaction replaced a more global mode when cell membranes formed at the time of gastrulation, then mutations that systematically increase nuclear density should either increase the number of segments or seriously affect determination of parts within segments. Apparently the first does not happen and the second need not: mutations that endow the embryo with half-size haploid nuclei delay cellularization by one cycle (i.e., to the end of the 15th cycle rather than the 14th) and thus result in an increased density of cells in a normal-sized blastoderm (Edgar et al., 1986). Although segmentation defects were common in these haploid embryos, especially in the head region, many haploid embryos did produce the normal number and pattern of thoracic and abdominal segments (Edgar et al., 1986). This implies that formation of a normal initial segmental pattern does not depend on having a normal number of cells within that segment.

Following this line of reasoning, one would predict that spatial patterns of expression of the late as well as the early acting segmentation genes should be independent of nuclear density. One might even anticipate that gene expression bands would appear even if cell membranes were selectively prevented from forming in mutants such as the one at the *shibire* locus studied by Swanson and Poodry (1981).

The more indirect line of evidence favoring the supracellular view is the existence of mutations that generate both whole-embryo and intrasegmental mirror-images. One set of mutations, at the *costal-2* locus, has been reported to produce *patched*-type segmental duplications and double-abdomen embryos in different offspring of single crosses (Grau and Simpson, 1987). These same mutations also generated duplications in adult cuticle formed from imaginal discs (Simpson and Grau, 1987). Owing to a genetic complication, namely the simultaneous presence of a second mutation *(Costal-1)* that is believed to be a dominant enhancer of *costal-2*, we cannot yet be certain that all of these effects resulted from the same altered gene action. The initial results, however, do ". . . lend some support to the idea that secondary fields rely on a mechanism of positional signaling similar to that used in primary embryonic organization" (Grau and Simpson, 1987, p. 200). Because primary embryonic organization is intracellular, any similar mechanism operating at a later stage may transcend the cell boundaries that have formed in the interim.

Morphallactic reorganization

In embryos of *Drosophila,* as in ciliates, mirror-image patterns are generated by internal reorganization, not by localized growth. This conclusion is uncontested for whole-embryo reversals (Section 11.3.4), whereas for intrasegmental reversals it is supported by the bulk of the evidence, although it has not yet been proven conclusively (Section 11.3.5). Thus, positional respecification in the embryo is primarily reorganizational (morphallactic) and in that respect resembles the respecification that occurs when mirror-image forms are produced in ciliates.

In drawing this conclusion for patterning in embryos, I am not challenging the clear-cut evidence that regeneration after exicisions of *Drosophila* imaginal discs is epimorphic (Kiehle and Schubiger, 1985; Bryant and Fraser, 1988). Nonetheless, the mechanisms by which imaginal discs generate mirror-image duplications under the influence of mutations that affect pattern formation is less obviously epimorphic. Of the two best-known mutations of this type, *engrailed* is cell-autonomous (Tokunaga, 1961) and *wingless* is nonautonomous. In principle, *wingless* could generate mirror-image wing–disc reversals by stimulating localized regeneration; clonal analysis, however, does not indicate formation of a localized blastema (Morata and Lawrence, 1977), and in the absence of extra cell death (James and Bryant, 1981) there is no obvious mechanism for stimulating such a blastema to form. In situ respecification, therefore, is likely in *wingless* as well as *engrailed* discs. Thus, the conclusion drawn by Maden (1981) for pattern formation in amphibian limbs probably applies to insect systems as well: ". . . the progression during embryonic development from the morphallactic behavior of the primary field to the epimorphic behavior of secondary fields is unlikely to involve a sudden and complete change of pattern-forming mechanisms. Rather, it is more likely to produce the pattern by the same process, but now in a system which is growing" (Maden, 1981, p. 166).

Continuity

A basic postulate that both the polar coordinate model and the cylindrical coordinate model (Chapter 10) share is that positional continuity is required for developmental stability. In *Drosophila* embryos, the continuity rule appears to be obeyed by all intrasegmental reversals, but *not* by all whole-embryo reversals. Inspection of Figure 11.5 shows that the symmetrical double-abdomen embryos (Fig. 11.5e) obey the rule of normal neighbors, but the asymmetrical classes (Fig. 11.5c, d) do not. At first glance, this might suggest that these reversals are brought about by "generative rules" different from those of *both* the PCM and the CCM, which hold positional continuity to be virtually sacred. The lack of continuity, however, in the asymmetrical mirror-image reversals could be an artifact derived from the rigid time schedule characteristic of specification of parts in embryos. Symmetrical double-abdomen embryos result when the disruption of the anteroposterior system of the embryo is most severe, whereas the asymmetrical double-abdomen embryos (including the inverted tail-regions at the anterior pole) occur after milder disturbance. If, in the milder cases, reorganization is stimulated less strongly than in the severe cases, that reorganization might still be incomplete at the time that segments become irreversibly specified. Interestingly, the descending order of incidence of asymmetrical double-abdomen forms, most common in *Drosophila* (Nüsslein-Volhard, 1977; Mohler and Wieschaus, 1986; Fröhnhofer et al., 1986), rarer in *Chironomus* (Percy et al., 1986), and virtually unknown in *Smittia* (Kalthoff, 1983), is parallel to the descending order of size of these embryos, with

Drosophila the largest and *Smittia* the smallest. Possibly, smaller size is associated with greater ease of reorganization.

Spacing

One distinctive feature of the ciliate CCM is its emphasis on spacing of positional values as a condition for stability. A central idea of the CCM is that when positional values become more closely spaced than usual relative to an independent structural repeat (the ciliary rows), the positional system tends to readjust itself to restore a normal spacing of positional values (see Section 10.2). Readjustment may be stimulated if the positional values become too *widely* spaced, but this was not included as a basic postulate of the CCM because it was not indispensable for explaining pattern readjustments in ciliates.

In the *Drosophila* embryo, the observation that mirror-image reorganization occurs when one-half or more of the positional values are missing can most easily be thought of as a consequence of a spacing rule. To make this more obvious, imagine that there were no limit to the degree that positional values can "stretch." Then we could ask, if the pattern specifications for more than half of a segment (or more than half of an embryo) were eliminated, why does the remainder of the pattern not simply expand to cover the available space? In the segmental case, one can answer this question by noting that the inevitable discontinuities between adjacent incomplete patterns themselves would trigger reverse-intercalation. For the whole *Drosophila* embryo, however, as Mohler and Wieschaus (1986) pointed out, no discontinuities would result; therefore, if there were no limit to the width of segments, then no mirror-image embryos should ever form. The fact that they do form suggests that insect embryos have a "spacing rule" of some kind.

In *Drosophila,* mirror-image reversals are observed only when positional values are expanded in relation to the length of the blastoderm and not when they are compressed. Embryos heterozygous for *Krüppel* mutations, however, bring about compression of the bands of molecular expression of segmentation genes *(ftz, even-skipped)* (Carroll and Scott, 1986; Frasch and Levine, 1987) and phenotypic defects (Wieschaus et al., 1984) in the same segments. Inappropriately closely spaced positional values thus may interfere with pattern formation in *Drosophila* embryos. The evidence for a "spacing rule" controlling intracellular organization in both ciliates and insect embryos suggests that some type of molecular network capable of dynamic readjustment might be of critical importance in governing supracellular pattern stability.

Generative rules

In the previous sections, I have argued that the intracellular-global-morphallactic conceptual framework for large-scale patterning in ciliates applies equally to the large-scale organization of *Drosophila* embryos. This viewpoint has at least three specific experimental consequences for *Drosophila*. First, it suggests that mutations that interfere with cellularization in a specific manner might not block at least some aspects of segmentation, including localized expression of segment polarity genes. Second, it implies that certain pleiotropic mutations that act repetitively (perhaps including *costal-2*) may be as important for understanding pattern formation as are mutations that code for specific localized determinants. Third, it suggests that intensive investigation of what is happening near the cell surface, both in the cyto-skeleton and the membrane, will be necessary for understanding diversification at both the whole-embryo and intrasegmental levels.

A further implication of the intracellular-global-morphallactic viewpoint is that activation of localized genomes may not be telling us all that there is to know about regional diversification. In ciliates, different positional values must be "inter-preted" without intervention of localized genomes, presumably through regionally differing influences on assembly processes. The precept that "interpretation" and pattern formation in multicellular organisms occur entirely through differential activation of genes is still a matter of faith rather than of proof. My most general prediction is that cloning of genes and study of the spatial and temporal distribu-tion of their products will not by itself yield a *complete* understanding of mechan-isms of pattern formation; a serious search for understanding of the generative rules that are at work at the periphery is necessary as well. The arguments made by Jaffe (1969) for a *centripetal* approach to development are, in my opinion, as valid now as they were two decades ago.

11.4 WHAT ARE THE MECHANISMS?

As I pointed out at the beginning of this book (Section 1.1), although the goal of comparing pattern formation in ciliates with that in whole multicellular organisms requires thinking in an organismic context, understanding the actual mechanisms that might operate in ciliates (and in embryos as well) requires a return to a cellular context. At present, we do not know how positional information is generated within ciliates, or in other cells. The analyses in ciliates that have been described earlier in this book tell us that the substratum of this information must be suffi-ciently stable for its longitudinal coordinates to be propagated during growth of the clonal cylinder, yet sufficiently labile to undergo a large-scale reorganization within a time scale of hours. A fixed structural scaffold, such as is represented by most aspects of the microscopically visible ciliary units considered in Chapter 2, is too static to meet these requirements; on the other hand, most of the endoplasm, undergoing cyclosis (Sikora, 1981), would be too unstable. The large-scale pattern-ing mechanism must use some aspect of surface organization that is potentially stable yet capable of dynamic readjustment over long intracellular distances.

There are three cellular regions in which we might locate that organization: the cell membrane, the underlying cytoskeleton (interpreting the latter in a broad sense, to include the membrane skeleton), and the outermost layer of the subcortical cyto-plasm that does not participate in bulk cyclotic flow (Aufderheide, 1977). Because no distinctive structural or chemical organization is currently known for the sub-cortical layer, I confine my attention to the cytoskeleton and the cell membrane. In a variety of other cellular systems, pharmacological inhibitors and mutations (in yeast) have been used to assess the roles of the two best understood components of the cytoskeleton, actin-containing microfilaments and microtubules, in the origin and maintenance of cell polarity. Studies of this kind have indicated that F-actin is involved in establishing or maintaining cell polarity in a great variety of systems, including the yeast *Saccharomyces* (Novick and Botstein, 1985), the uncleaved egg of the nematode *Caenorhabdites* (Strome and Wood, 1983), the blastula cells of the sea urchin *Lytechinus* (Nelson and McClay, 1988), and the zygote of the brown algae *Fucus* and *Pelvetia* (Brawley and Robinson, 1985); in the first three of these systems disruption of microtubules has been shown to have no effect on cell polarity (*Saccharomyces:* Jacobs et al., 1988; *Caenorhabdites:* Strome and Wood, 1983; *Lytechinus:* Nelson and McClay, 1988). The negative findings with inhibitors of microtubule polymerization have a clear ciliate counterpart in the natural experiment that *Oxytricha* carries out when it dismantles all of its cortical microtubules during encystment yet maintains its large-scale organization (see Section 4.3.3). The positive results obtained with inhibitors of actin polymerization may have limited application to ciliates, because the quantity of orthodox actin in ciliates is low and because most of the actin that is present probably is associated with formation of food vacuoles (see Section 2.2.1). Ciliates are not the only organisms that might maintain cell polarity without involvement of actin, as the amoeboid sperm cell of *Caenorhabdites* manages to maintain a high degree of internal asymmetry with negligible quantities of both actin and tubulin (Ward, 1986).

Even this brief survey indicates that no persuasive generalizations can be made on the basis of our current knowledge of specific cytoskeletal components. However, we should be aware that the roles of only the most abundant cytoskeletal components, for which specific inhibitors are available, have as yet been assessed. In the case of ciliates, the molecular analysis of the most promising part of the cytoskeleton, the membrane skeleton or "epiplasm" (see Section 2.2.1), is still in its infancy.

Our knowledge of the role of specific membrane components in intracellular patterning is even more primitive. However, there is one suggestive precedent that indicates that a large-scale redistribution of position-specific molecules within the cell membrane of ciliates is conceivable. Ciliates are "swimming sensory cells" (Machemer and de Peyer, 1977) that carry a large variety of ion channels. Conductances of depolarizing calcium-sensitive and hyperpolarizing potassium-sensitive mechanoreceptor channels are distributed in anteroposterior gradients of opposite slope both in *Paramecium* (Fig. 11.8) (Ogura and Machemer, 1980) and in *Stylonychia* (de Peyer and Machemer, 1978); in *Stentor,* only the depolarizing gradient is present (Wood, 1982). This distribution must change when the equatorial region of the parent cell becomes transformed into the anterior and posterior ends of the future daughters. Redistribution appears to take place before division in *Stylonychia,* so that ". . . de- and hyperpolarizing receptor potentials can be observed not only at the anterior and posterior end of the mother cell, but also at the prospective anterior and posterior of the daughter cells near the furrow" (Machemer and Deitmer, 1987, p. 262).

If the spatial distribution of ion-sensitive conductances reflects the distributions of the channels responsible for these conductances, as is likely (Machemer and Deitmer, 1985), then the redistribution of conductances implies either localized removal and insertion of ion channels, especially in the equatorial zone, or else a lateral movement of these channels within the membrane. Other well-studied cases support the latter hypothesis (McCloskey and Poo, 1984). Although there is no evidence that the mechanoreceptor channels of ciliates have any causal relationship to pattern formation, their probable redistribution before fission implies that other

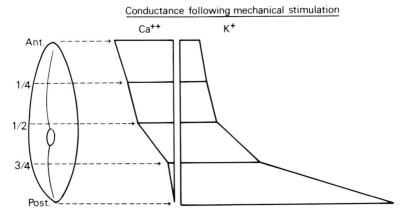

Figure 11.8 Gradients of calcium (Ca^{++})-mediated and potassium (K^{+})-mediated ionic conductances as observed in deciliated *Paramecium caudatum* cells given a standardized mechanical stimulus at each of the indicated positions along the anteroposterior axis. Slightly modified from Figure 19 of Aufderheide et al. (1980), in turn taken from Table 1 of Ogura and Machemer (1980), with permission.

membrane macromolecules that might be related to pattern formation also could be subject to such redistribution. The active modulation of lateral diffusion by mechanisms that might include anchorage to the underlying membrane skeleton (Schlessinger, 1983; Jacobson et al., 1987) could provide the combination of stability and capacity for redistribution that is required by the large-scale patterning system.

The reader will probably be correct in viewing this as a speculative house of cards. In reality, the molecular mechanism for large-scale patterning in ciliates—and other organisms—is unknown, and, as Pringle and his colleagues have pointed out for the equally unsolved problems of localization of the budding site in yeast, ". . . mechanisms [may be] involved in morphogenesis that are as yet undreamt of in the philosophies of biochemists and cell biologists" (Pringle et al., 1986, p. 59). In a condition of uncertainty with regard to mechanism, research should proceed along as broad a front as possible and should include a search for new theories (Holliday, 1988). For ciliates, a sample of approaches that might prove valuable in addition to the genetic and microsurgical methods that have been pursued thus far might include cloning of genes (such as the *janus* genes) whose mutant alleles perturb pattern, a search for "position-specific antigens" (Leptin and Wilcox, 1986) using monoclonal antibodies, a search for localized ion currents using the vibrating probe (Nucitelli, 1983), and an attempt to replace the cyclindrical-coordinate-model with a dynamically and empirically sufficient model (see Section 10.5). Each approach has both technical and conceptual problems, which can be illustrated by two examples that represent the antipodes of contemporary biological thought. Gene cloning in ciliates has the technical problem that methods that allow relatively straightforward isolation of genes with unknown products in yeast (shuttle vectors and allelic replacement) are not (yet) feasible in ciliates, and the conceptual problem that even if and when they do become feasible, the deduced sequence of a particular gene product may not tell one very much about the process that it is influencing. Application of the vibrating probe, which can detect tiny voltage differences in a restricted area, has the technical problem that one must get the ciliate to "sit still" by a method that does not perturb the ionic currents that one is measuring, and the conceptual problem that ionic currents may or may not be actual causes of cell polarity (Jaffe, 1981; Harold, 1986). My own view is that the true mechanisms are not yet understood, not even in principle, and that answers will be found by scientists who are willing to think and work unconventionally.

References

Adlakha, R. C., Sahasrabuddhe, C. G., Wright, D. A., and Rao, P. U. (1983). Evidence for the presence of inhibitors of mitotic factors during G1 period in mammalian cells. *J. Cell Biol.* 97, 1707–1713.

Adoutte, A. (1988). Exocytosis: Biogenesis, transport, and secretion of trichocysts. In H.-D. Görtz (ed.), *Paramecium*. Berlin, Springer Verlag, pp. 324–362.

Adoutte, A., Ramanathan, R., Lewis, R. M., Dute, R. R., Ling, K.-Y., Kung, C., and Nelson, D. L. (1980). Biochemical studies of the excitable membrane of *Paramecium tetraurelia*. III. Proteins of cilia and ciliary membranes. *J. Cell Biol.* 84, 717–738.

Albrecht-Buehler, G. (1977). Daughter 3T3 cells: Are they mirror images of each other? *J. Cell Biol.* 72, 595–603.

Allen, R. D. (1967). Fine structure, reconstruction, and possible functions of components of the cortex of *Tetrahymena pyriformis*. *J. Protozool.* 14, 553–565.

Allen, R. D. (1969). The morphogenesis of basal bodies and accessory structures of the cortex of the ciliated protozoan *Tetrahymena pyriformis*. *J. Cell. Biol.* 40, 716–733.

Allen, R. D. (1971). Fine structure of membranous and microfibrillar systems in the cortex of *Paramecium caudatum*. *J. Cell. Biol.* 49, 1–20.

Allen, R. D. (1974). Food vacuole membrane growth with microtubule-associated membrane transport in *Paramecium*. *J. Cell. Biol.* 63, 904–922.

Allen, R. D. (1978a). Membranes of ciliates: Ultrastructure, biochemistry, and fusion. In G. Poste and G. L. Nicolson (eds.), *Membrane Fusion, Cell Surface Reviews* 5, Amsterdam, North Holland, pp. 658–753.

Allen, R. D. (1978b). Particle arrays in the surface membrane of *Paramecium:* Junctional and possible sensory sites. *J. Ultrastr. Res.* 63, 64–78.

Allen, R. D., and Wolf, R. W. (1979). Membrane recycling at the cytoproct in *Tetrahymena*. *J. Cell Sci.* 35, 217–227.

Ammermann, D. (1985). Species characterization and speciation in the *Stylonychia/Oxytricha* group (Ciliata, Hypotrichida, Oxytrichidae). *Atti Soc. Tosc. Sci. Nat. Mem.,* Serie B, 92, 15–27.

Andersen, H. A. (1972). Requirements for DNA replication preceding cell division in *Tetrahymena pyriformis*. *Exp. Cell Res.* 75, 89–94.

Andersen, H. A. (1977). Replication and functions of macronuclear DNA in synchronously growing populations of *Tetrahymena pyriformis*. *Carlsberg Res. Commun.* 42, 225–248.

Andersen, H. A., and Zeuthen, E. (1971). DNA replication sequence in *Tetrahymena* is not repeated from generation to generation. *Exp. Cell Res.* 68, 309–314.

Anderson, H., and French, V. (1985). Cell division during intercalary regeneration in the cockroach leg. *J. Embryol. Exp. Morph.* 90, 57–78.

Anderson, R.G.W., and Brenner, R. M. (1971). The formation of basal bodies (centrioles) in the rhesus monkey oviduct. *J. Cell Biol.* 50, 10–34.

Antipa, G. A. (1980). A temporal analysis of cell cycle and morphogenetic events in *Tetrahymena pyriformis*. *Acta Protozool.* 19, 1–14.

Aufderheide, K. J. (1977). Saltatory motility of uninserted trichocysts and mitochondria in *Paramecium aurelia*. *Science* 198, 299–300.

Aufderheide, K. J. (1979). Mitochondrial associations with specific microtubular components of the cortex of *Tetrahymena thermophila*. I. Cortical patterning of mitochondria. *J. Cell Sci.* 39, 299–312.

Aufderheide, K. J. (1980). Mitochondrial associations with specific microtubular components of the cortex of *Tetrahymena thermophila*. II. Response of the mitochondrial pattern to changes in the microtubule pattern. *J. Cell Sci.* 42, 247–260.

Aufderheide, K. J. (1986). Identification of basal bodies and kinetodesmal fibers in living cells of *Paramecium tetraurelia* Sonneborn, 1975 and *Paramecium sonneborni* Aufderheide, Daggett & Nerad, 1983. *J. Protozool.* 33, 77–80.

Aufderheide, K. J., Frankel, J., and Williams, N. E. (1980). Formation and positioning of surface-related structures in protozoa. *Microbiol. Rev.* 44, 252–302.

Baker, N. E. (1987). Embryonic and imaginal requirements for *wingless*, a segment-polarity gene in *Drosophila*. *Dev. Biol.* 125, 96–108.

Bakowska, J. (1980). Size dependent regulation of serially repeated structures of a protozoan *Paraurostyla weissei*. *Acta Protozool.* 19, 297–314.

Bakowska, J. (1981). The ultrastructural analysis of the regulation of frontal cirri in *Paraurostyla weissei*. *Acta Protozool.* 20, 25–38.

Bakowska, J., and Jerka-Dziadosz, M. (1978). Ultrastructural analysis of the infraciliature of the oral apparatus of *Paraurostyla weissei* (Hypotricha). *Acta Protozool.* 17, 285–301.

Bakowska, J., and Jerka-Dziadosz, M. (1980). Ultrastructural aspect of size-dependent regulation of surface pattern of complex ciliary organelle in a protozoan ciliate. *J. Embryol. Exp. Morph.* 59, 355–375.

Bakowska, J., Frankel, J., and Nelsen, E. M. (1982a). Regulation of the pattern of basal bodies within the oral apparatus of *Tetrahymena thermophila*. *J. Embryol. Exp. Morph.* 69, 83–105.

Bakowska, J., Nelsen, E. M., and Frankel, J. (1982b). Development of the ciliary pattern of the oral apparatus of *Tetrahymena thermophila*. *J. Protozool.* 29, 366–382.

Banchetti, R., Nobili, R., Ricci, N., and Esposito, F. (1980). Mating-type dependent cell pairing in *Oxytricha hymenostoma*. III. Chromatin mixing after macronuclear fusion. *J. Protozool.* 27, 459–466.

Bardele, C. F. (1977). Comparative study of axopodial microtubule patterns and possible mechanisms of pattern control in the centrohelidian Heliozoa *Acanthocystis, Raphidiophrys,* and *Heterophrys*. *J. Cell Sci.* 25, 205–232.

Bardele, C. F. (1983). Mapping of highly ordered membrane domains in the plasma membrane of the ciliate *Cyclidium glaucoma*. *J. Cell Sci.* 61, 1–30.

Baroin, A., Perasso, R., Qu, L.-H., Brugerolle, G., Bachelliere, J.-P., and Adoutte, A. (1988). Partial phylogeny of the unicellular eukaryotes based on rapid sequencing of a portion of 28S ribosomal RNA. *Proc. Natl. Acad. Sci. U.S.* 85, 3474–3478.

Bart, A. (1988). Proximodistal leg regeneration in *Carausius morosus:* Growth, proximalization, and distalization. *Development* 102, 71–84.

Bartnik, E., Osborn, M., and Weber, K. (1985). Intermediate filaments in non-neuronal cells of invertebrates: Isolation and biochemical characterization of intermediate filaments from the esophageal epithelium of the mollusc *Helix promatia*. *J. Cell Biol.* 101, 427–440.

Bateson, W. (1892). On numerical variation in teeth, with a discussion of the conception of homology. *Proc. Zool. Soc. London* 1892, 102–115.

Bateson, W. (1894). *Materials for the Study of Variation, treated with especial regard to Discontinuity in the Origin of Species*. London, Macmillan.

Batson, B. S. (1983). *Tetrahymena dimorpha* Sp. nov. (Hymenostomatida, Tetrahyminidae), a new ciliate parasite of Simuliidae (Diptera) with potential as a model for the study of ciliate morphogenesis. *Phil. Trans. R. Soc. Lond. B* 301, 345–363.

Behnke, O. (1970). Microtubules in disc-shaped blood cells. *Int. Rev. Exp. Pathol.* 9, 1–92.

Beisson, J., and Sonneborn, T. M. (1965). Cytoplasmic inheritance of the organization of the cell cortex of *Paramecium aurelia. Proc. Natl. Acad. Sci. U.S.* 53, 275–282.

Beisson, J., Lefort-Tran, M., Pouphile, M., Rossignol, M., and Satir, B. (1976a). Genetic analysis of membrane differentiation in *Paramecium.* Freeze-fracture study of the trichocyst cycle in wild-type and mutant cells. *J. Cell Biol.* 69, 126–143.

Beisson, J., Rossignol, M., Ruiz, F., Adoutte, A., and Grandchamp, S. (1976b). Genetic analysis of morphogenetic processes in *Paramecium:* A mutation affecting cortical pattern and nuclear reorganization. *J. Protozool.* 23 (Suppl.), 3A–4A.

Berger, J. D. (1976). Gene expression and phenotypic change in *Paramecium tetraurelia* exconjugants. *Genet. Res. Cambr.* 27, 123–134.

Berger, J. D. (1979). Regulation of macronuclear DNA content in *Paramecium tetraurelia. J. Protozool.* 26, 18–28.

Berger, J. D. (1984). The ciliate cell cycle. In P. Nurse and E. Streiblova (eds.), *The Microbial Cell Cycle,* Boca Raton, Fla., CRC Press, pp. 191–208.

Berger, J. D. (1988). The cell cycle and regulation of cell mass and macronuclear DNA content in *Paramecium.* In H.-D. Görtz (ed.), *Paramecium,* Berlin, Springer Verlag, pp. 97–119.

Berger, J. D., and Ching, A.S.-L. (1988). The timing of initiation of DNA synthesis in *Paramecium tetraurelia* is established during the preceding cell cycle as cells become committed to cell division. *Exp. Cell Res.* 174, 355–366.

Berger, J. D., and Morton, G. T. (1980). Studies on the macronuclei of doublet *Paramecium tetraurelia:* Distribution of macronuclei and DNA at fission. *J. Protozool.* 27, 443–450.

Berger, J. D., and Schmidt, H. J. (1978). Regulation of macronuclear DNA content in *Paramecium tetraurelia. J. Cell Biol.* 76, 116–126.

Bernard, F., and Bohatier, J. (1981). Ultrastructure et mise en place des organelles buccaux au cours de la régénération orale chez *Stentor coeruleus* (Cilié Hétérotriche). *Can. J. Zool.* 59, 2306–2318.

Berrill, N. J. (1961). *Growth, Development, and Pattern.* San Francisco, Freeman.

Blackburn, E. H., and Gall, J. G. (1978). A tandemly repeated sequence at the termini of the extrachromosomal ribosomal RNA genes in *Tetrahymena. J. Mol. Biol.* 120, 33–53.

Blackburn, E. H., and Karrer, K. M. (1986). Genomic reorganization in ciliated protozoans. *Ann Rev. Genet.* 20, 501–521.

Blakemore, R. P., and Frankel, R. B. (1981). Magnetic navigation in bacteria. *Sci. Am.* 245, (6), 58–65.

Bohatier, J. (1979). Morphogenese de régénération chez la cilié *Condylostoma magnum* (Spiegel): Etude ultrastructurale. *J. Protozool.* 26, 404–414.

Bohn, H. (1970). Interkalare Regeneration und segmentale Gradienten bei den Extremitäten von *Leucophaea*-larven (Blattaria). I. Femur und Tibia. *Wilh. Roux Arch Entwickl.-Mech. Org.* 165, 303–341.

Bohn, H. (1971). Interkalare Regeneration und segmentale Gradienten bei den Extremitäten von *Leucophaea*-larven (Blattaria). III. Die Herkunft des interkalaren Regenerates. *Wilh. Roux Arch. Entwickl.-Mech. Org.* 167, 209–221.

Bohn, H. (1976). Regeneration of proximal tissues from more distal amputation level in the insect leg (*Blaberus craniifer,* Blattaria). *Dev. Biol.* 53, 285–293.

Borror, A. C. (1972). Revision of the order Hypotrichida (Ciliophora, Protozoa). *J. Protozool.* 19, 1–23.

Brawley, S. H., and Robinson, K. R. (1985). Cytochalasin treatment disrupts the endogenous currents associated with cell polarization in fucoid zygotes: Studies of the role of F-actin in embryogenesis. *J. Cell Biol.* 100, 1173–1184.

Bray, D., Heath, J., and Moss, D. (1986). The membrane-associated "cortex" of animal cells: Its structure and mechanical properties. In A. V. Grimstone, H. Harris, and R. T. Johnson (eds.), *Prospects in Cell Biology, J. Cell Sci.* 4 (Suppl.), 71–88.

Brinkley, B. R., Cox, S. M., Pepper, D. A., Wible, L., Brenner, S. L., and Pardue, R. L. (1981). Tubulin assembly sites and the organization of cytoplasmic microtubules in cultured mammalian cells. *J. Cell Biol.* 90, 554–562.

Browne, E. N. (1909). The production of new hydranths in *Hydra* by the insertion of small grafts. *J. Exp. Zool.* 7, 1–23.

Bruns, P. J., and Brussard, T. B. (1974). Positive selection for mating with functional heterokaryons in *Tetrahymena pyriformis. Genetics* 78, 831–841.

Bryant, P. J., and Fraser, S. E. (1988). Wound healing, cell communication, and DNA synthesis during imaginal disc regeneration in *Drosophila. Dev. Biol.* 127, 197–208.

Bryant, S. V., French, V., and Bryant, P. J. (1981). Distal regeneration and symmetry. *Science* 212, 993–1002.

Buhse, H. R. (1966a). Oral morphogenesis during transformation from microstome to macrostome and macrostome to microstome in *Tetrahymena vorax* Strain V₂, type S. *Trans. Amer. Micr. Soc.* 85, 305–313

Buhse, H. R. (1966b). An analysis of macrostome production in *Tetrahymena vorax* strain V₂S type. *J. Protozool.* 13, 429–435.

Bulliére, D. (1971). Utilisation de la régénération intercalaire pour l'étude de la détermination cellulaire au cours de la morphogenenese chez *Blabera craniifer* (Insects Dictyoptere). *Dev. Biol.* 25, 672–709.

Burchill, B. R. (1968). Synthesis of RNA and protein in relation to oral regeneration in the ciliate *Stentor coeruleus. J. Exp. Zool.* 167, 427–438.

Bussers, J.-C. (1984). Les appareils protecteurs des infusoires ciliés. *Traité de Zoologie* 2 (fasc. 1). Paris, Masson, pp. 439–455.

Byrne, G., and Cox, E. C. (1987). Genesis of spatial pattern in the cellular slime mold *Polysphondylium pallidum. Proc. Natl. Acad. Sci. U.S.* 84, 4140–4144.

Calkins, G. N. (1911). Effects produced by cutting *Paramecium* cells. *Biol. Bull.* 21, 36–72.

Calvo, P., Martin, J., Delgado, P., and Torres, A. (1988). Cortical morphogenesis during excystment in *Histriculus similis* (Hypotricha, Oxytrichidae). *J. Protozool.* 35, 177–181.

Campos-Ortega, J. A., and Hartenstein, V. (1985). *The Embryonic Development of Drosophila melanogaster.* Berlin, Springer Verlag.

Carroll, S. B., and Scott, M. P. (1985). Localization of the *fushi tarazu* protein during *Drosophila* embryogenesis. *Cell* 43, 47–57.

Carroll, S. B., and Scott, M. P. (1986). Zygotically active genes that affect the spatial expression of the *fushi tarazu* segmentation gene during early *Drosophila* embryogenesis. *Cell* 45, 113–126.

Chandebois, R. (1976). *Histogenesis and Morphogenesis in Planarian Regeneration.* Basel, S. Karger.

Chandebois, R. (1984). Intercalary regeneration and level interactions in the fresh-water planarian *Dugesia lugubris.* I. The anteroposterior system. *Roux's Arch. Dev. Biol.* 193, 149–157.

Chatton, E. (1921). Réversion de la scission chez les Ciliés. Réalisation d'individus distomes et polyenergides de *Glaucoma scintillans* se multipliant indéfiniment par scissiparité. *Compt. Rend. Acad. Sci.* 173, 393–395.

Chatton, E., and Lwoff, A. (1935). La constitution primitive de la strié ciliare des infusoires. La desmodexie. *Compt. Rend. Soc. Biol.* 118, 1068–1071.

Chatton, E., and Villaneuve, S. (1937). *Gregarella fabrearum* Chatton et Brachon, protiste parasite du Cilié *Fabrea salina* Henneguy. La notion de dépolarisation chez les Flagellés et la conception des apomastigines. *Arch. Zool. Exp. et Gen.* 78 (no. 4), 216–237.

Chen-Shan, L. (1969). Cortical morphogenesis in *Paramecium aurelia* following amputation of the posterior region. *J. Exp. Zool.* 170, 205–228.

Chen-Shan, L. (1970). Cortical morphogenesis in *Paramecium aurelia* following amputation of the anterior region. *J. Exp. Zool.* 174, 463–478.

Child, C. M. (1915). *Individuality in Organisms.* Chicago, University of Chicago Press.

Ching, A.S.-L., and Berger, J. D. (1986a). The timing of initiation of macronuclear DNA synthesis is set during the preceding cell cycle in *Paramecium tetraurelia.* Analysis of the effects of abrupt changes in nutrient level. *Exp. Cell Res.* 167, 177–190.

Ching, A.S.-L., and Berger, J. D. (1986b). Control of cell division in *Paramecium tetraurelia.* Effects of abrupt changes in nutrient level on accumulation of macronuclear DNA and cell mass. *Exp. Cell Res.* 167, 191–202.

Cho, P. L. (1971). Cortical pattern in two syngens of *Glaucoma. J. Protozool.* 18, 180–183.

Cleffmann, G. (1968). Regulieurng der DNS-Menge im Makronukleus von *Tetrahymean. Exp. Cell Res.* 50, 193–207.

Cleffmann, G. (1980). Chromatin elimination and the genetic organization of the macronucleus in *Tetrahymena thermophila. Chromosoma* 78, 313–325.

Cleveland, L. R. (1957). Types and life cycles of centrioles of flagellates. *J. Protozool.* 4, 230–241.

Cleveland, L. R. (1960). The centrioles of *Trichonympha* from termites and their functions in reproduction. *J. Protozool.* 7, 326–341.

Cohen, J., and Beisson, J. (1980). Genetic analysis of the relationships between the cell surface and the nuclei in *Paramecium tetraurelia. Genetics* 95, 797–818.

Cohen, J., and Beisson, J. (1988). The cytoskeleton. In H.-D. Görtz (ed.), *Paramecium.* Berlin, Springer Verlag, pp. 363–392.

Cohen, J., Beisson, J., and Tucker, J. B. (1980). Abnormal microtubule deployment during defective macronuclear division in a *Paramecium* mutant. *J. Cell Sci.* 44, 153–167.

Cohen, J., Adoutte, A., Grandchamp, S., Houdebine, L.-M., and Beisson, J. (1982). Immunochemical study of microtubular structures throughout the cell cycle of *Paramecium. Biol. Cell* 44, 35–44.

Cohen, J., Garreau de Loubresse, N., and Beisson, J. (1984). Actin microfilaments in *Paramecium:* Localization and role in intracellular movements. *Cell Motility* 4, 443–468.

Cohen, J., Garreau de Loubresse, N., Klotz, C., Ruiz, F., Bordes, N., Sandoz, D., Bornens, M., and Beisson, J. (1987). Organization and dynamics of a cortical fibrous network of *Paramecium:* The outer lattice. *Cell Motility* 7, 315–324.

Cole, E. S., Frankel, J., and Jenkins, L. M. (1987). *bcd:* A mutation affecting the width of organelle domains in the cortex of *Tetrahymena thermophila. Roux's Arch. Dev. Biol.* 196, 421–433.

Cole, E. S., Frankel, J., and Jenkins, L. M. (1988). Interactions between the *janus* and *bcd* cortical pattern mutants in *Tetrahymena thermophila:* An investigation into global intracellular patterning mechanisms using double-mutant analysis. *Roux's Arch. Dev. Biol.,* 197, 476–489.

Collins, T., Baker, R. L., Wilhelm, J. M., and Olmstead, J. B. (1980). A cortical scaffold in the ciliate *Tetrahymena. J. Ultrastr. Res.* 70, 92–103.

Cooke, J. (1981). Scale of body pattern adjusts to available cell number in amphibian embryos. *Nature* 290, 775–778.

Cooke, J. (1985). The system specifying body position in the early development of *Xenopus,* and its response to early perturbations. *J. Embryol. Exp. Morph.* 89 (Suppl.), 69–87.

Cooke, J. (1987). Dynamics of the control of body pattern in the development of *Xenopus laevis.* IV. Timing and pattern in the development of twinned bodies after reorientation of eggs in gravity. *Development* 99, 417–427.

Cooke, J., and Webber, J. A. (1985). Dynamics and control of body pattern in the development of *Xenopus laevis.* I. Timing and pattern in the development of dorsoanterior and posterior blastomere pairs, isolated at the 4-cell stage. *J. Embryol. Exp. Morph.* 88, 85–112.

Corliss, J. O. (1979). *The Ciliated Protozoa: Characterization, Classification, and Guide to the Literature.* 2nd ed. Oxford, Pergamon Press.

Corliss, J. O. (1984). The kingdom Protista and its 45 phyla. *Biosystems* 17, 87–126.

Corliss, J. O., and Lom, J. (1985). An annotated glossary of protozoological terms. In J. J. Lee, S. H. Hutner, and E. C. Bovee (eds.), *An Illustrated Guide to the Protozoa.* Lawrence, Kan., Society of Protozoologists, pp. 576–602.

Coss, R. A. (1974). Mitosis in *Chlamydomonas reinhardtii:* Basal bodies and the mitotic apparatus. *J. Cell Biol.* 63, 325–329.

Crick, F.H.C. (1970). Diffusion in embryogenesis. *Nature* 225, 420–422.

Crick, F.H.C., and Watson, D. (1956). The structure of small viruses. *Nature* 177, 473–475.

Dalcq, A. M. (1938). *Form and Causality in Early Development.* Cambridge, Cambridge University Press.

Dale, L., and Slack, J.M.W. (1987). Regional specification within the mesoderm of early embryos of *Xenopus laevis. Development* 100, 279–295.

Dawson, J. A. (1920). An experimental study of an amicronucleate *Oxytricha.* II. The formation of double animals or twins. *J. Exp. Zool.* 30, 129–157.

Dembowska, W. S. (1938). Körperreorganisation von *Stylonychia mytilus* beim hungern. *Arch. Protistenk.* 91, 89–105.

de Peyer, J. E., and Machemer, H. (1978). Hyperpolarizing and depolarizing mechanoreceptor potentials in *Stylonychia. J. Comp. Physiol.* 127, 255–266.

de Puytorac, P., Savoie, A., and Roque, M. (1973). Observations cytologiques et biologiques sur le Cilié polymorphe *Glaucoma ferox* Savoie et de Puytorac, 1971. *Protistologica* 9, 45–63.

de Terra, N. (1964). Nucleocytoplasmic interactions during the differentiation of oral structures in *Stentor coeruleus. Dev. Biol.* 10, 269–288.

de Terra, N. (1966). Leucine incorporation into the membranellar bands of regenerating and nonregenerating *Stentor. Science* 153, 543–545.

de Terra, N. (1969). Differential growth in the cortical fibrillar system as the trigger for oral differentiation and cell division in *Stentor. Exp. Cell Res.* 56, 142–153.

de Terra, N. (1970). Cytoplasmic control of macronuclear events in the cell cycle of *Stentor.* In *Control of Organelle Development (Society for Experimental Biology Symposium* 24). Cambridge, Cambridge University Press, pp. 345–368.

de Terra, N. (1972). Kinetosome production in *Condylostoma* occurs during cell division. *J. Protozool.* 19, 602–603.

de Terra, N. (1975). Evidence for cell surface control of macronuclear DNA synthesis in *Stentor. Nature* 258, 300–303.

de Terra, N. (1977). The role of cortical pattern in timing of cell division and morphogenesis in *Stentor. J. Exp. Zool.* 200, 237–242.

de Terra, N. (1978). Some regulatory interactions between cell structures at the supramolecular level. *Biol. Rev.* 53, 427–463.

de Terra. N. (1979). Dependence of oral development and cleavage on cell size in the ciliate *Stentor. J. Exp. Zool.* 209, 57–64.

de Terra, N. (1981). Cortical control of macronuclear positioning in the ciliate *Stentor. J. Exp. Zool.* 216, 367–376.

de Terra, N. (1985a). Does the oral apparatus of the ciliate *Stentor* inhibit oral development by release of a diffusible substance? *J. Embryol. Exp. Morph.* 87, 241–247.

de Terra, N. (1985b). Cytoskeletal discontinuities in the cell body cortex initiate basal body assembly and oral development in the ciliate *Stentor. J. Embryol. Exp. Morph.* 87, 249–257.

Diener, D. R., Burchill, B. R., and Burton, P. R. (1983). Microtubules and filaments beneath the fission furrow of *Stentor coeruleus. J. Protozool.* 30, 83–90.

DiNardo, S., Sher, E., Heemskerk-Jongens, J., Kassis, J. A., and O'Farrell, P. H. (1988). Two-tiered regulation of spatially patterned *engrailed* gene expression during *Drosophila* embryogenesis. *Nature* 332, 604–609.

Dippell, R. V. (1968). The development of basal bodies in *Paramecium. Proc. Natl. Acad. Sci. U.S.* 61, 461–468.

Dippell, R. V. (1976). Effects of nuclease and protease digestion on the ultrastructure of *Paramecium* basal bodies. *J. Cell Biol.* 69, 622–637.

Doerder, F. P. (1979). Regulation of macronuclear DNA content in *Tetrahymena thermophila. J. Protozool.* 26, 28–35.

Doerder, F. P., and Berkowitz, M. S. (1986). Purification and partial characterization of the H immobilization antigens of *Tetrahymena thermophila. J. Protozool.* 33, 204–208.

Doerder, F. P., and DeBault, L. E. (1978). Life cycle variation and regulation of macronuclear DNA content in *Tetrahymena thermophila. Chromosoma* 69, 1–19.

Doerder, F. P., Frankel, J., Jenkins, L. M., and DeBault, L. E. (1975). Form and pattern in ciliated protozoa: Analysis of a genic mutant with altered cell shape in *Tetrahymena pyriformis,* syngen 1. *J. Exp. Zool.* 192, 237–258.

Driever, W., and Nüsslein-Volhard, C. (1988a). A gradient of *bicoid* protein in *Drosophila* embryos. *Cell* 54, 83–93.

Driever, W., and Nüsslein-Volhard, C. (1988b). The *bicoid* protein determines position in the *Drosophila* embryo in a concentration-dependent manner. *Cell* 54, 95–104.

Dubielecka, B., and Jerka-Dziadosz, M. (1989). Defective spatial control in patterning of microtubular structures in mutants of the ciliate *Paraurostyla.* I. Morphogenesis in multi-left-marginal cirri. *Eur. J. Protistol.,* in press.

Dubielecka, B., and Kaczanowska, J. (1984). Experimental studies on positioning and sizing of the cytoproct in *Paramecium tetraurelia. J. Exp. Zool.* 229, 349–359.

Dubochet, C.-F., Peck, R. K., and de Haller, G. (1979). Morphogenesis in the heterotrich ciliate *Climacostomum virens.* I. Oral development during cell division. *J. Protozool.* 26, 218–226.

Dustin, P. (1984). *Microtubules,* 2nd ed. Berlin, Springer.

Eberhardt, R. (1962). Untersuchungen zur morphogenese von *Blepharisma* und *Spirostomum. Arch. Protistenk.* 106, 241–341.

Edgar, B. A., Kiehle, C. P., and Schubiger, G. (1986). Cell cycle control by the nucleo-cytoplasmic ratio in early *Drosophila* development. *Cell* 44, 365–372.

Edgar, B. A., Odell, G. M., and Schubiger, G. (1987). Cytoarchitecture and the patterning of *fushi tarazu* expression in the *Drosophila* blastoderm. *Genes and Development* 1, 1226–1237.

Elwood, H. J., Olsen, G. J., and Sogin, M. L. (1985). The small-subunit ribosomal RNA sequences from the hypotrichous ciliates *Oxytricha nova* and *Stylonychia pustulata. Molec. Biol. Evol.* 2, 399–410.

Esposito, F., Banchetti, R., and Ricci, N. (1978). Mating-type dependent cell pairing in *Oxytricha hymenostoma:* Macronuclear fusion and lack of meiosis. *J. Protozool.* 25, 238–241.

Fahrni, J. F. (1985). Morphogenesis in the heterotrich ciliate *Climacostomum virens.* II. Oral development during regeneration. *J. Protozool.* 32, 460–473.

Fantes, P. A. (1981). Isolation of cell size mutants of a fission yeast by a new selective method: Characterization of mutants and implications for division control mechanism. *J. Bact.* 146, 746–754.

Fantes, P. A. (1983). Control of timing of cell cycle events by the *wee 1⁺* gene. *Nature* 302, 153–155.

Fantes, P. A., and Nurse, P. (1977). Control of cell size at division in fission yeast by a growth modulated size control over nuclear division. *Exp. Cell Res.* 107, 377–386.

Fauré-Fremiet, E. (1932). Division et morphogénèse chez *Folliculina ampulla* O. F. Muller. *Bull. Biol. France Belg.* 66, 77–110.

Fauré-Fremiet, E. (1945). Symétrie et polarité chez les Ciliés bi ou multicomposites. *Bull. Biol. France Belg.* 79, 106–150.

Fauré-Fremiet, E. (1948). Doublets homopolaires et régulation morphogénétique chez le cilié *Leucophrys patula*. *Arch. Anat. Microscop. Morphol. Exp.* 37, 183–203.

Fauré-Fremiet, E. (1962). Le genre *Paranassula* Kahl (Ciliata Cyrtophorina). *Cah. Biol. Mar.* 3, 61–77.

Fauré-Fremiet, E., and André, J. (1968). Structure fine de l'*Euplotes eurystomus* Wrz. *Arch. Anat. Microscop.* 57, 53–78.

Fernández-Galiano, D. (1978). Le comportement des cinétodesmes pendant la division de *Paramecium caudatum*. *Protistologica* 14, 291–294.

Field, K. G., Olsen, G. J., Lane, D. J., Giovannoni, S. J., Ghiselin, M. T., Raff, E. C., Pace, N. R., and Raff, R. A. (1988). Molecular phylogeny of the animal kingdom. *Science* 239, 748–753.

Firby, P. A., and Gardiner, C. F. (1982). *Surface Topology.* Chichester, Ellis Horwood.

Fleury, A., Iftode, F., Deroux, G., Fryd-Versavel, G., and Génermont, J. (1985). Relation entre caractérès ultrastructuraux et morphogénétiques chez les Hypotriches (Protozoa, Ciliata): Définition des Euhypotrichina n. S. O. et des Pseudohypotrichina n. S. O. *Compt. Rend. Acad. Sci., Paris* 300, 567–570.

Foe, V. E., and Alberts, B. M. (1983). Studies of nuclear and cytoplasmic behaviour during the five mitotic cycles that precede gastrulation in *Drosophila* embryogenesis. *J. Cell Sci.* 61, 31–70.

Foissner, W., and Adam, H. (1983). Morphologie und Morphogenese des Bodenciliates *Oxytricha granulifera* sp. n. (Ciliophora, Oxytrichidae). *Zoologica Scripta* 12, 1–11.

Frankel, J. (1961). Spontaneous astomy: Loss of oral areas in *Glaucoma chattoni*. *J. Protozool.* 8, 250–256.

Frankel, J. (1962). The effects of heat, cold, and *p*-fluorophenylalanine on morphogenesis in synchronized *Tetrahymena pyriformis* GL. *Compt. Rend. Trav. Lab. Carlsberg.* 33, 1–52.

Frankel, J. (1964). The effects of high temperature on the pattern of oral development in *Tetrahymena pyriformis* GL. *J. Exp. Zool.* 155, 403–436.

Frankel, J. (1965). The effects of nucleic acid antagonists on cell division and oral organelle development in *Tetrahymena pyriformis*. *J. Exp. Zool.* 159, 113–148.

Frankel, J. (1967a). Studies on the maintenance of development in *Tetrahymena pyriformis* GL-C. I. An analysis of the mechanism of resorption of developing oral structures. *J. Exp. Zool.* 164, 435–460.

Frankel, J. (1967b). Studies on the maintenance of oral development in *Tetrahymena pyriformis* GL-C. II. The relationship of protein synthesis to cell division and oral organelle development. *J. Cell Biol.* 34, 841–858.

Frankel, J. (1967c). Critical phases of oral primordium development in *Tetrahymena pyriformis* GL-C: An analysis employing low temperature treatments. *J. Protozool.* 14, 639–649.

Frankel, J. (1969a). Participation of the undulating membrane in the formation of the oral replacement primordium of *Tetrahymena pyriformis. J. Protozool.* 16, 26–35.

Frankel, J. (1969b). The relationship of protein synthesis to cell division and oral development in synchronized *Tetrahymena pyriformis* GL-C: An analysis employing cycloheximide. *J. Cell Physiol.* 74, 135–148.

Frankel, J. (1970). The synchronization of oral development without cell division in *Tetrahymena pyriformis* GL-C. *J. Exp. Zool.* 173, 79–100.

Frankel, J. (1972). The stability of cortical phenotypes in continuously growing cultures of *Tetrahymena pyriformis. J. Protozool.* 19, 648–654.

Frankel, J. (1973a). A genically determined abnormality in the number and arrangement of basal bodies in a ciliate. *Dev. Biol.* 30, 336–365.

Frankel, J. (1973b). Dimensions of control of cortical patterns in *Euplotes:* The role of preexisting structure, the clonal life cycle, and the genotype. *J. Exp. Zool.* 183, 71–94.

Frankel, J. (1974). Positional information in unicellular organisms. *J. Theoret. Biol.* 47, 439–481.

Frankel, J. (1975a). An analysis of the spatial distribution of ciliary units in the ciliate, *Euplotes minuta. J. Embryol. Exp. Morph.* 33, 553–580.

Frankel, J. (1975b). Pattern formation in ciliary organelle systems of ciliated protozoa. In *Cell Patterning (Ciba Foundation Symposium* 29). Amsterdam, Elsevier/Excerpta Medica/North Holland, pp. 25–49.

Frankel, J. (1979). An analysis of cell-surface patterning in *Tetrahymena.* In S. Subtelny and I. R. Konigsberg (eds.), *Determinants of Spatial Organization (Society for Developmental Biology Symposium* 37). New York, Academic Press, pp. 215–246.

Frankel, J. (1980). Propagation of cortical differences in *Tetrahymena. Genetics* 94, 607–623.

Frankel, J. (1982). Global patterning in single cells. *J. Theor. Biol.* 99, 119–134.

Frankel, J. (1983). What are the developmental underpinnings of evolutionary changes in protozoan morphology? In B. C. Goodwin, N. Holder, and C. G. Wylie (eds.), *Development and Evolution (British Society for Developmental Biology Symposium* 6). Cambridge, Cambridge University Press, pp. 279–314.

Frankel, J. (1984). Pattern formation in ciliated protozoa. In G. M. Malacinski and S. V. Bryant (eds.), *Pattern Formation: A Primer in Developmental Biology.* New York, Macmillan, pp. 163–196.

Frankel, J., and Jenkins, L. M. (1979). A mutant of *Tetrahymena thermophila* with a partial mirror-image duplication of cell surface pattern. II. Nature of genic control. *J. Embryol. Exp. Morph.* 49, 203–227.

Frankel, J., and Nelsen, E. M. (1981). Discontinuties and overlaps in patterning within single cells. *Phil. Trans. Roy. Soc. Lond.* B 295, 525–538.

Frankel, J., and Nelsen, E. M. (1986a). Intracellular pattern reversal in *Tetrahymena thermophila.* II. Transient expression of a *janus* phenocopy in balanced doublets. *Dev. Biol.* 114, 72–86.

Frankel, J., and Nelsen, E. M. (1986b). How the mirror-image pattern specified by a *janus* mutation of *Tetrahymena* comes to expression. *Dev. Genet.* 6, 213–238.

Frankel, J., and Nelsen, E. M. (1987). Positional reorganization in compound *janus* cells of *Tetrahymena thermophila. Development* 99, 51–68.

Frankel, J., and Williams, N. E. (1973). Cortical development in *Tetrahymena.* In A. M. Elliott (ed.), *The Biology of Tetrahymena.* Stroudsburg, Pa., Dowden, Hutchinson, and Ross, pp. 375–409.

Frankel, J., Jenkins, L. M., and DeBault, L. E. (1976). Causal relations among cell cycle processes in *Tetrahymena pyriformis:* An analysis employing temperature-sensitive mutants. *J. Cell Biol.* 71, 242–260.

Frankel, J., Nelsen, E. M., and Jenkins, L. M. (1977). Mutations affecting cell division in *Tetrahymena pyriformis,* syngen 1. II. Phenotypes of single and double homozygotes. *Dev. Biol.* 58, 255–275.

Frankel, J., Mohler, J., and Frankel, A.W.K. (1980a). Temperature-sensitive periods of mutations affecting cell division in *Tetrahymena thermophila. J. Cell Sci.* 43, 59–74.

Frankel, J., Mohler, J., and Frankel, A. W. K. (1980b). The relationship between the excess-delay phenomenon and temperature-sensitive periods in *Tetrahymena thermophila. J. Cell Sci.* 43, 75–91.

Frankel, J., Nelsen, E. M., and Martel, E. (1981). Development of the ciliature of *Tetrahymena thermophila.* II. Spatial subdivision prior to cytokinesis. *Dev. Biol.* 88, 39–54.

Frankel, J., Jenkins, L. M., and Bakowska, J. (1984a). Selective mirror-image reversal of ciliary patterns in *Tetrahymena thermophila* homozygous for a *janus* mutation. *Roux's Arch. Dev. Biol.* 194, 107–120.

Frankel, J., Jenkins, L. M., Bakowska, J., and Nelsen, E. M. (1984b). Mutational analysis of patterning of oral structures of *Tetrahymena.* I. Effects of increased size on organization. *J. Embryol. Exp. Morph.* 82, 41–66.

Frankel, J., Nelsen, E. M., Bakowska, J., and Jenkins, L. M. (1984c). Mutational analysis of patterning of oral structures in *Tetrahymena.* II. A graded basis for the individuality of intracellular structural arrays. *J. Embryol. Exp. Morph.* 82, 67–95.

Frankel, J., Nelsen, E. M., and Jenkins, L. M. (1987). Intracellular pattern reversal in *Tetrahymena thermophila: janus* mutants and their geometrical phenocopies. In W. F. Loomis (ed.), *Genetic Regulation of Development* (*Society for Developmental Biology Symposium* 45). New York, Alan R. Liss, pp. 219–244.

Frasch, M., and Levine, M. (1987). Complementary patterns of *even-skipped* and *fushi tarazu* expression involve their differential regulation by a common set of segmentation genes in *Drosophila. Genes and Development* 1, 881–995.

French, V., Bryant, P. J., and Bryant, S. V. (1976). Pattern regulation in epimorphic fields. *Science* 193, 969–981.

Fröhnhofer, H. G., and Nüsslein-Volhard, C. (1986). Organization of anterior pattern in the *Drosophila* embryo by the maternal-effect gene *bicoid. Nature* 324, 120–125.

Fröhnhofer, H. G., and Nüsslein-Volhard, C. (1987). Maternal genes required for the anterior localization of *bicoid* activity in the embryo of *Drosophila. Genes and Development* 1, 880–890.

Fröhnhofer, H. G., Lehmann, R., and Nüsslein-Volhard, C. (1986). Manipulating the antero-posterior pattern in the *Drosophila* embryo. In C. C. Wylie (ed.), *Determinative Mechanisms in Early Development, J. Embryol. Exp. Morph.* 97 (Suppl.), 169–179.

Fulton, C. (1971). Centrioles. In J. Reinhardt and H. Ursprung (eds.), *Origin and Continuity of Cell Organelles. (Results and Problems in Cell Differentiation. Vol. 2).* Berlin, Springer-Verlag, pp. 170–221.

Gabe, P. R., and DeBault, L. E. (1973). Macromolecular synthesis related to the reproductive cyst of *Tetrahymena patula. J. Cell Biol.* 59, 615–623.

Gabe, P. R., and Williams, N. E. (1982). The formation and fate of *Tetrahymena patula* cryptostomes. *J. Protozool.* 29, 2–7.

Gall, J. G. (ed.) (1986). *The Molecular Biology of Ciliated Protozoa.* New York, Academic Press.

Garreau de Loubresse, N., Keryer, G., Vigues, B., and Beisson, J. (1988). A contractile cytoskeletal network of *Paramecium:* The infraciliary lattice. *J. Cell Sci.* 90, 351–364.

Gavin, R. H. (1965). The effects of heat and cold on cellular development in *Tetrahymena pyriformis* WH-6. *J. Protozool.* 12, 307–318.

Gavin, R. H. (1977). The oral apparatus of *Tetrahymena pyriformis,* strain WH-6. IV. Observations on the organization of microtubules and filaments in the isolated oral apparatus and the differential effect of potassium chloride on the stability of oral apparatus microtubules. *J. Morph.* 151, 239–258.

Gavin, R. H. (1984). *In vitro* reassembly of basal body components. *J. Cell Sci.* 66, 147–154.

Gavin, R. H., and Frankel, J. (1966). The effects of mercaptoethanol on cellular development in *Tetrahymena pyriformis. J. Exp. Zool.* 161, 63–82.

Gavin, R. H., and Frankel, J. (1969). Macromolecular synthesis, differentiation, and cell division in *Tetrahymena pyriformis* mating type I variety 1. *J. Cell Physiol.* 74, 123–134.

Gehring, W. J. (1978). Imaginal discs: Determination. In M. Ashburner and T.R.F. Wright (eds.), *The Genetics and Biology of Drosophila* 2c. London, Academic Press, pp. 511–554.

Gergen, J. P., and Wieschaus, E. (1985). The localized requirements for a gene affecting segmentation in *Drosophila:* Analysis of larvae mosaic for *runt. Dev. Biol.* 109, 321–335.

Gergen, J. P., and Wieschaus, E. (1986a). Localized requirements for gene activity in segmentation of *Drosophila* embryos: Analysis of *armadillo, fused, giant,* and *unpaired* mutations in mosaic embryos. *Roux's Arch. Dev. Biol.* 195, 49–62.

Gergen, J. P., and Wieschaus, E. (1986b). Dosage requirements for *runt* in the segmentation of *Drosophila* embryos. *Cell* 45, 289–299.

Gergen, J. P., Coulter, D., and Wieschaus, E. (1986). Segmental pattern and blastoderm cell identities. In J. G. Gall (ed.), *Gametogenesis and the Early Embryo (Society for Developmental Biology Symposium* 44). New York, Alan R. Liss, pp. 195–220.

Gerhart, J. C. (1980). Mechanisms regulating pattern formation in the amphibian egg and early embryo. In R. F. Goldberger (ed.), *Molecular Organization and Cell Function, (Biological Regulation and Development Vol. 2).* New York, Plenum Press, pp. 133–316.

Gierer, A., and Meinhardt, H. (1972). A theory of biological pattern formation. *Kybernetik* 12, 30–39.

Giese, A. C. (1973). Encystment and Excystment. In A. C. Giese (ed.), *Blepharisma.* Stanford, Stanford University Press, pp. 247–265.

Gillies, C. G., and Hanson, E. D. (1968). Morphogenesis in *Paramecium trichium. Acta Protozool.* 6, 15–31.

Glass, L. (1977). Patterns of supernumerary limb regeneration. *Science* 198, 321–322.

Golinska, K. (1986). Modifications of size and pattern of microtubular organelles in overfed cells of a ciliate *Dileptus. J. Embryol. Exp. Morph.* 93, 85–104.

Golinska, K., and Jerka-Dziadosz, M. (1973). The relationship between cell size and capacity for division in *Dileptus anser* and *Urostyla cristata. Acta Protozool.* 12, 1–21.

Golinska, K., and Kink, J. (1977). The regrowth of oral structures in *Dileptus cygnus* after partial excision. *Acta Protozool.* 15, 143–163.

Goodenough, U. W., and St. Clair, H. S. (1975). *Bald-2:* A mutation affecting the formation of doublet and triplet sets of microtubules in *Chlamydomonas reihardtii. J. Cell Biol.* 66, 480–491.

Goodwin, B. C. (1976). *Analytical Physiology of Cells and Developing Organisms.* London, Academic Press.

Goodwin, B. C. (1980). Pattern formation and its regeneration in the protozoa. In G. W. Gooday, D. Lloyd, and A.P.J. Trinci (eds.), *The Eukaryotic Microbial Cell (Society for General Microbiology Symposium* 30). Cambridge, Cambridge University Press, pp. 377–404.

Goodwin, B. C. (1985). What are the causes of morphogenesis? *Bioessays* 3, 32–36.

Goodwin, B. C. (1988). Morphogenesis and heredity. In M.-W. Ho and S. W. Fox (eds.), *Evolutionary Processes and Metaphors.* New York, Wiley, pp. 145–162.

Goodwin, B. C. (1989). Unicellular Morphogenesis. In W. D. Stein and F. Bronner (eds.),

Cell Shape: Determinants, Regulation, and Regulatory Role, New York Academic Press, in press.

Goodwin, B. C., and Trainor, L.E.H. (1983). The ontogeny and phylogeny of the pentadactyl limb. In B. C. Goodwin, N. Holder, and C. C. Wylie (eds.), *Development and Evolution* (*British Society for Developmental Biology Symposium* 6). Cambridge, Cambridge University Press, pp. 75–98.

Görtz, H.-D. (1982). The behavior and fine structure of the dorsal bristles of *Euplotes minuta, E. aediculatus,* and *Stylonychia mytilus. J. Protozool.* 29, 353–359.

Grain, J. (1984). Cinétosome, cil, systemes fibrillaires en relation avec le cinétosome. *Traité de Zoologie,* 2 (fasc. 1), Paris, Masson, pp. 35–179.

Grain, J. (1986). The cytoskeleton of protists: Nature, structure, and functions. *Int. Rev. Cytol.* 104, 153–249.

Grain, J., and Bohatier, J. (1977). La régénération chez les protozoaires ciliés. *Ann. Biol.* 16, 193–240.

Grain, J., and Golinska, K. (1969). Structure et ultrastructure de *Dileptus cygnus* Claparede et Lachmann, 1859, Cilié Holotriche Gymnostome. *Protistologica* 5, 269–291.

Grandchamp, S., and Beisson, J. (1981). Positional control of nuclear differentiation in *Paramecium. Dev. Biol.* 81, 336–341.

Grau, Y., and Simpson, P. (1987). The segment polarity gene *costal-2* in *Drosophila.* I. The organization of both primary and secondary embryonic fields may be affected. *Dev. Biol.* 122, 186–200.

Green, P. B. (1984). Shifts in plant axiality: Histogenetic influences on cellulose orientation in the succulent, *Graptopetalum. Dev. Biol.* 103, 18–27.

Green, P. B. (1987). Inheritance of pattern: Analysis from phenotype to gene. *Am. Zool.* 27, 657–673.

Grell, K. B. (1973). *Protozoology.* Berlin, Springer Verlag.

Grim, J. N. (1972). Fine structure of the surface and infraciliature of *Gastrostyla steini. J. Protozool.* 19, 113–126.

Grim, J. N. (1982). Subpellicular microtubules of *Euplotes eurystomus:* Their geometry relative to cell form, surface contours, and ciliary organelles. *J. Cell Sci.* 56, 471–484.

Grimes, G. W. (1972). Cortical structure in nondividing and cortical morphogenesis in dividing *Oxytricha fallax. J. Protozool.* 19, 428–445.

Grimes, G. W. (1973a). Morphological discontinuity of kinetosomes during the life cycle of *Oxytricha fallax. J. Cell Biol.* 57, 229–232.

Grimes, G. W. (1973b). Differentiation during encystment and excystment in *Oxytricha fallax. J. Protozool.* 20, 92–104.

Grimes, G. W. (1973c). Origin and development of kinetosomes in *Oxytricha fallax. J. Cell Sci.* 13, 43–53.

Grimes, G. W. (1973d). An analysis of the determinative difference between singlets and doublets of *Oxytricha fallax. Genet. Res. Cambr.* 21, 57–66.

Grimes, G. W. (1976). Laser microbeam induction of incomplete doublets of *Oxytricha fallax. Genet. Res., Cambr.* 27, 213–226.

Grimes, G. W. (1982). Pattern determination in hypotrich ciliates. *Am. Zool.* 22, 35–46.

Grimes, G. W. (1989). Inheritance of cortical patterns in ciliated protozoa. In G. M. Malacinski (ed.), *Cytoplasmic Organization Systems, Primers in Developmental Biology,* Volume IV. New York, McGraw-Hill, in press.

Grimes, G. W., and Adler, J. A. (1976). The structure and development of the dorsal bristle complex of *Oxytricha fallax* and *Stylonychia pustulata. J. Protozool.* 23, 135–143.

Grimes, G. W., and Adler, J. A. (1978). Regeneration of ciliary pattern in longitudinal fragments of the hypotrichous ciliate, *Stylonychia. J. Exp. Zool.* 204, 57–80.

Grimes, G. W., and Gavin, R. H. (1987). Ciliary protein conservation during development in the ciliated protozoan, *Oxytricha. J. Cell Biol.* 105, 2855–2859.

Grimes, G. W., and Goldsmith-Spoegler, C. M. (1990a). Structure and morphogenesis of mirror-image doublets of hypotrichs. I. Asexual reproduction. Forthcoming.

Grimes, G. W. and Goldsmith-Spoegler, C. M. (1990b). Structure and morphogenesis of mirror-image doublets of hypotrichs. II. Regenerative morphogenesis after longitudinal separation. Forthcoming.

Grimes, G. W., and Hammersmith, R. L. (1980). Analysis of the effects of encystment and excystment on incomplete doublets of *Oxytricha fallax. J. Embryol. Exp. Morph.* 59, 19–26.

Grimes, G. W., and L'Hernault, S. W. (1979). Cytogeometrical determination of ciliary pattern formation in the hypotrich ciliate *Stylonychia mytilus. Dev. Biol.* 70, 372–395.

Grimes, G. W., McKenna, M. E., Goldsmith-Spoegler, C. M., and Knaupp, E. A. (1980). Patterning and assembly of ciliature are independent processes of hypotrich ciliates. *Science* 209, 281–283.

Grimes, G. W., Knaupp-Waldvogel, E. A., and Goldsmith-Spoegler, C. M. (1981). Cytogeometrical determination of ciliary pattern formation in the hypotrich ciliate *Stylonychia mytilus.* II. Stability and field regulation. *Dev. Biol.* 84, 477–480.

Grimstone, A. V. (1961). Fine structure and morphogenesis in protozoa. *Biol. Rev. Camb.* 36, 97–150.

Groliere, C. A., and Dupy-Blanc, J. (1985). Action du thiram (T.M.T.D.) sur le cortex de *Tetrahymena pyriformis* au cours de la stomotogenèse. *Protistologica* 21, 525–540.

Gubb, D., and Garcia-Bellido, A. (1982). A genetic analysis of the determination of cuticular polarity during development in *Drosophila melanogaster. J. Embryol. Exp. Morph.* 68, 37–57.

Hammersmith, R. L. (1976a). Differential cortical degradation in the two members of early conjugant pairs of *Oxytricha fallax. J. Exp. Zool.* 196, 45–70.

Hammersmith, R. L. (1976b). The redevelopment of heteropolar doublets and monster cells of *Oxytricha fallax* after cystment. *J. Cell Sci.* 22, 563–573.

Hammersmith, R. L., and Grimes, G. W. (1981). Effects of cystment on cells of *Oxytricha fallax* possessing supernumerary dorsal bristle rows. *J. Embryol. Exp. Morph.* 63, 17–27.

Hanson, E. D. (1962). Morphogenesis and regeneration of oral structures in *Paramecium aurelia:* An analysis of intracellular development. *J. Exp. Zool.* 150, 45–68.

Hanson, E. D., and Ungerlieder, R. M. (1973). The formation of the feeding organelle in *Paramecium aurelia. J. Exp. Zool.* 185, 175–188.

Harold, F. M. (1986). Transcellular ion currents in tip-growing organisms: Where are they taking us? In R. Nucitelli (ed.), *Ionic Currents in Development.* New York, Alan R. Liss, pp. 359–366.

Harper, D.S., and Jahn, C.L. (1989). Differential use of termination codons in ciliated protozoa. *Proc. Natl. Acad. Sci. U.S.* 86, 3252–3256.

Harrison, L. G. (1982). An overview of kinetic theory in developmental modelling. In S. Subtelny and P. B. Green (eds.), *Developmental Order: Its Origin and Regulation (Society for Developmental Biology Symposium* 40). New York, Alan R. Liss, pp. 3–35.

Harrison, L. G. (1987). What is the status of reaction-diffusion theory thirty-four years after Turing? *J. Theoret. Biol.* 125, 369–384.

Harrison, L. G., and Hillier, N. A. (1985). Quantitative control of *Acetabularia* morphogenesis by extracellular calcium: A test of kinetic theory. *J. Theoret. Biol.* 114, 177–192.

Hartman, H., Puma, J. P., and Gurney, T. L. Jr. (1974). Evidence for the association of RNA with the ciliary basal bodies of *Paramecium. J. Cell Sci.* 16, 241–259.

Hashimoto, K. (1961). Stomatogenesis and formation of cirri in fragments of *Oxytricha fallax* Stein. *J. Protozool.* 8, 433–442.

Hashimoto, K. (1962). Relationships between feeding organelles and encystment in *Oxytricha fallax* Stein. *J. Protozool.* 9, 161–169.

Hashimoto, K. (1963). Formation of ciliature in excystment and induced re-encystment of *Oxytricha fallax* Stein. *J. Protozool.* 10, 156–166.

Hashimoto, K. (1964). Localization of ciliary primordia in induced abnormal cysts of *Oxytricha falla*. *J.Protozool.* 11, 75–84.

Hausmann, K. (1978). Extrusive organelles of protists. *Int. Rev. Cytol.* 52, 197–276.

Hausmann, K., and Kaiser, J. (1979). Arrangement and structure of plates in the cortical alveoli of the hypotrich ciliate, *Euplotes vannus*. *J. Ultrastr. Res.* 67, 15–22.

Heath, I. B. (1980). Variant mitoses in lower eukaryotes: Indicators of the evolution of mitosis? *Int. Rev. Cytol.* 64, 1–80.

Heath, I. B. (1986). Nuclear division: A marker for protist phylogeny? *Progr. in Protistology* 1. Bristol, Biopress Ltd., pp. 115–162.

Heckmann, K. (1965). Totale Konjugation bei *Urostyla hologama* n. sp. *Arch. Protistenk.* 108, 55–62.

Heckmann, K., and Frankel, J. (1968). Genic control of cortical pattern in *Euplotes*. *J. Exp. Zool.* 168, 11–38.

Hedges, R. W. (1985). Inheritance of magnetosome polarity in magnetotropic bacteria. *J. Theoret. Biol.* 112, 607–608.

Heidemann, S. R., Sander, G., and Kirschner, M. W. (1977). Evidence for a functional role of RNA in centrioles. *Cell* 10, 337–350.

Held, L. I., Duarte, C. M., and Derakhshanian, K. (1986). Extra tarsal joints and abnormal cuticular polarities in various mutants of *Drosophila melanogaster*. *Roux's Arch. Dev. Biol.* 195, 145–157.

Hendrix, R. (1985). Shape determination in virus assembly: The bacteriophage example. In S. Casjens (ed.), *Virus Structure and Assembly*. Boston, Jones and Bartlett, pp. 169–203.

Hershey, A. D. (1970). Genes and hereditary characteristics. *Nature* 226, 697–700.

Hirono, M., Endoh, H., Okada, N., Numata, O., and Watanabe, Y. (1987a). *Tetrahymena* actin: Cloning and sequencing the *Tetrahymena* actin gene and identification of its gene product. *J. Mol. Biol.* 194, 181–192.

Hirono, M., Nakamura,M.,Tsunemoto, M., Yasuda, T., Ohba, H., Numata, O., and Watanabe, Y. (1987b). *Tetrahymena* actin: Localization and possible biological roles of actin in *Tetrahymena* cells. *J. Biochem. (Tokyo)*, 102, 537–545.

Hjelm, K. K. (1983). Relative daughter cell volume and position of the division furrow in *Tetrahymena*. *J. Cell Sci.* 61, 273–287.

Hjelm, K. K. (1986). Is non-genic inheritance involved in carcinogenesis? A cytotactic model of transformation. *J. Theor. Biol.* 119, 89–101.

Holliday, R. (1988). Successes and limitations of molecular biology. J. Theoret. Biol. 132, 253–262.

Holtfreter, J. (1943). A study of the mechanics of gastrulation. Part I. J. Exp. Zool. 94, 261–318.

Holtfreter, J., and Hamburger, V. (1955). Amphibians. In B. H. Willier, P. A. Weiss, and V. Hamburger (eds.), Analysis of Development. Philadelphia, WB Saunders, pp. 230–296.

Honda, H., and Miyake, A. (1976). Cell-to-cell contact by locally differentiated surfaces in conjugation in *Blepharisma*. *Dev. Biol.* 52, 221–230.

Hoops, H. J., and Witman, G. B. (1985). Basal bodies and associated structures are not required for normal flagellar motion or phototaxis in the green alga *Chlorogonium elongatum. J. Cell Biol.* 100, 297–309.

Horio, T., and Hotani, H. (1986). Visualization of the dynamic instability of individual microtubules by dark-field microscopy. *Nature* 321, 605–607.

Hörstadius, S. (1973). *Experimental Embryology of Sea Urchins.* Oxford, Clarendon Press.

Huang, B., and Pitelka, D. R. (1973). The contractile process in the ciliate, *Stentor coeruleus.* I. The role of microtubules and filaments. *J. Cell Biol.* 57, 704–728.

Huang, B., Ramanis, Z., Dutcher, S. K., and Luck, D.J.L. (1982). Uniflagellar mutants of *Chlamydomonas:* Evidence for the role of basal bodies in transmission of positional information. *Cell* 29, 745–753.

Hufnagel, L. A. (1969). Cortical ultrastructure of *Paramecium aurelia.* Studies on isolated pellicles. *J. Cell Biol.* 40, 779–801.

Hufnagel, L. A. (1981). Particle assemblies in the plasma membrane of *Tetrahymena:* Relationship to cell surface topography and cellular morphogenesis. *J. Protozool.* 28, 192–203.

Hufnagel, L. A. (1983). Freeze-fracture analysis of membrane events during early neogenesis of cilia in *Tetrahymena:* Changes in fairy-ring morphology and membrane topography. *J. Cell Sci.* 60, 137–156.

Hufnagel, L. A., and Torch, R. (1967). Intraclonal dimorphism of caudal cirri in *Euplotes vannus:* cortical determination. *J. Protozool.* 14, 429–439.

Huxley, J. S., and DeBeer, G. R. (1934). *The Elements of Experimental Embryology.* Cambridge, Cambridge University Press.

Iftode, F., Cohen, J., Ruiz, F., Torres-Rueda, A., Chen-Shan, L., Adoutte, A., and Beisson, J. (1989). Development of surface pattern during division of *Paramecium.* I. Mapping of duplication and reorganization of cortical cytoskeletal structures in the wild-type. *Development,* 105, 191–211.

Jacobs, C. W., Adams, A.E.M., Szaniszlo, P. J., and Pringle, J. R. (1988). Functions of microtubules in the *Saccharomyces cerevisiae* cell cycle. *J. Cell Biol.,* 107, 1409–1426.

Jacobson, K., Ishihara, A., and Inman, R. (1987). Lateral diffusion of proteins in membranes. *Ann. Rev. Physiol.* 49, 163–175.

Jaeckel-Williams, R. (1978). Nuclear divisions with reduced numbers of microtubules in *Tetrahymena. J. Cell Sci.* 34, 303–319.

Jaffe, L. F. (1969). On the centripetal course of development, the *Fucus* egg, and self-electrophoresis. In A. Lang (ed.), *Communication in Development (Society for Developmental Biology Symposium* 28). New York, Academic Press, pp. 83–111.

Jaffe, L. F. (1981). The role of ionic currents in establishing developmental patterns. *Phil. Trans. R. Soc. Lond. B* 295, 553–566.

James, A. A., and Bryant, P. J. (1981). Mutations causing pattern deficiencies and duplications in the imaginal wing disc of *Drosophila melanogaster. Dev. Biol.* 85, 39–54.

James, E. A. (1967). Regeneration and division in *Stentor coeruleus:* The effects of microinjected and externally applied actinomycin D and puromycin. *Dev. Biol.* 16, 577–593.

Jareno, M. A. (1984). Macronuclear events and some morphogenetic details during the excystment of *Onychodromus acuminatus* (Ciliophora, Hypotrichida). *J. Protozool.* 31, 489–492.

Jareno, M. A., and Tuffrau, M. (1979). Le positionnement des primordiums cinétosomiens et la morphogenese sous kyste chez le cilié hypotriche *Onychodromus acuminatus. Protistologica* 15, 597–605.

Jeffery, W. R., Stuart, K. D., and Frankel, J. (1970). The relationship between deoxyribonucleic acid replication and cell division in heat synchronized *Tetrahymena. J. Cell Biol.* 46, 533–543.

Jeffery, W. R., Frankel, J., DeBault, L. E., and Jenkins, L. M. (1973). Analysis of the schedule of DNA replication in heat-synchronized *Tetrahymena. J. Cell Biol.* 59, 1–11.

Jenkins, R. A. (1973). Fine Structure. In A. C. Giese (ed.), *Blepharisma.* Stanford, Stanford University Press, pp. 39–93.

Jerka-Dziadosz, M. (1964). Localization of the organization area in course of regeneration of *Urostyla grandis* Ehrbg. *Acta Protozool.* 2, 129–136.

Jerka-Dziadosz, M. (1972). Cortical development in *Urostyla.* I. Comparative study of morphogenesis in *U. cristata* and *U. grandis. Acta Protozool.* 10, 73–96.

Jerka-Dziadosz, M. (1974). Cortical development in *Urostyla.* II. The role of positional information and preformed structures in formation of cortical pattern. *Acta. Protozool.* 12, 239–274.

Jerka-Dziadosz, M. (1976). The proportional regulation of cortical structure in a hypotrich ciliate *Paraurostyla weissei. J. Exp. Zool.* 195, 1–14.

Jerka-Dziadosz, M. (1980). Ultrastructural study on development of the hypotrich ciliate *Paraurostyla weissei.* I. Formation and morphogenetic movements of ventral ciliary primordia. *Protistologica* 16, 571–589.

Jerka-Dziadosz, M. (1981a). Ultrastructural study on development of the hypotrich ciliate *Paraurostyla weissei.* II. Formation of the adoral zone of membranelles and its bearing on problems of ciliate morphogenesis. *Protistologica* 17, 67–81.

Jerka-Dziadosz, M. (1981b). Ultrastructural study on development of the hypotrich ciliate *Paraurostyla weissei.* III. Formation of the paroral membranelles and an essay on comparative morphogenesis. *Protistologica* 17, 83–97.

Jerka-Dziadosz, M. (1981c). Cytoskeleton-related structures in *Tetrahymena thermophila:* Microfilaments at the apical and division-furrow rings. *J. Cell Sci.* 51, 241–253.

Jerka-Dziadosz, M. (1981d). Patterning of ciliary structures in *janus* mutant of *Tetrahymena* with mirror-image cortical duplications. An ultrastructural study. *Acta Protozool.* 20, 337–356.

Jerka-Dziadosz, M. (1982). Ultrastructural study on development of the hypotrich ciliate *Paraurostyla weissei.* IV. Morphogenesis of dorsal bristles and caudal cirri. *Protistologica* 18, 237–251.

Jerka-Dziadosz, M. (1983). The origin of mirror-image symmetry doublet cells in the hypotrich ciliate *Paraurostyla weissei. Roux's Arch. Dev. Biol.* 192, 179–188.

Jerka-Dziadosz, M. (1985). Mirror-image configuration in the cortical pattern causes modifications in propagation of microtubular structures in the hypotrich ciliate *Paraurostyla weissei. Roux's Arch. Dev. Biol.* 194, 311–324.

Jerka-Dziadosz, M. (1989). Defective spatial control in patterning of microtubular structures in mutants of the ciliate *Paraurostyla.* II. Spatial coordinates in a double-recessive mutant. *Eur. J. Protistol,* in press.

Jerka-Dziadosz, M., and Banaczyk, I. A. (1983). Cell shape, growth rate, and cortical pattern aberrations in an abnormal strain of the hypotrich ciliate *Paraurostyla weissei. Acta Protozool.* 22, 139–156.

Jerka-Dziadosz, M., and Dubielecka, B. (1985). Transmission of a genetic trait through total conjugation in a hypotrich ciliate *Paraurostyla weissei.* Genetic basis of the multi-left-marginal mutant. *Genet. Res., Cambr.* 46, 263–271.

Jerka-Dziadosz, M., and Frankel, J. (1969). An analysis of the formation of ciliary primordia in the hypotrich ciliate *Urostyla weissei. J. Protozool.* 16, 612–638.

Jerka-Dziadosz, M., and Frankel, J. (1970). The control of DNA synthesis in macronuclei and micronuclei of a hypotrich ciliate: A comparison of normal and regenerating cells. *J. Exp. Zool.* 173, 1–22.

Jerka-Dziadosz, M., and Frankel, J. (1979). A mutant of *Tetrahymena thermophila* with a partial mirror-image duplication of cell surface pattern. I. Analysis of the phenotype. *J. Embryol. Exp. Morph.* 49, 167–202.

Jerka-Dziadosz, M., and Golinska, K. (1977). Regulation of ciliary pattern in ciliates. *J. Protozool.* 24, 19–26.

Jerka-Dziadosz, M., and Janus, I. (1972). Localization of primordia during cortical development of *Keronopsis rubra* (Ehrbg 1838) *(Hypotrichida). Acta Protozool.* 10, 249–262.

Jerka-Dziadosz, M., and Janus, I. (1975). Discontinuity of cortical pattern during total conjugation of a hypotrich ciliate *Paraurostyla weissei. Acta Protozool.* 13, 309–332.

Jerka-Dziadosz, M., Dosche, C., Kuhlmann, H.-W., and Heckmann, K. (1987). Signal-induced reorganization of the microtubular cytoskeleton in the ciliated protozoan *Euplotes octocarinatus. J. Cell Sci.* 87, 555–564.

Johnson, U. G., and Porter, K. R. (1968). Fine structure and cell division in *Chlamydomonas reinhardtii. J. Cell Biol.* 38, 403–425.

Jones, W. R. (1976). Oral morphogenesis during asexual reproduction in *Paramecium tetraurelia. Genet. Res. Cambr.* 27, 187–204.

Jurand, A., and Ng, S. F. (1988). Ultrastructural features of the oral region of amicronucleate *Paramecium tetraurelia* in autogamy. *J. Protozool.* 35, 256–259.

Jurand, A., and Selman, C. G. (1969). *The Anatomy of Paramecium aurelia.* London, MacMillan.

Jürgens, G. (1987). Segmental organization of the tail region in the embryo of *Drosophila melanogaster.* A blastoderm fate map of the cuticle structures of the larval tail region. *Roux's Arch. Dev. Biol.* 196, 141–157.

Jürgens, G., Lehmann, R., Schardin, M., and Nüsslein-Volhard, C. (1986). Segmental organization of the head in the embryo of *Drosophila melanogaster.* A blastoderm fate map of the cuticular structures of the larval head. *Roux's Arch. Dev. Biol.* 195, 359–377.

Kaczanowska, J. (1971). Topography of cortical organelles in early dividers of *Chilodonella cucullulus* O.F.M. *Acta Protozool.* 8, 231–250.

Kaczanowska, J. (1974). The pattern of morphogenetic control in *Chilodonella cucullulus. J. Exp. Zool.* 187, 47–62.

Kaczanowska, J. (1975). Shape and pattern regulation in regenerants of *Chilodonella cucullulus* O.F.M. *Acta Protozool.* 13, 343–360.

Kaczanowska, J. (1981). Polymorphism and specificity of positioning of contractile vacuole pores in a ciliate, *Chilodonella steini. J. Embryol. Exp. Morph.* 65, 57–71.

Kaczanowska, J., and Dubielecka, B. (1983). Pattern determination and pattern regulation in *Paramecium tetraurelia. J. Embryol. Exp. Morph.* 74, 47–68.

Kaczanowska, J., and Kowalska, D. (1969). Studies on the topography of cortical organelles of *Chilodonella cucullulus* O.F.M. I. The cortical organelles and intraclonal dimorphism. *Acta Protozool.* 7, 1–15.

Kaczanowska, J., and Moracewski, J. (1981). Short-range positioning of contractile vacuole systems in a ciliate *Chilodonella steini. Acta Protozool.* 20, 39–50.

Kaczanowska, J., Wychowaniec, L., and Ostrowski, M. (1982). A densitometrical method for the study of pattern formation in a ciliate *Chilodonella. Roux's Arch. Dev. Biol.* 191, 325–330.

Kaczanowski, A. (1975). A single-gene-dependent abnormality of adoral membranelles in *Tetrahymena pyriformis,* species 1. *Genetics* 81, 631–639.

Kaczanowski, A. (1976). An analysis of MP gene affected morphogenesis in *Tetrahymena pyriformis,* syngen 1 (species 1) ciliates. *J. Exp. Zool.* 196, 215–230.

Kaczanowski, A. (1978). Gradients of proliferation of ciliary basal bodies and the determination of the position of the oral primordium in *Tetrahymena. J. Exp. Zool.* 204, 417–430.

Kalnins, V. I., and Porter, K. R. (1969). Centriole replication during ciliogenesis in the chick tracheal epithelium. *Z. Zellf.* 100, 1–30.

Kalthoff, K. (1979). Analysis of a morphogenetic determinant in an insect embryo *(Smittia Spec., Chironomidia, Diptera).* In S. Subtelny and I. R. Konigsberg (eds.), *Determinants of Spatial Organization* (*Society for Developmental Biology Symposium* 37). New York, Academic Press, pp. 97–126.

Kalthoff, K. (1983). Cytoplasmic determinants in dipteran eggs. In W. R. Jeffery and R. A. Raff (eds.), *Time, Space, and Pattern in Embryonic Development.* New York, Alan R. Liss, pp. 313–348.

Kalthoff, K., and Sander, K. (1968). Der Entwicklungsgang der Missbildung "Doppelabdomen" im partiell UV-bestrahlten Ei von *Smittia parthenogenetica* (Dipt., Chironomidae). *Wilhelm Roux' Arch.* 161, 129–146.

Kaneda, M., and Hanson, E. D. (1974). Growth patterns and morphogenetic events in the cell cycle of *Paramecium aurelia.* In W. J. Van Wagtendonck (ed.), *Paramecium. A Current Survey.* Amsterdam, Elsevier, pp. 219–262.

Katsura, I. (1987). Determination of bacteriophage lambda tail length by a protein ruler. *Nature* 327, 73–75.

Kiehle, C. P., and Schubiger, G. (1985). Cell proliferation changes during pattern regulation in imaginal leg discs of *Drosophila melanogaster. Dev. Biol* 109, 336–346.

Kikuchi, Y., and King, J. (1975). Genetic control of bacteriophage T4 baseplate morphogenesis. III. Formation of the central plug and the overall assembly pathway. *J. Mol. Biol.* 99, 695–716.

King, J., Hall, C., and Casjens, S. (1978). Control of the synthesis of phage P22 scaffolding protein is coupled to capsid assembly. *Cell* 15, 551–560.

King, R. L. (1954). Origin and morphogenetic movements of the pores of the contractile vacuoles in *Paramecium aurelia. J. Protozool.* 1, 121–130.

Kink, J. (1976). A localized region of basal body proliferation in growing cells of *Dileptus visscheri* (Ciliata, Gymnostomata). *J. Cell Sci.* 20, 115–133.

Kirschner, M., and Mitchison, T. (1986). Beyond self-assembly: From microtubules to morphogenesis. *Cell* 45, 329–342.

Klobutcher, L. A., and Prescott, D. M. (1986). The special case of hypotrichs. In J. G. Gall (ed.), *The Molecular Biology of Ciliated Protozoa.* New York, Academic Press, pp. 111–154.

Klug, S. H. (1968). Cortical studies on *Glaucoma. J. Protozool.* 15, 321–327.

Kowalska, D., and Kaczanowska, J. (1970). Studies on the topography of cortical organelles of *Chilodonella cucullulus.* O.F.M. II. Topographical relations of the total number of kineties to the disposition of CVPs. *Acta Protozool.* 7, 181–192.

Kozloff, E. N. (1956). A comparison of the parasitic phase of *Tetrahymena limacis* (Warren) with clones in culture with particular reference to variability in the number of primary ciliary meridians. *J.Protozool.* 3, 20–28.

Kubai, D. F. (1973). Unorthodox mitosis in *Trichonympha agilis:* Kinetochore differentiation and chromosome movement. *J. Cell Sci.* 13, 511–552.

Kuhn, K. L., Percy, J., Laurel, M., and Kalthoff, K. (1987). Instability of the anteroposterior axis in spontaneous double abdomen (*sda*), a genetic variant of *Chironomus samoaensis. Development* 101, 591–603.

Kumazawa, H. (1979). Homopolar grafting in *Blepharisma japonicum. J. Exp. Zool.* 207, 1–16.

Lacalli, T. C. (1981). Dissipative structures and morphogenetic pattern in unicellular algae. *Phil. Trans. R. Soc. Lond. (Ser. B)* 294, 547–588.

Laemmli, U. K. (1970). Cleavage of structural proteins during the assembly of the head of bacteriophage T4. *Nature* 227, 680–685.

Laloë, F. (1979). Contribution a l'étude de la variabilité intraspécifique pour le nombre de cinéties dorsales et de cirres caudaux chez une espece du complexe *Euplotes vannus* (ciliés Hypotriches). *Arch. Zool. Exp. Gen.* 120, 109–120.

Lansing, T. J., Frankel, J., and Jenkins, L. M. (1985). Oral ultrastructure and oral development in the misaligned undulating membrane mutant of *Tetrahymena thermophila*. *J. Protozool.* 32, 126–139.

Lawrence, P. A. (1966). Gradients in the insect segment: The orientation of hairs in the milkweed bug *Oncopeltus fasciatus. J. Exp. Biol.* 44, 607–620.

Lawrence, P. A. (1985). Molecular development: Is there a light burning in the hall? *Cell* 40, 221.

Lawrence, P. A., and Green, S. M. (1975). The anatomy of a compartment border. The intersegmental boundary in *Oncopeltus. J. Cell Biol.* 65, 373–382.

Lehmann, R., and Nüsslein-Volhard, C. (1986). Abdominal segmentation, pole cell formation, and embryonic polarity require the localized activity of *oskar,* a maternal gene in *Drosophila. Cell* 47, 141–152.

Lehmann, R., and Nüsslein-Volhard, C. (1987). *hunchback,* a gene required for segmentation of an anterior and posterior region of the *Drosophila* embryo. *Dev. Biol.* 119, 402–417.

Leptin, M., and Wilcox, M. (1986). The *Drosophila* position-specific antigens. Clues to their morphogenetic role. *BioEssays* 5, 204–207.

Lewis, J. (1981). Simpler rules for epimorphic regeneration: The polar coordinate model without polar coordinates. *J. Theor. Biol.* 88, 371–392.

Lewis, J. (1982). Continuity and discontinuity in pattern formation. In S. Subtelny and P. B. Green (eds.), *Developmental Order: Its Origin and Regulation (Society for Developmental Biology Symposium* 40). New York, Alan R. Liss, pp. 511–531.

Lewis, J., and Wolpert, L. (1976). The principle of nonequivalence in development. *J. Theoret. Biol.* 62, 479–490.

Lewontin, R. (1974). *The Genetic Basis of Evolutionary Change.* New York, Columbia University Press.

Lillie, F. R. (1896). On the smallest parts of *Stentor* capable of regeneration: A contribution on the limits of divisibility of living matter. *J. Morph.* 12, 239–249.

Lillie, F. R. (1906). Observations and experiments concerning the elementary phenomena of development in *Chaetopterus. J. Exp. Zool.* 3, 153–267.

Locke, M. (1960). The cuticular pattern in an insect—the intersegmental membranes. *J. Exp. Biol.* 37, 398–406.

Lwoff, A. (1950). *Problems of Morphogenesis in Ciliates: The Kinetosomes in Development, Reproduction, and Evolution.* New York, Wiley.

Lynn, D. H. (1977). Proportional control of organelle position by a mechanism which similarly monitors cell size of wild type and conical form-mutant *Tetrahymena. J. Embryol. Exp. Morph.* 42, 261–274.

Lynn, D. H. (1981). The organization and evolution of microtubular organelles in ciliated protozoa. *Biol. Rev.* 56, 243–292.

Lynn, D. H. (1988). Cytoterminology of cortical components of ciliates: somatic and oral kinetids. *BioSystems* 21, 299–307.

Lynn, D. H. (1990). *A Biology of Ciliates.* London, Chapman, Routledge, and Hall, Forthcoming.

Lynn, D. H., and Sogin, M. L. (1988). Assessment of phylogenetic relationships among ciliated protists using partial ribosomal RNA sequences derived from reverse transcripts. *BioSystems* 21, 249–254.

Lynn, D. H., and Tucker, J. B. (1976). Cell size and proportional distance assessment during determination of organelle position in the cortex of the ciliate *Tetrahymena. J. Cell Sci.* 21, 35–46.

Mabuchi, I. (1986). Biochemical aspects of cytokinesis. *Int. Rev. Cytol.* 101, 175–213.

Machelon, V., Génermont, J., and Dattée, Y. (1984). A biometrical analysis of morphological variation within a section of genus *Euplotes* (Ciliata, Hypotrichida), with special reference to the *E. vannus* complex of sibling species. *Origins of Life* 13, 249–267.

Machemer, H., and de Peyer, J. E. (1977). Swimming sensory cells: Electrical membrane parameters, receptor properties and motor control in ciliated protozoa. *Verh. Dtsch. Zool. Ges.* 1977, 86–110.

Machemer, H., and Deitmer, J. W. (1985). Mechanoreception in ciliates. In D. Ottoson (ed.), *Progress in Sensory Physiology* 5. Berlin, Springer Verlag, pp. 81–118.

Machemer, H., and Deitmer, J. W. (1987). From structure to behavior: *Stylonychia* as a model system for cellular physiology. *Progress in Protistology* 2. Bristol, Biopress Ltd., pp. 213–330.

Maden, M. (1981). Morphallaxis in an epimorphic system: Size growth control, and pattern formation during amphibian limb regeneration. In R. M. Gaze (ed.), *Growth and Development of Pattern, J. Embryol. Exp. Morph.* 65 (Suppl.), 151–167.

Martin, J. (1982). Evolution des patrons morphogénétiques et phylogenese dans le sous-ordre des Sporadotrichina (Ciliophora, Hypotrichida). *Protistologica* 17, 431–447.

Martin, J., Fedriani, C., and Perez-Silva, J. (1983). Morphogenetic pattern in *Laurentiella acuminata* (Ciliophora, Hypotrichida): Its significance for the comprehension of ontogeny and phylogeny of hypotrichous ciliates. *J. Protozool.* 30, 519–529.

Martinez-Arias, A. (1985). Development of *fused* embryos of *Drosophila melanogaster. J. Embryol. Exp. Morph.* 87, 99–114.

Martinez-Arias, A., Baker, N. E., and Ingham, P. W. (1988). Role of segment polarity genes in the definition and maintenance of cell states in the *Drosophila* embryo. *Development* 103, 157–170.

Mayo, K. A., and Orias, E. (1985). Lack of expression of micronuclear genes determining two different enzymatic activities in *Tetrahymena thermophila. Differentiation* 28, 217–224.

McCloskey, M., and Poo, M.-m. (1984). Protein diffusion in cell membranes: Some biological implications. *Int. Rev. Cytol.* 87, 19–81.

McCoy, J. W. (1974). New features of the tetrahyminid cortex revealed by protargol staining. *Acta Protozool.* 13, 155–159.

McDonald, B. B. (1966). The exchange of RNA and protein during conjugation in *Tetrahymena. J. Protozool.* 13, 277–285.

McIntosh, J. R. (1979). Cell Division. In K. Roberts and J. S. Hyams (eds.), *Microtubules.* London, Academic Press, pp. 381–441.

McIntosh, J. R. (1983). The centrosome as the organizer of the cytoskeleton. In B. H. Satir (ed.), *Modern Cell Biology* 2. New York, Alan R. Liss, pp. 115–142.

McKanna, J. A. (1973). Fine structure of the contractile vacuole pore in *Paramecium. J. Protozool.* 20, 631–638.

Meedel, T. H., Crowther, R. J., and Whittaker, J. R. (1987). Determinative properties of muscle lineage in ascidian embryos. *Development* 100, 245–260.

Meinhardt, H. (1982). *Models of Biological Pattern Formation.* London, Academic Press.

Méténier, G. (1979). A propos de la régulation de la teneur en ADN macronucléaire pendant le cycle cellulaire du Cilié *Tetrahymena paravorax. Compt. Rend. Acad. Sci. Paris.* 289, 497–500.

Méténier, G. (1981). Chronologie de la synthese du DNA macronucléaire chez les formes prédatrices de *Tetrahymena paravorax. Eur. J. Cell Biol.* 24, 252–258.

Méténier, G. (1984). Actin in *Tetrahymena paravorax:* Ultrastructural localization of HMM-binding filaments in glycerinated cells. *J. Protozool.* 31, 205–215.

Méténier, G., and Groliere, C.-A. (1979). Précisions temporelles sur la processus de la stomatogenèse de *Tetrahymena paravorax* analysés en microscopie photonique. *J. Protozool.* 26, 75–82.

Mikami, K. (1980). Differentiation of somatic and germinal nuclei correlated with intracellular localization in *Paramecium caudatum* exconjugants. *Dev. Biol.* 80, 46–55.

Miller, K. G., Karr, T. L., Kellogg, D. R., Mohr, I. J., Walter, M., and Alberts, B. M. (1985).
Studies on cytoplasmic organization in early *Drosophila* embryos. *Molecular Biology
of Development* (*Cold Spring Harbor Symposium of Quantitative Biology* 50). Cold
Spring Harbor, Cold Spring Harbor Press, pp. 79–90.

Mitchison, J. M. (1971). *The Biology of the Cell Cycle.* Cambridge, Cambridge University
Press.

Mittenthal, J. E. (1981). The rule of normal neighbors: A hypothesis for morphogenetic pattern regulation. *Dev. Biol.* 88, 15–26.

Miyake, A. (1981). Cell interactions by gamones in *Blepharisma.* In D. H. O'Day and P.A.
Horgen (eds.), *Sexual Interactions in Eukaryotic Microbes.* New York, Academic
Press, pp. 95–129.

Mohler, J., and Wieschaus, E. F. (1985). Bicaudal mutations of *Drosophila melanogaster:*
Alteration of blastoderm cell fate. *Molecular Biology of Development* (*Cold Spring
Harbor Symposium of Quantitative Biology* 50). Cold Spring Harbor, Cold Spring
Harbor Press, pp. 105–111.

Mohler, J., and Wieschaus, E. F. (1986). Dominant maternal-effect mutations of *Drosophila
melanogaster* causing the production of double abdomen embryos. *Genetics* 112, 803–
822.

Moore, K. C. (1972). Pressure-induced regression of oral apparatus microtubules in synchronized *Tetrahymena. J. Ultrastr. Res.* 41, 499–518.

Morata, G., and Lawrence, P. A. (1977). The development of *wingless,* a homoeotic mutation
of *Drosophila. Dev. Biol.* 56, 227–240.

Morgan, T. H. (1901). Regeneration of proportionate structures in *Stentor. Biol. Bull.* 2, 311–
328.

Morris, W. (ed.) (1976). *The American Heritage Dictionary of the English Language.* Boston,
Houghton Mifflin.

Morton, G. T., and Berger, J. D. (1978). Comparison of singlet and doublet *Paramecium
tetraurelia:* DNA content, protein content, and the cell cycle. *J.Protozool.* 25, 203–207.

Mulisch, M., and Hausmann, K. (1983). Lorica construction in *Eufolliculina* sp. (Ciliophora,
Heterotrichida). *J. Protozool.* 30, 97–104.

Mulisch, M., and Hausmann, K. (1984). Structure and ultrastructure of the oral apparatus of
Eufolliculina uhligi Mulisch and Patterson 1983. *Protistologica* 20, 415–429.

Nanney, D. L. (1966a). Cortical integration in *Tetrahymena:* An exercise in cytogeometry. *J.
Exp. Zool.* 161, 307–318.

Nanney, D. L. (1966b). Corticotypes in *Tetrahymena pyriformis. Am. Nat.* 100, 303–318.

Nanney, D. L. (1966c). Corticotype transmission in *Tetrahymena. Genetics* 54, 955–968.

Nanney, D. L. (1967a). Cortical slippage in *Tetrahymena. J. Exp. Zool.* 166, 163–170.

Nanney, D. L. (1967b). Comparative corticotype analyses in *Tetrahymena. J. Protozool.* 14,
690–697.

Nanney, D. L. (1968a). Patterns of cortical stability in *Tetrahymena. J. Protozool.* 15, 109–
112.

Nanney, D. L. (1968b). Cortical patterns in cellular morphogenesis. *Science* 160, 496–502.

Nanney, D. L. (1971a). Cortical characteristics of strains of syngens 10, 11, and 12 of *Tetrahymena pyriformis. J. Protozool.* 18, 33–37.

Nanney, D. L. (1971b). The constancy of cortical units in *Tetrahymena* with varying numbers of ciliary rows. *J. Exp. Zool.* 178, 177–182.

Nanney, D. L. (1975). Patterns of basal body addition in ciliary rows in *Tetrahymena. J. Cell Biol.* 65, 503–512.

Nanney, D. L. (1977). Cell–cell interactions in ciliates: Evolutionary and genetic constraints. In J. L. Reissig (ed.), *Microbial Interactions.* London, Chapman and Hall, pp. 353–397.

Nanney, D. L. (1985). Heredity without genes: Ciliate explorations of clonal heredity. *Trends in Genetics* 1, 295–298.

Nanney, D. L., and McCoy, J. W. (1976). Characterization of the species of the *Tetrahymena pyriformis* complex. *Trans. Am. Micros. Soc.* 95, 664–682.

Nanney, D. L. Chow, M., and Wozencraft, B. (1975). Considerations of symmetry in the cortical integration of *Tetrahymena* doublets. *J. Exp. Zool.* 193, 1–14.

Nanney, D. L., Chen, S. S., and Meyer, E. B. (1978). Scalar constraints in *Tetrahymena* evolution. Quantitative basal body variations within and between species. J. Cell Biol. 79, 727–736.

Nanney, D. L., Nyberg, D., Chen, S. S., and Meyer, E. B. (1980). Cytogeometric constraints in *Tetrahymena* evolution: Contractile vacuole pore positions in nineteen species of the *Tetrahymena pyriformis* complex. *Am. Nat.* 115, 705–717.

Nardi, J. B., and Kafatos, F. C. (1976). Polarity and gradients in lepidopteran wing epidermis. II. The differential adhesiveness model: Gradient of a non-diffusible cell-surface parameter. *J. Embryol. Exp. Morph.* 36, 489–512.

Nardi, J. B., and Stocum, D. L. (1983). Surface properties of regenerating limb cells: Evidence for gradation along the proximodistal axis. *Differentiation* 25, 27–31.

Nelsen, E. M. (1970). Division delays and abnormal oral development produced by colchicine in *Tetrahymena. J. Exp. Zool.* 175, 69–84.

Nelsen, E. M. (1978). Transformation in *Tetrayhmena thermophila:* Development of an inducible phenotype. *Dev. Biol.* 66, 17–31.

Nelsen, E. M. (1981). The undulating membrane of *Tetrahymena:* formation and reconstruction. *Trans. Amer. Micr. Soc.* 100, 285–295.

Nelsen, E. M., and DeBault, L. E. (1978). Transformation in *Tetrahymena pyriformis:* Description of an inducible phenotype. *J. Protozool.* 25, 113–119.

Nelsen, E. M., and Frankel, J. (1979). Regulation of corticotype through kinety insertion in *Tetrahymena. J. Exp. Zool.* 210, 277–288.

Nelsen, E. M., and Frankel, J. (1986). Intracellular pattern reversal in *Tetrahymena thermophila.* I. Evidence for reverse intercalation in unbalanced doublets. *Dev. Biol.* 114, 53–71.

Nelsen, E. M., and Frankel, J. (1989). Maintenance and regulation of cellular handedness in *Tetrahymena. Development,* 105, 457–471.

Nelsen, E. M., Frankel, J., and Martel, E. (1981). Development of the ciliature of *Tetrahymena thermophila.* I. Temporal coordination with oral development. *Dev. Biol.* 88, 27–38.

Nelsen, E. M., Frankel, J., and Jenkins, L. M. (1989a). Non-genic inheritance of cellular handedness. *Development,* 105, 447–456.

Nelsen, E. M., Frankel, J. and Williams, N. E. (1989b). Effects of cellular handedness on oral assembly in *Tetrahymena. J. Protozool,* in press.

Nelson, S. H., and McClay, D. R. (1988). Cell polarity in sea urchin embryos: Reorientation of cells occurs quickly in aggregates. *Dev. Biol.* 127, 235–247.

Ng, S. F. (1976a). Extra cytoproct mutant in *Paramecium tetraurelia:* Phenotype and biometrical analysis. *Protistologica* 12, 69–86.

Ng, S. F. (1976b). Extra cytoproct mutant in *Paramecium tetraurelia:* Morphogenetical analysis of proters and opisthes. *J. Exp. Zool.* 196, 167–182.

Ng, S. F. (1977). Analysis of contractile vacuole pore morphogenesis in *Tetrahymena pyriformis* by 180° rotation of ciliary meridians. *J. Cell Sci.* 25, 233–246.

Ng, S. F. (1978). Directionality of microtubule assembly: An *in vivo* study with the ciliate *Tetrahymena. J. Cell Sci.* 33, 227–234.

Ng, S. F. (1979a). Origin and inheritance of an extra band of microtubules in *Tetrahymena* cortex. *Protistologica* 15, 5–15.

Ng, S. F. (1979b). The precise site of origin of the contractile vacuole pore in *Tetrahymena* and its morphogenetic implications. *Acta Protozool.* 18, 305–312.

Ng, S. F. (1979c). Unidirectional regeneration is an intrinsic property of longitudinual microtubules in *Tetrahymena*—An *in vivo* study. *J. Cell Sci.* 36, 109–119.

Ng, S. F. (1979d). Interdependence of cell cycle events in *Tetrahymena thermophila:* Formation of contractile vacuole pore and fission gap in ciliary meridians. *J. Protozool.* 26, 583–586.

Ng, S. F. (1986). The somatic function of the micronucleus of ciliated protozoa. *Progress in Protistology* 1. Bristol, Biopress Ltd., pp. 215–286.

Ng, S. F., and Frankel, J. (1977). 180° rotation of ciliary rows and its morphogenetic implications in *Tetrahymena pyriformis. Proc. Natl. Acad. Sci. U.S.* 74, 1115–1119.

Ng, S. F., and Mikami, K. (1981). Morphogenetic role of the germ nucleus in *Paramecium tetraurelia. Protistologica* 17, 497–509.

Ng, S. F., and Newman, A. (1984a). The role of the micronucleus in stomatogenesis in sexual reproduction of *Paramecium tetraurelia:* Micronuclear and stomatogenic events. *Protistologica* 20, 43–64.

Ng, S. F., and Newman, A. (1984b). The role of the micronucleus in stomatogenesis in sexual reproduction of *Paramecium tetraurelia:* Conjugation of amicronucleates. *Protistologica* 20, 517–523.

Ng, S. F., and Newman, A. (1985). The macronuclear anlage does not play an essential role in stomatogenesis in conjugation in *Paramecium tetraurelia. Protistologica* 21, 391–398.

Ng, S. F., and Newman, A. (1987). Formation of the oral apparatus in the absence of macronuclear anlage differentiation during sexual reproduction in *Paramecium tetraurelia. Acta Protozool.* 26, 205–211.

Ng, S. F., and Williams, R. J. (1977). An ultrastructural investigation of 180°-rotated ciliary meridians of *Tetrahymena pyriformis. J. Protozool.* 24, 257–263.

Nieto, J. J., Calvo, P., Torres, A., and Perez-Silva, J. (1981). Régénération chez *Gastrostyla steini. Acta Protozool.* 20, 373–384.

Nilsson, J. R., and Williams, N. E. (1966). An electron microscope study of the oral apparatus of *Tetrahymena pyriformis. Compt. Rend. Trav. Lab. Carlsberg.* 35, 119–141.

Novick, P., and Botstein, D. (1985). Phenotypic analysis of temperature-sensitive yeast actin mutants. *Cell* 40, 405–416.

Nübler-Jung, K. (1974). Cell migration during pattern reconstitution in the insect segment *(Dysdercus intermedius). Nature* 248, 610–611.

Nübler-Jung, K. (1977). Pattern stability in the insect segment. I. Pattern reconstitution by intercalary regeneration and cell sorting in *Dysdercus intermedius. Roux's Arch. Dev. Biol.* 183, 17–40.

Nübler-Jung, K. (1979). Pattern stability in the insect segment. II. The intersegmental region. *Roux's Arch. Dev. Biol.* 186, 211–233.

Nübler-Jung, K. (1987). Tissue polarity in an insect segment: Denticle patterns resemble spontaneously forming fibroblast patterns. *Development* 100, 171–177.

Nübler-Jung, K., and Grau, V. (1987). Pattern control in insect segments: Superimposed features of the pattern may be subject to different control mechanisms. *Roux's Arch. Dev. Biol.* 196, 290–294.

Nübler-Jung, K., Bonitz, R., and Sonnenschein, M. (1987). Cell polarity during wound healing in an insect epidermis. *Development* 100, 163–170.

Nuccitelli, R. (1983). Transcellular ion currents: Signals and effectors of cell polarity. In J. R. McIntosh, (ed.), *Spatial Organization of Eukaryotic Cells* (*Modern Cell Biology* 2). New York, Alan R. Liss, pp. 451–481.

Numata, O., and Watanabe, Y. (1982). *In vitro* assembly and disassembly of 14-nm filament from *Tetrahymena pyriformis*. The protein component of 14-nm filament is a 49,000-dalton protein. *J. Biochem (Tokyo)* 91, 1563–1573.

Numata, O., Hirono, M., and Watanabe, Y. (1983). Involvement of *Tetrahymena* intermediate filament protein, a 49K protein, in the oral morphogenesis. *Exp. Cell Res.* 148, 207–220.

Nurse, P., and Thuriaux, P. (1980). Regulatory genes controlling mitosis in the fission yeast *Schizosaccharomyces pombe*. *Genetics* 96, 627–637.

Nüsslein-Volhard, C. (1977). Genetic analysis of pattern formation in the embryo of *Drosophila melanogaster*. Characterization of the maternal effect mutation *bicaudal*. *Roux's Arch. Dev. Biol.* 183, 249–268.

Nüsslein-Volhard, C., and Wieschaus, E. (1980). Mutations affecting segment number and polarity in *Drosophila*. *Nature* 287, 795–801.

Nüsslein-Volhard, C., Wieschaus, E., and Jürgens, C. (1982). Segmentierung bei *Drosophila*—eine Genetische Analyse. *Verh. Deutsche Zool. Ges.* 75, 91–104.

Nüsslein-Volhard, C., Kluding, H., and Jürgens, G. (1985). Genes affecting the segmental subdivision of the *Drosophila* embryo. *Molecular Biology of Development, (Cold Spring Harbor Symposium of Quantitative Biology* 50), 145–154.

Nüsslein-Volhard, C., Fröhnhofer, G., and Lehmann, R. (1987). Determination of anteroposterior polarity in *Drosophila*. *Science* 238, 1675–1681.

Ogura, A., and Machemer, H. (1980). Distribution of mechanoreceptor channels in the *Paramecium* surface membrane. *J. Comp. Physiol* 135, 233–242.

Ohba, H., Ohmori, I., Numata, O., and Watanabe, Y. (1986). Purification and immunofluorescence localization of the mutant gene product of a *Tetrahymena cdaAl* mutant affecting cell division. *J. Biochem. (Tokyo)* 100, 797–808.

Okada, Y. K., and Sugino, H. (1937). Transplantation experiments in *Planaria gonocephala* Dugés. *Japan. J. Zool.* 7, 373–439.

Orias, E., and Pollock, N. A. (1975). Heat-sensitive development of the phagocytotic organelle in a *Tetrahymena* mutant. *Exp. Cell Res.* 90, 345–356.

Orias, E., and Rasmussen, L. (1976). Dual capacity for nutrient uptake in *Tetrahymena*. IV. Growth without food vacuoles and its implications. *Exp. Cell Res.* 102, 127–137.

Osmani, S. A., Engle, D. B., Doonan, J. H., and Morris, J. R. (1988). Spindle formation and chromation condensation in cells blocked at interphase by mutation of a negative cell cycle control gene. *Cell* 52, 241–251.

Pang, Y.-b., Zou, S.-f, and Tchang, T.-r. (Zhang, Z.-r.) (1984). A study on regulation of jumelles *Stylonychia mytilus* to normal singlets. *J. East China Normal Univ., Nat. Sci.* 3, 93–99 (in Chinese, with English summary).

Parducz, B. (1962). On a new concept of cortical organization in *Paramecium*. *Acta Biol. Acad. Sci. Hungaricae* 13, 299–322.

Parker, J. W., and Giese, A. C. (1966). Nuclear activity during regeneration in *Blepharisma intermedium* Bhandary. *J. Protozool.* 13, 617–622.

Patterson, D. J. (1981). On the origin of the postoral microtubules in *Paramecium putrinum* (Hymenostomatida, Ciliophora). *Protistologica* 17, 525–531.

Paulin, J. J., and Bussey, J. (1971). Oral regeneration in the ciliate *Stentor coeruleus:* A scanning and transmission electron optical study. *J. Protozool.* 18, 201–213.

Pearson, P. J., and Tucker, J. B. (1977). Control of shape and pattern during the assembly of a large microtubule bundle. Evidence for a microtubule-nucleating-template. *Cell Tiss. Res.* 180, 241–252.

Peck, R. K. (1977). The ultrastructure of the somatic cortex of *Pseudomicrothorax dubius:* Structure and function of the epiplasm in ciliated protozoa. *J. Cell Sci.* 25, 367–385.

Peck, R. K., Pelvat, B., Bolivar, I., and de Haller, G. (1975). Light and electron microscopic observations on the heterotrich ciliate *Climacostomum virens. J. Protozool.* 22, 368–385.

Pelvat, B. (1985). Observations sur l'ultrastructure de l'appareil buccal chez le cilié hétérotriche *Stentor coeruleus. Protistologica* 21, 61–80.

Pelvat, B., and de Haller, G. (1979). La régénération de l'appariel oral chez *Stentor coeruleus:* Étude au protargol et essai de morphogénèse comparée. *Protistologica* 15, 369–386.

Percy, J., Kuhn, K. L., and Kalthoff, K. (1986). Scanning electron microscopic analysis of the spontaneous and UV-induced abnormal segment patterns in *Chironomus samoaensis.* (Diptera, Chironomidae). *Roux's Arch. Dev. Biol.* 195, 92–102.

Perez-Paniagua, F., Camacho-Fumanal, R., and de Puytorac, P. (1988). Observations sur la stomatogenèse de bipartition du cilié *Paramecium tetraurelia. Ann. des Sci. Nat., Zool. Paris, 13ᵉ ser* 9, 13–20.

Perrimon, N., and Mahowald, A. P. (1987). Multiple functions of segment polarity genes in *Drosophila. Dev. Biol.* 119, 587–600.

Peterson, E. L., and Berger, J. D. (1976). Mutational blockage of DNA synthesis in *Paramecium tetraurelia. Can. J. Zool.* 54, 2089–2097.

Peterson, S. P., and Berns, M. W. (1978). Evidence for a centriolar region RNA functioning in spindle formation in dividing PTK_2 cells. *J. Cell Sci.* 34, 289–301.

Pitelka, D. R. (1968). Fibrillar systems in protozoa. In T. T. Chen (ed.), *Research in Protozoology* 3. Oxford, Pergamon Press, pp. 280–388.

Plattner, H., Westphal, C., and Tiggemann, R. (1982). Cytoskeleton—secretory vesicle interactions during the docking of secretory vesicles at the cell membrane of *Paramecium tetraurelia* cells. *J. Cell Biol.* 92, 368–377.

Polyani, M. (1968). Life's irreducible structure. *Science* 160, 1308–1312.

Pouphile, M., Lefort-Tran, M., Plattner, H., Rossignol, M., and Beisson, J. (1986). Genetic dissection of the morphogenesis of exocytosis sites in *Paramecium. Biology of the Cell* 56, 151–161.

Poyton, R. O. (1983). Memory and membranes: The expression of genetic and spatial memory during the assembly of organelle macrocompartments. In B. H. Satir (ed.), *Modern Cell Biology* 2. New York, Alan R. Liss, pp. 15–72.

Preer, J. R., Jr. (1986). Surface antigens in *Paramecium.* In J. G. Gall (ed.), *The Molecular Biology of Ciliated Protozoa.* New York, Academic Press, pp. 310–339.

Pringle, J. R., Lillie, S. H., Adams, A.E.M., Jacobs, C. W., Haarer, B. K., Coleman, K. G., Robinson, J. S., Bloom, L., and Preston, R. A. (1986). Cellular morphogenesis in the yeast cell cycle. In J. Hicks (ed.), *Yeast Cell Biology.* New York, A. R. Liss, pp. 47–80.

Rasmussen, C. D., Ching, A. S.-L., and Berger, J. D. (1985). The full schedule of macronuclear DNA synthesis is not required for cell division in *Paramecium aurelia. J. Protozool.* 32, 366–368.

Rasmussen, L. (1966). Effects of DL-*p*-fluorophenylalanine on *Paramecium aurelia* during the cell generation cycle. *Exp. Cell. Res.* 45, 501–504.

Rasmussen, L. (1967). Effects of metabolic inhibitors on *Paramecium aurelia* during the cell generation cycle. *Exp. Cell Res.* 48, 132–139.

Reverberi, G. (1961). The embryology of ascidians. *Adv. Morphogen.* 1, 55–101.

Reverberi, G., and Ortolani, C. (1962). Twin larvae from halves of the same egg in ascidians. *Dev. Biol.* 5, 84–100.

Ricci, N. (1981). Preconjugant cell interactions in *Oxytricha bifaria* (Ciliata, Hypotrichida): A two-step recognition process leading to cell fusion and induction of meiosis. In D. H. O'Day and P. A. Horgen (eds.), *Sexual Interactions in Eukaryotic Microbes.* New York, Academic Press, pp. 319–350.

Ricci, N., Banchetti, R., and Cetera, R. (1980). Initiation of meiosis and other nuclear changes in two species of *Oxytricha. Protistologica* 16, 413–417.

Rickoll, W. L., and Counce, S. J. (1980). Morphogenesis in the embryo of *Drosophila melanogaster*—Germ band extension. *Roux's Arch. Dev. Biol.* 188, 163–177.

Roberts, L. (1988). Chromosomes: The ends in view. *Science* 240, 982–983.

Roque, M. (1961). Recherches sur les infusoires ciliés: Les hyménostomes péniculiens. *Bull. Biol. France Belg.* 95, 432–519.

Roth, L. E., Pihlaja, D. J., and Shigenaka, Y. (1970). Microtubules in the heliozoan axopodium. I. The gradion hypothesis of allosterism in structural proteins. *J. Ultrastr. Res.* 30, 7–37.

Ruffolo, J. J., Jr. (1970). Regulation of cortical proteins in *Euplotes. J. Protozool.* 17, 115–124.

Ruffolo, J. J., Jr. (1976a). Fine structure of the dorsal bristle complex and pellicle of *Euplotes. J. Morph.* 148, 469–488.

Ruffolo, J. J., Jr. (1976b). Cortical morphogenesis during the cell division cycle in *Euplotes:* An integrated study using light optical, scanning electron and transmission electron microscopy. *J. Morph.* 148, 489–528.

Ruiz, F., Garreau de Loubresse, N., and Beisson, J. (1987). A mutation affecting basal body duplication and cell shape in *Paramecium. J. Cell Biol.* 104, 417–430.

Russell, M. A. (1985). Positional information in insect segments. *Dev. Biol.* 108, 269–283.

Russell, P., and Nurse, P. (1987). Negative regulation of mitosis by *wee1*⁺, a gene encoding a protein kinase homolog. *Cell* 49, 559–567.

Saló, E., and Baguñá, J. (1985). Proximal and distal transformation during intercalary regeneration in the planarian *Dugesia (S) mediterranea.* Evidence using a chromosomal marker. *Roux's Arch Dev. Biol.* 194, 364–368.

Sander, K. (1959). Analyse der ooplasmatischen Reaktionssystems von *Euscelis plebejus* Fall (Cicadina) durch Isolieren und Kombinieren von Keimteilen. I. Die Differenzierungsleistungen vorderer und hinterer Eiteile *Wilh. Roux' Arch. Entwickl.-Mech. Org.* 151, 430–497.

Sander, K. (1971). Pattern formation in longitudinal halves of leaf hopper eggs (Homoptera) and some remarks on the definition of "embryonic regulation". *Wilh. Roux' Arch Entwickl.-Mech. Org.* 167, 336–352.

Sander, K. (1976). Specification of the basic body pattern in insect embryogenesis. *Adv. Insect Physiol.* 12, 125–238.

Sander, K. (1981). Pattern generation and pattern conservation in insect embryogenesis—Problems, data, and models. *Fortschr. der Zool.* 26, 101–119.

Sapra, G. R., and Ammerman, D. (1974). An analysis of the developmental program in relation to RNA metabolism in the ciliate *Stylonychia mytilus. Dev. Biol.* 36, 105–112.

Satir, B., Schooley, C., and Satir, P. (1972). Membrane reorganization during secretion in *Tetrahymena. Nature* 235, 53–54.

Sattler, C. A., and Staehelin, L. A. (1976). Reconstruction of oral cavity of *Tetrahymena pyriformis* using high voltage electron microscopy. *Tissue and Cell* 8, 1–18.

Sattler, C. A., and Staehelin, L. A. (1979). Oral cavity of *Tetrahymena pyriformis*. A freeze-fracture and high-voltage electron microscopy study of the oral ribs, cytostome, and forming food vacuole. *J. Ultrastr. Res.* 66, 132–150.

Sawyer, H. R., Jr., and Jenkins, R. A. (1977). Stomatogenic events accompanying binary fission in *Blepharisma*. *J. Protozool.* 24, 140–149.

Schäfer, E., and Cleffmann, G. (1982). Division and growth kinetics of the division mutant *conical* of *Tetrahymena*. A contribution to the regulation of generation time. *Exp. Cell Res.* 137, 277–281.

Schlessinger, J. (1983). Mobilities of cell-membrane proteins: How are they modulated by the cytoskeleton? *Trends in Neurosciences* 6, 360–363.

Schroeder, T. E. (1975). Dynamics of the contractile ring. In S. Inoue and R. E. Stephens (eds.), *Molecules and Cell Movement* (*Society for General Physiology Symposium* 30). New York, Raven Press, pp. 305–334.

Schubiger, G., and Wood, W. J. (1977). Determination during early embryogenesis in *Drosophila melanogaster*. *Am. Zool.* 17, 565–576.

Schubiger, G., Mosely, R. C., and Wood, W. J. (1977). Interaction of different egg parts in determination of various body regions in *Drosophila melanogaster*. *Proc. Natl. Acad. Sci.* 74, 2050–2053.

Schulte, H., and Schwartz, V. (1970). Regulatives Zellwachstum und Determination der Teilungslinie bei *Stentor coeruleus* Ehrbg. *Arch. Protistenk.* 112, 305–311.

Schüpbach, T., and Wieschaus, E. (1986). Maternal-effect mutations altering the anterior-posterior patterns of the *Drosophila* embryo. *Roux's Arch Dev. Biol.* 195, 302–317.

Schwartz, V. (1935). Versuche über Regeneration und Kerndimorphismus bei *Stentor coeruleus* Ehrbg. *Archiv f. Protistenk.* 85, 100–139.

Schwartz, V. (1963). Die Sicherung der Arttypischen Zellform bei Ciliaten. *Naturwiss.* 20, 631–640.

Shay, J. W., Peters, T. T., and Fuseler, J. W. (1978). Cytoplasmic transfer of microtubule organizing centers in mouse tissue culture cells. *Cell* 14, 835–842.

Shi, X.-b (1980a). The morphology and morphogenesis of the buccal apparatus of *Paramecium* and their phylogenetic implications. I. Morphology of the buccal apparatus. *Acta Zool. Sinica* 26, 205–213 (In Chinese, with English summary).

Shi, X.-b. (1980b). The morphology and morphogenesis of the buccal apparatus of *Paramecium* and their phylogenetic implications. II. Stomatogenesis. *Acta Zool. Sinica* 26, 289–300 (In Chinese, with English summary).

Shi, X.-b, and Frankel, J. (1989a). Morphology and development of mirror-image doublet *Stylonychia mytilus*. *J. Protozool.*, submitted.

Shi, X.-b. and Frankel, J. (1989b). Morphology and development of left-handed singlets derived from mirror-image doublets of *Stylonychia mytilus*. *J. Protozool.*, submitted.

Shi, X.-b., Qiu, Z:j. Lu, L., and Frankel, J. (1990a). Microsurgically generated discontinuities provoke heritable changes in cellular handedness in a ciliate, *Stylonychia mytilus*. Forthcoming.

Shi, X.-b., He, W., and Frankel, J. (1990b). Creation of mirror-image doublets of *Stylonychia mytilus* with neighboring oral structures. Forthcoming.

Shinbrot, M. (1966). Fixed point theorems. *Sci. Am.* 214 (1), 105–110.

Siegel, R. W. (1970). Organellar damage and revision as a possible basis for intraclonal variation in *Paramecium*. *Genetics* 66, 305–314.

Sikora, J. (1981). Cytoplasmic streaming in *Paramecium*. *Protoplasma* 109, 57–77.

Simanis, V., and Nurse, P. (1986). The cell cycle control gene $cdc2^+$ of fission yeast encodes a protein kinase potentially regulated by phosphorylation. *Cell* 45, 261–268.

Simpson, P., and Grau, Y. (1987). The segment polarity gene *costal-2* in *Drosophila*. II. The origin of imaginal pattern duplications. *Dev. Biol.* 122, 201–209.

Simpson, R. E., and Williams, N. E. (1970). The effects of pressure on cell division and oral morphogenesis in *Tetrahymena*. *J. Exp. Zool.* 175, 85–98.

Slack, J.M.W. (1983). *From Egg to Embryo: Determinative Events in Early Development.* Cambridge, Cambridge University Press.

Slack, J.M.W. (1987). We have a morphogen! *Nature* 327, 553–554.

Slack, J.M.W., and Forman, D. (1980). An interaction between dorsal and ventral regions of the marginal zone in early amphibian embryos. *J. Embryol. Exp. Morph.* 56, 283–299.

Small, E. B., and Lynn, D. H. (1981). A new macrosystem for the phylum Ciliophora Doflein 1901. *BioSystems* 14, 387–401.

Small, E. B., and Lynn, D. H. (1985). Phylum Ciliophora Doflein 1901. In J. J. Lee, S. H. Hutner, and E. B. Bovee (ed.), *An Illustrated Guide to the Protozoa.* Lawrence, Kan., Society of Protozoologists, pp. 393–575.

Smith, H. E. (1982). Oral apparatus structure in the microstomial form of *Tetrahymena vorax. Trans. Am. Micr. Soc.* 101, 36–58.

Smith, J. C., and Slack, J.M.W. (1983). Dorsalization and neural induction: Properties of the organizer in *Xenopus laevis. J. Embryol. Exp. Morph.* 78, 299–317.

Smith, L. D., and Ecker, R. E. (1970). Uterine suppression of biochemical and morphogenetic events in *Rana pipiens. Dev. Biol.* 22, 622–637.

Smith-Sonneborn, J., and Plaut, W. (1967). Evidence for the presence of DNA in the pellicle of *Paramecium. J. Cell Sci.* 2, 225–234.

Sogin, M. L., and Elwood, H. J. (1986). Primary structure of the *Paramecium aurelia* small-subunit rRNA coding region: Phylogenetic relationships within the Ciliophora: *J. Mol. Evol.* 23, 53–60.

Sogin, M. L., Swanton, M. T., Gunderson, J. H., and Elwood, H. J. (1986a). Sequence of the small subunit ribosomal RNA gene from the hypotrichous ciliate *Euplotes aedicula-tus. J. Protozool.* 33, 26–29.

Sogin, M. L., Elwood, H. J., and Gunderson, J. H. (1986b). Evolutionary diversity of eukaryotic small-subunit rRNA genes. *Proc. Natl. Acad. Sci. U.S.* 83, 1383–1387.

Sogin, M. L., Ingold, A., Karlok, M., Nielsen, H., and Engberg J. (1986c). Phylogenetic evidence for the acquisition of ribosomal RNA introns subsequent to the divergence of some major *Tetrahymena* groups. *EMBO J.* 5, 3625–3630.

Solomon, F. (1981). Specification of cell morphology by endogenous determinants. *J. Cell Biol.* 90, 547–553.

Sonneborn, T. M. (1930). Genetic studies in *Stenostomum incaudatum.* II. The effects of lead acetate on the hereditary constitution. *J. Exp. Zool.* 57, 409–439.

Sonneborn, T. M. (1954). Patterns of nucleocytoplasmic integration in *Paramecium.* In *Proc. Intl. Congr. Genet. (Caryologia)* 9 (Suppl.), 307–325.

Sonneborn, T. M. (1963). Does preformed cell structure play an essential role in cell heredity? In J. M. Allen (ed.), *The Nature of Biological Diversity.* New York, McGraw-Hill, pp. 165–221.

Sonneborn, T. M. (1964). The differentiation of cells. *Proc. Natl. Acad. Sci. U.S.* 51, 915–929.

Sonneborn, T. M. (1970). Gene action in development. *Proc. Roy. Soc. Lond. B* 176, 347–366.

Sonneborn, T. M. (1974). Ciliate morphogenesis and its bearing on cellular morphogenesis. *Actualites Protozoologiques, Proc. IV Intl. Congr. Protozool.* 1, 327–355.

Sonneborn, T. M. (1975a). Positional information and nearest neighbor interactions in relation to spatial patterns in ciliates. *Ann. Biol.* 14, 565–584.

Sonneborn, T. M. (1975b). The *Paramecium aurelia* complex of fourteen sibling species. *Trans. Am. Micr. Soc.* 94, 155–178.

Sonneborn, T. M. (1977). Local differentiation of the cell surface of ciliates: Their determination, effects, and genetics. In G. Poste and G. L. Nicolson (eds.), *The Synthesis, Assembly, and Turnover of Cell Surface Components* (*Cell Surface Reviews* 4). Amsterdam, Elsevier/North Holland, pp. 829–856.

Spemann, H., and Mangold, H. (1964). Induction of embryonic primordia by implantation of organizers from a different species. In B. H. Willier and J. M. Oppenheimer (eds.), *Foundations of Experimental Embryology*. Englewood Cliffs, N.J., Prentice Hall, pp. 144–184. [Translation of 1924 original].

Stein-Gaven, S., Wells, J. M., and Karrer, K. M. (1987). A germ line specific DNA sequence is transcribed in *Tetrahymena*. *Dev. Biol.* 120, 259–269.

Steinberg, M. (1978). Cell–cell recognition in multicellular assembly: Levels of specificity. *Cell-Cell Interactions* (*Society for Experimental Biology Symposium* 32). Cambridge, Cambridge University Press, pp. 25–49.

Stent, G. (1985). Thinking in one dimension: The impact of molecular biology on development. *Cell* 40, 1–2.

Stern, C. (1954). Two or three bristles. *Am. Sci* 42, 213–247.

Stern, C. (1968). *Genetic Mosaics and Other Essays*. Cambridge, Mass., Harvard University Press.

Sternberg, P. N. (1988). Lateral inhibition during vulval induction in *Caenorhabdites elegans*. *Nature,* 335, 551–554.

Sternberg, P. N., and Horvitz, H. R. (1986). Pattern formation during vulval development in *C. elegans*. *Cell* 44, 761–772.

Stocum, D. L. (1984). The urodele limb regeneration blastema: Determination and organization of the morphogenetic field. *Differentiation* 27, 13–28.

Stout, J. D. (1954). The ecology, life history, and parasitism of *Tetrahymena* (*Paraglaucoma*) *rostrata* (Kahl) Corliss. *J. Protozool.* 1, 211–215.

Strome, S., and Wood, W. B. (1983). Generation of asymmetry and segregation of germ-line granules in early *C. elegans* embryos. *Cell* 35, 15–25.

Stumpf, H. F. (1966). Über gefälleabhängige Bildungen des Insektensegmentes. *J. Insect Physiol.* 12, 601–617.

Stumpf, H. F. (1967). Über die Lagebestimmung der Kutikularzonen innerhalb eines Segmentes von *Galleria melonella*. *Dev. Biol.* 16, 144–167.

Stumpf, H. F. (1968). Further studies on gradient-dependent diversification in the pupal cuticle of *Galleria mellonella*. *J. Exp. Biol.* 49, 49–60.

Suganuma, Y., Shimode, C., and Yamamoto, H. (1984). Conjugation in *Tetrahymena:* Formation of a special junction area for conjugation during the co-stimulation period. *J. Electron Microsc.* 33, 10–18.

Sugino. H. (1941). Homopolar union in *Planaria gonocephala*. *Japan J. Zool.* 9, 175–183.

Suhama, M. (1971). The formation of cortical structures in the cell cycle of *Paramecium trichium*. I. Proliferation of cortical units, *J. Sci. Hiroshima Univ., Ser B., Div. 1,* 23, 105–119.

Suhama, M. (1973). The formation of cortical structures in the cell cycle of *Paramecium trichium*. II. Morphogenetic movement of the contractile vacuole pores. *J. Sci. Hiroshima Univ., Ser. B. Div. 1,* 24, 165–181.

Suhama, M. (1975). Changes of proliferation of cortical units in *Paramecium trichium*, caused by excision of the anterior region. *J. Sci Hiroshima Univ., Ser. B. Div. 1,* 26, 37–51.

Suhama, M. (1982). Homopolar doublets of the ciliate *Glaucoma scintillans* with a reversed oral apparatus. I. Development of the oral primordium. *J. Sci. Hiroshima Univ., Ser. B. Div. 1,* 30, 51–65.

Suhama, M. (1983). The location of dividing micronucleus respecting cortical structures in binary fission of *Glaucoma scintillans* (Ciliophora). *J. Sci. Hiroshima Univ. Ser. B. Div. 1* 31, 49–62.

Suhama, M. (1984). The position of the contractile vacuole pore in doublets of the ciliate *Glaucoma scintillans,* manifesting pattern reversal. *Zool. Sci.* (Tokyo) 1, 888 (abstr.).

Suhama, M. (1985). Reproducing singlets with an inverted oral apparatus in *Glaucoma scintillans* (Ciliophora, Hymenostomatida). *J. Protozool.* 32, 454–459.

Suhr-Jessen, P. B., Stewart, J. M., and Rasmussen, L. (1977). Timing and regulation of nuclear and cortical events in the cell cycle of *Tetrahymena pyriformis. J. Protozool.* 24, 299–303.

Sullivan, W. (1987). Independence of *fushi tarazu* expression with respect to cellular density in *Drosophila* embryos. *Nature* 327, 164–167.

Sulston, J. E., and White, J. G. (1980). Regulation and cell autonomy during postembryonic development of *Caenorhabdites elegans. Dev. Biol.* 78, 577–597.

Sundararaman, V., and Hanson, E. D. (1976). Longitudinal microtubules and their functions during asexual reproduction in *Paramecium tetraurelia. Genet. Res. Cambr.* 27, 205–211.

Suzuki, S. (1957). Morphogenesis in the regeneration of *Blepharisma undulans japonicus* Suzuki. *Bull. Yamagata Univ. Nat. Sci.* 4, 85–192.

Suzuki, S. (1973a). General Morphology. In A. C. Giese (ed.), *Blepharisma.* Stanford, Stanford University Press, pp. 5–17.

Suzuki, S. (1973b). Nuclear Behavior. In A. C. Giese (ed.), *Blepharisma.* Stanford, Stanford University Press, pp. 18–38.

Suzuki, S. (1973c). Morphogenesis. In A. C. Giese (ed.), *Blepharisma.* Stanford, Stanford University Press, pp. 172–214.

Swan, J. A., and Solomon, F. (1984). Reformation of the marginal band of avian erythrocytes in vitro using calf-brain tubulin: Peripheral determinants of microtubule form. *J. Cell Biol.* 99, 2108–2113.

Swanson, M. M., and Poodry, C. A. (1981). The *Shibire*[ts] mutant of *Drosophila:* A probe for the study of embryonic development. *Dev. Biol.* 84, 465–470.

Szabad, J., Schüpbach, T., and Wieschaus, E. (1979). Cell lineage and development in the larval epidermis of *Drosophila melanogaster. Dev. Biol.* 73, 256–271.

Szablewski, L. (1985). Adaptation of stomatogenesis and cell division in *Tetrahymena pyriformis* GL to the continuous presence of colistin in the medium. *Acta Protozool.* 24, 23–35.

Tam, L.-W., and Ng, S. F. (1987). Genetic analysis of heterokaryons in search of active micronuclear genes in stomatogenesis of *Paramecium tetraurelia. Europ. J. Protistol.* 23, 43–50.

Tamm, S., and Tamm, S. L. (1980). Origin and development of free kinetosomes in the flagellates *Deltotrichonympha* and *Koruga. J. Cell Sci.* 42, 189–205.

Tamm, S., and Tamm, S. L. (1988). Development of macrociliary cells in Beroë. *J. Cell Sci.* 89, 67–80.

Tamm, S. L., Sonneborn, T. M., and Dippell, R. V. (1975). The role of cortical orientation in the control of the direction of ciliary beat in *Paramecium. J. Cell Biol.* 64, 98–112.

Tamura, S., Toyoshima, Y., and Watanabe, Y. (1966). Mechanism of temperature-induced synchrony in *Tetrahymena pyriformis.* Analysis of the leading cause of synchronization. *Jap. J. Med. Sci. Biol.* 19, 85–96.

Tartar, V. (1954a). Anomalies of regeneration of *Paramecium. J. Protozool.* 1, 11–17.

Tartar, V. (1954b). Reactions of *Stentor coeruleus* to homoplastic grafting. *J. Exp. Zool.* 127, 511–576.

Tartar, V. (1956a). Grafting experiments concerning primordium formation in *Stentor coeruleus*. *J. Exp. Zool.* 131, 75–122.

Tartar, V. (1956b). Further experiments correlating primordium sites with cytoplasmic pattern in *Stentor coeruleus*. *J. Exp. Zool.* 132, 269–298.

Tartar, V. (1956c). Pattern and substance in *Stentor*. In D. Rudnick (ed.), *Cellular Mechanisms of Differentiation and Growth* (*Society for Developmental Biology Symposium* 14). Princeton, Princeton University Press, pp. 73–100.

Tartar, V. (1957). Deletion experiments on the oral primordium of *Stentor coeruleus*. *J. Exp. Zool.* 136, 53–73.

Tartar, V. (1958a). Induced resorption of oral primordia in regenerating *Stentor coeruleus*. *J. Exp. Zool.* 139, 1–31.

Tartar, V. (1958b). Specific inhibition of the oral primordium by formed oral structures in *Stentor coeruleus*. *J. Exp. Zool.* 139, 479–505.

Tartar, V. (1959). Equational division of carbohydrate reserves in *Stentor coeruleus*. *J. Exp. Zool.* 140, 269–280.

Tartar, V. (1960). Reconstitution of minced *Stentor coeruleus*. *J. Exp. Zool.* 144, 187–207.

Tartar, V. (1961). *The Biology of Stentor*. Oxford, Pergamon Press.

Tartar, V. (1962). Morphogenesis in *Stentor*. *Adv. Morphogen.* 2, 1–26.

Tartar, V. (1964). Morphogenesis in homopolar tandem grafted *Stentor coeruleus*. *J. Exp. Zool.* 156, 243–252.

Tartar, V. (1966a). Synchronization of oral primordia in *Stentor coeruleus*. *J. Exp. Zool.* 161, 53–62.

Tartar, V. (1966b). Fission after division primordium removal in the ciliate *Stentor coeruleus* and comparable experiments on reorganizers. *Exp. Cell Res.* 42, 357–370.

Tartar, V. (1966c). Induced division and division regression by cell fusion in *Stentor*. *J. Exp. Zool.* 163, 297–310.

Tartar, V. (1966d). Stentors in dilemmas. *Z. Allg. Mikrobiol.* 6, 125–134.

Tartar, V. (1967a). Cell division after removal of the division line in *Stentor*. *Nature* 216, 695–697.

Tartar, V. (1967b). Morphogenesis in protozoa. In T. T. Chen (ed.), *Research in Protozoology* 2. Oxford, Pergamon Press, pp. 1–116.

Tartar, V. (1968). Micrurgical experiments on cytokinesis in *Stentor coeruleus*. *J. Exp. Zool.* 167, 21–36.

Tartar, V. (1979). Bipolar forms of *Condylostoma magnum*. *J. Protozool.* 26, 26A.

Tchang, T.-r., and Pang, Y.-b. (1965). Conditions for the artificial induction of monster jumelles of *Stylonychia mytilus* which are capable of reproduction. *Scientia Sinica* 14, 1331–1338.

Tchang, T.-r., and Pang, Y.-b. (1977). The cytoplasmic differentiation of jumelle *Stylonychia*. *Scientia Sinica* 20, 234–243.

Tchang, T.-r., and Pang, Y.-b. (1979). Phenomenon of macronuclear regulation of artificial dorsi-conjugant *Stylonychia* and its relation to the cytoplasm. *Scientia Sinica* 22, 467–473.

Tchang, T.-r., and Pang, Y.-b. (1981). The "ciliary pattern" of jumelle *Stylonychia* and its genetic behavior. *Scientia Sinica* 24, 122–129.

Tchang, T.-r. (Zhang, Z.-r.), and Pang, Y.-b. (1982). Formation of jumelles and dorsal connected doublets in hypotrich ciliates. *Kexue Tongbao* (foreign language edition) 27, 1213–1217.

Tchang, T.-r., Shi, X.-b., and Pang, Y.-b. (1964). An induced monster ciliate transmitted through three hundred and more generations. *Scientia Sinica* 13, 850–853.

Thormar, H. (1959). Delayed division in *Tetrahymena pyriformis* induced by temperature changes. *Compt. Rend. Trav. Lab. Carlsberg* 31, 207–225.

Tiwari, S. C., Wick, S. M., Williamson, R. E., and Gunning, B.E.S. (1984). Cytoskeleton and integration of cellular function in cells of higher organisms. *J. Cell. Biol.* 99, 63s–69s.

Tokunaga, C. (1961). The differentiation of the secondary sexcomb under the influence of the gene *engrailed* in *Drosophila melanogaster. Genetics* 46, 157–176.

Tokunaga, C., and Gerhart, J. C. (1976). The effect of growth and joint formation on the bristle pattern of *Drosophila melanogaster. J. Exp. Zool.* 198, 79–95.

Torres, A., Fedriani, C., Morenza, C., and Gutierrez-Navarro, A. M. (1980). Cell cycle alterations of *Laurentia acuminata* induced by cortical damage. *Protistologica* 16, 227–232.

Totwen-Nowakowska, I. (1965). Doublets of *Stylonychia mytilus* (O.F.M.) evoked by action of thermic shocks. *Acta Protozool.* 3, 355–361.

Townes, P. L., and Holtfreter, J. (1955). Directed movements and selective adhesion of embryonic amphibian cells. *J. Exp. Zool.* 128, 53–120.

Truby, P. R. (1986). The growth of supernumerary legs in the cockroach. *J. Embryol. Exp. Morph.* 92, 115–131.

Tsunemoto, M., Numata, O., Sugai, T., and Watanabe, Y. (1988). Analysis of oral replacement by scanning electron microscopy and immunofluorescence microscopy in *Tetrahymena thermophila* during conjugation. *Zool. Sci.* (Tokyo) 5, 119–131.

Tucker, J. B. (1968). Fine structure and function of the cytopharyngeal basket in the ciliate *Nassula. J. Cell. Sci.* 3, 493–514.

Tucker, J. B. (1970). Morphogenesis of a large microtubular organelle and its association with basal bodies in the ciliate *Nassula. J. Cell Sci.* 6, 385–429.

Tucker, J. B. (1971a). Microtubules and a contractile ring of microfilaments associated with the cleavage furrow. *J. Cell Sci.* 8, 557–571.

Tucker, J. B. (1971b). Development and deployment of cilia, basal bodies, and other microtubular organelles in the cortex of the ciliate *Nassula. J. Cell. Sci.* 9, 539–567.

Tucker, J. B. (1977). Shape and pattern specification during microtubule bundle assembly. *Nature* 266, 22–26.

Tucker, J. B. (1979). Spatial organization of microtubules. In K. Roberts and S. Hyams (eds.), *Microtubules.* London, Academic Press, pp. 315–357.

Tucker, J. B. (1981). Cytoskeletal coordination and intercellular signalling during metazoan embryogenesis. *J. Embryol. Exp. Morph.* 65, 1–25.

Tucker, J. B., Beisson, J., Roche, D.L.J., and Cohen, J. (1980). Microtubules and control of macronuclear 'amitosis' in *Paramecium. J. Cell Sci.* 44, 135–151.

Tuffrau, M. (1969). L'origine du primordium buccal chez les ciliés hypotriches. *Protistologica* 5, 227–237.

Tuffrau, M. (1984). Les aspects généraux de la morphogenese chez les ciliés. *Traité de Zoologie,* 2 (fasc. 1). Paris, Masson, pp. 537–580.

Tuffrau, M. (1987). Proposition d'une classification nouvelle de l'ordre Hypotrichida (Protozoa, Ciliophora), fondée sur quelques données récentes. *Ann. des Sci. Nat., Zool., Paris, 13e ser* 8, 111–117.

Tuffrau, M., and Totwen-Nowakowska, I. (1988). Différents types de doublets chez *Stylonychia mytilus* (cilié Hypotriche): Genese, morphologie, reorganisation, polarité. *Ann. des Sci. Nat., Zool., Paris, 13e ser.* 9, 21–36.

Tuffrau, M., Fryd-Versavel, G., and Tuffrau, H. (1981). La réorganisation infraciliaire au cours de la conjugaison chez *Stylonychia mytilus. Protistologica* 17, 387–396.

Turing, A. M. (1952). The chemical basis of morphogenesis. *Phil. Trans. Roy. Soc. Lond. B* 237, 37–72.

Turner, F. R., and Mahowald, A. P. (1977). Scanning electron microscopy of *Drosophila melanogaster* embryogenesis. II. Gastrulation and segmentation. *Dev. Biol.* 57, 403–416.

Uhlig, G. (1959). Polaritätsabhängige Anlagenentwicklung bei *Stentor coeruleus. Z. Naturf.* 14b, 353–354.

Uhlig, G. (1960). Entwicklungsphysiologische Untersuchungen zur Morphogenese von *Stentor coeruleus* Ehrbg. *Arch. Protistenk.* 105, 1–109.

Uhlig, G. (1963a). Untersuchungen über die Folliculiniden *(Ciliata, Heterotrichida)* der Deutschen Bucht, *Veröff. Inst. Meeresforsch. Bremerhaven.* Sonderband, 115–121.

Uhlig, G. (1963b). Der Gehäusebau be *Metafolliculina andrewsi* (Ciliata; Heterotricha). *Zool Anz.* (Suppl.) 27, 498–507.

Valbonesi, A., Ortensi, C., and Luporini, P. (1988). An integrated study of the species problem in the *Euplotes crassus-minuta-vannus* group. *J. Protozool.,* 35, 38–45.

Van Bell, C. T. (1985). The 5S and 5.8S ribosomal RNA sequences of *Tetrahymena thermophila* and *T. pyriformis. J. Protozool.* 32, 640–644.

Vaudaux, P. (1976). Isolation and identification of specific cortical proteins in *Tetrahymena pyriformis* strain GL. *J. Protozool.* 23, 458–464.

Villaneuve-Brachon, S. (1940). Recerces sur les ciliés hétérotriches. Cinétome, argyrome, myoneme. *Arch. Zool. Exp. et Gen.* 82 (fasc. 1), 1–180.

Vinson, C. R., and Adler, P. N. (1987) Directional non-cell autonomy and the transmission of polarity information by the *frizzled* gene of *Drosophila. Nature* 329, 549–551.

Walker, G. K., Maugel, T. K., and Goode, D. (1980). Encystment and excystment in hypotrich ciliates. I. *Gastrostyla steini. Protistologica* 16, 511–524.

Ward, S. (1986). Asymmetric localization of gene products during the development of *Caenorhabdites elegans* spermatozoa. In J. G. Gall, (ed.), *Gametogenesis and the Early Embryo* (*Society for Developmental Biology Symposium* 44). New York, Alan R. Liss, pp. 55–75.

Weber, K., and Osborn, M. (1981). Microtubule and intermediate filament networks in cells viewed by immunofluorescent microscopy. In G. Poste and G. L. Nicolson (eds.), *Cytoskeletal Elements and Plasma Membrane Organization* (*Cell Surface Reviews* 8). Amsterdam, North Holland, pp. 1–53.

Webster, G. (1971). Morphogenesis and pattern formation in *Hydra. Biol. Rev.* 46, 1–64.

Whitman, C. O. (1893). The inadequacy of the cell theory of development. *J. Morph.* 8, 639–658.

Whitson, G. L. (1964). Temperature sensitivity and its relation to changes in growth, control of cell division, and stability of morphogenesis in *Paramecium aurelia* syngen 4, stock 51. *J. Cell. Comp. Physiol.* 64, 455–464.

Whitten, J. M. (1973). Kinetosomes in insect epidermal cells and their orientation with respect to cell symmetry and intracellular patterning. *Science* 181, 1066–1067.

Wieschaus, E., and Riggleman, R. (1987). Autonomous requirements for the segment polarity gene *armadillo* during *Drosophila* embryogenesis. *Cell* 49, 177–184.

Wieschaus, E., Nüsslein-Volhard, C., and Kluding, H. (1984). *Krüppel,* a gene whose activity is required early in the zygotic genome for normal embryonic segmentation. *Dev. Biol.* 104, 172–186.

Wille, J. J., Jr. (1966). Induction of altered patterns of cortical morphogenesis and inheritance in *Paramecium aurelia. J. Exp. Zool.* 163, 191–214.

Williams, N. E. (1960). The polymorphic life history of *Tetrahymena patula. J. Protozool.* 7, 10–17.

Williams, N. E. (1964). Relations between temperature sensitivity and morphogenesis in *Tetrahymena pyriformis* GL. *J. Protozool.* 11, 566–572.

Williams, N. E. (1975). Regulation of microtubules in *Tetrahymena. Int. Rev. Cytol.* 41, 59–86.

Williams, N. E. (1984). An apparent disjunction between the evolution of form and substance in the genus *Tetrahymena. Evolution* 38, 25–33.

Williams, N. E. (1986a). Evolutionary change in cytoskeletal proteins and cell architecture in lower eukaryotes. *Progr. in Protistology* 1, 309–324. Biopress Ltd., Bristol.

Williams, N. E. (1986b). The nature and organization of filaments in the oral apparatus of *Tetrahymena. J. Protozool.* 33, 352–358.

Williams, N. E. (1989). Structure, turnover, and assembly of ciliary membranes in *Tetrahymena.* In R. A. Bloodgood (ed.), *Structure and Function of Ciliary and Flagellar Surfaces.* New York, Plenum, (in press).

Williams, N. E., and Bakowska, J. (1982). Scanning electron microscopy of cytoskeletal elements in the oral apparatus of *Tetrahymena. J. Protozool.* 29, 382–389.

Williams, N. E., and Frankel, J. (1973). Regulation of microtubules in *Tetrahymena.* I. Electron microscopy of oral replacement. *J. Cell Biol.* 56, 441–457.

Williams, N. E., and Honts, J. E. (1987). The assembly and positioning of cytoskeletal elements in *Tetrahymena. Development* 100, 23–30.

Williams, N. E., and Luft, J. H. (1968). Use of a nitrogen mustard derivative in fixation for electron microscopy and observations on the ultrastructure of *Tetrahymena. J. Ultrastr. Res.* 25, 271–292.

Williams, N. E., Vaudaux, P. E., and Skriver, L. (1979). Cytoskeletal proteins of the cell surface in *Tetrahymena.* I. Identification and localization of major proteins. *Exp. Cell Res.* 123, 311–320.

Williams, N. E., Honts, J. E., and Graeff, R. W. (1986). Oral filament proteins and their regulation in *Tetrahymena pyriformis. Exp. Cell Res.* 164, 295–310.

Williams, N. E., Honts, J. E., and Jaeckel-Williams, R. F. (1987). Regional differentiation of the membrane skeleton in *Tetrahymena. J. Cell Sci.* 87, 457–463.

Williams, N. E., Honts, J. E., and Stuart, K. R. (1989a). Properties of microtubule-free cortical residues isolated from *Paramecium tetraurelia. J. Cell Sci.* 92, 427–432.

Williams, N. E., Honts, J. E., Lu, Q., Olson, C., and Moore, K. (1989b). Identification and localization of major cortical proteins in the ciliated protozoan, *Euplotes eurystomus. J. Cell Sci.,* 92, 433–439.

Winfree, A. T. (1980). *The Geometry of Biological Time.* Berlin, Springer Verlag.

Wirnsberger, E., Foissner, W., and Adam, H. (1985). Morphological, biometric, and morphogenetic comparison of two closely related species, *Stylonychia vorax* and *S. pustulata* (Ciliophora, Oxytrichidae). *J. Protozool.* 32, 261–268.

Wirnsberger, E., Foissner, W., and Adam, H. (1986). Biometric and morphogenetic comparison of the sibling species *Stylonychia mytilus* and *S. lemnae,* including a phylogenetic system for the Oxytrichids. *Arch. Protistenk.* 133, 167–185.

Wolfe, J., and Grimes, G. W. (1979). Tip transformation in *Tetrahymena:* A morphogenetic response to interactions between mating types. *J. Protozool* 26, 82–89.

Wolff, E. (1962). Recent researches on the regeneration of Planaria. In D. Rudnick (ed.), *Regeneration (Society for Developmental Biology Symposium* 20). New York, Ronald Press, pp. 53–84.

Wolpert, L. (1969). Positional information and the spatial pattern of cellular differentiation *J. Theoret. Biol.* 25, 1–47.

Wolpert, L. (1971). Positional information and pattern formation. *Curr. Top. Dev. Biol.* 6, 183–224.

Wood, D. C. (1982). Membrane permeabilities determining resting, action, and mechanoreceptor potentials in *Stentor coeruleus. J. Comp. Physiol.* 146, 537–550.

Wood, W. B. (1980). Bacteriophage T4 morphogenesis as a model for assembly of subcellular structure. *Quart. Rev. Biol.* 55, 353–367.

Wood, W. B., and Henninger, M. (1969). Attachment of tail fibers in bacteriophage T4 tail assembly: Some properties of the reaction in vitro and its genetic control. *J. Mol. Biol.* 39, 603–618.

Worthington, D. H., Salamone, M., and Nachtwey, D. S. (1976). Nucleocytoplasmic ratio requirements for initiation of DNA replication and fission in *Tetrahymena*. *Cell Tissue Kinet.* 9, 119–130.

Wright, D. A., and Lawrence, P. A. (1981). Regeneration of the segment boundary in *Oncopeltus*. *Dev. Biol.* 85, 317–327.

Wright, R. L., Chojnacki, B., and Jarvik, J. W. (1983). Abnormal basal-body number, location, and orientation in a striated fiber-defective mutant of *Chlamydomonas reinhardtii*. *J. Cell Biol.* 96, 1697–1707.

Yagiu, R. (1951). Studies on *Condylostoma spatiosum* Ozaki & Yagiu. IV. The relationship between an injury and the power of division. *J. Sci. Hiroshima Univ. Ser. B, Div 1* 12, 131–139.

Yagiu, R. (1952). Studies on *Condylostoma spatiosum* Ozaki and Yagiu. V. Abnormal phenomenon caused by being kept in fresh water. *J. Sci. Hiroshima Univ. Ser B, Div. 1,* 13, 92–109.

Yamada, T. (1940). Beeinflussung der Differenzierungsleistungen des isolierten Mesoderms von Molchkeimen durch zegefügtes Chorda- und Neuralmaterial. *Okajimas Folia Anat. Japon.* 19, 131–197.

Yamana, K., and Kageura, H. (1987). Reexamination of the "regulative development" of amphibian embryos. *Cell Differentiation* 20, 3–10.

Yanagi, A. (1987). Positional control of the fates of nuclei produced after meiosis in *Paramecium caudatum:* Analysis by nuclear transplantation. *Dev. Biol.* 122, 535–539.

Yanagi, A., and Hiwatashi, K. (1985). Intracellular positional control of survival or degeneration of nuclei during conjugation in *Paramecium caudatum*. *J. Cell Sci.* 79, 237–246.

Yano, J. (1985a). Mating types and conjugant fusion with macronuclear union in *Stylonychia pustulata* (Ciliophora). *J. Sci. Hiroshima Univ., Ser. B, Div. 1* 32, 157–175.

Yano, J. (1985b). Degeneration of the cortical organelles and nuclear changes in the split members from the early conjugating pairs in *Stylonychia pustulata* (Ciliophora). *J. Sci. Hiroshima Univ. Ser. B. Div. 1* 32, 193–207.

Yano, J. (1987). Morphogenesis of mirror-image symmetry doublet cells in the ciliate *Stylonychia pustulata*. *Zool. Sci.* (Tokyo). 4, 993 (abstr.).

Yasuda, T., Numata, O., Ohnishi, K., and Watanabe, Y. (1980). A contractile ring and cortical changes found in the dividing *Tetrahymena pyriformis*. *Exp. Cell Res.* 128, 407–417.

Yusa, A. (1957). The morphology and morphogenesis of the buccal organelles in *Paramecium* with particular reference to their systematic significance. *J. Protozool.* 4, 128–142.

Yusa, A. (1963) An electron microscope study on regeneration of trichocysts in *Paramecium caudatum*. *J. Protozool.* 10, 253–262.

Zeuthen, E. (1978). Induced reversal of order of cell division and DNA replication in *Tetrahymena*. *Exp. Cell Res.* 116, 39–46.

Zeuthen, E., and Rasmussen, L. (1972). Synchronized cell division in protozoa. In T. T. Chen (ed.), *Research in Protozoology* 4. Oxford, Pergamon Press, pp. 9–145.

Index